Introduction to the theory of molecular vibrations and vibrational spectroscopy

L. A. WOODWARD

Introduction to the theory of molecular vibrations and vibrational spectroscopy 1972

OXFORD · AT THE CLARENDON PRESS

1972

Oxford University Press, Ely House, London W. 1

GLASGOW NEW YORK TORONTO MELBOURNE WELLINGTON
CAPE TOWN IBADAN NAIROBI DAR ES SALAAM LUSAKA ADDIS ABABA
DELHI BOMBAY CALCUTTA MADRAS KARACHI LAHORE DACCA
KUALA LUMPUR SINGAPORE HONG KONG TOKYO

QC
454
W86

PRINTED IN GREAT BRITAIN
AT THE UNIVERSITY PRESS, OXFORD
BY VIVIAN RIDLER
PRINTER TO THE UNIVERSITY

Preface

As an Oxford Fellow and Tutor for more than a quarter of a century I have had the pleasure of initiating and supervising the experimental researches of a large number of young graduates in the field of molecular vibrational spectroscopy. In consequence I have learned where difficulties with the theory most often arise, and have had considerable experience in helping to deal with them. During sabbatical leave in 1968 as Visiting Professor at the University of Tennessee, I gave a series of about thirty talks on molecular vibrations to a class of doctorate candidates from the faculties of physics and of chemistry. This afforded me the opportunity to assemble and organize the relevant subject matter in note form. The present book is based upon these notes and the discussions to which the talks gave rise.

Nobody undertaking an exposition of this kind could fail to owe an incalculable debt to those two authoritative and extensive works, *Infra-red and Raman Spectra of Polyatomic Molecules* by Herzberg (Van Nostrand, 1945), and *Molecular Vibrations* by Wilson, Decius, and Cross (McGraw-Hill, 1955). I have made frequent use of them ever since their publication. It seemed to me, however, that there was a need for a book on a more modest scale, which would not only expound the essential features of vibrational theory but at the same time would explain and develop the special mathematical ideas and methods involved. The present volume is the result of my attempts to fill this need. It is divided into two related parts, the first of which deals with the classical treatment of the vibrations of molecular models, and the second with the quantum-mechanical theory of vibrational infra-red and Raman spectra. In the course of the exposition the relevant properties and uses of matrices and vectors are explained, and symmetry theory is progressively developed. No attempt is made to deal with a

v

broad range of molecular species; on the contrary, all the principal theoretical themes are illustrated by the same few simple examples, which are thus subjected to full and detailed treatment. It is hoped that the book will be found useful, not only by young graduates in physics and chemistry with an interest in vibrational spectroscopic research, but also by experienced workers in related fields and by Honours students.

It is a pleasure to express my gratitude to two Oxford colleagues: to Dr. C. J. Bradley of Jesus College for helpful discussions on symmetry theory, and to Dr. S. Altmann of Brasenose College for kindly reading through the completed work in typescript and sending me his comments. My thanks are also due to the U.S. National Science Foundation for their invitation to spend six months in the United States in 1968 as a Senior Foreign Scientist Fellow, and to the staff of the Chemistry Department of the University of Tennessee at Knoxville (and in particular to Professor W. H. Fletcher) who made me very welcome and gave me facilities for the compilation of the notes which formed the basis for this book. The main work of writing it has been done during the tenure of an Emeritus Fellowship, for which I am much indebted to the Leverhulme Foundation.

L. A. W.

Jesus College, Oxford
1971

Contents

Page numbers of the text are printed at the bottom outer corners of the pages. The pairs of numbers in square brackets at the top outer corners are those of the sections into which each chapter is subdivided.

Contents

Contents

Contents

Contents

PART II

THE QUANTUM-MECHANICAL TREATMENT OF MOLECULAR VIBRATIONAL SPECTROSCOPY

Contents

Contents

Contents

Part I

The treatment of molecular vibrations by classical mechanics

1 A simple example to illustrate the application of classical mechanics

1.1. Reasons for undertaking a classical treatment

THE vibrations of molecules are subject to quantum mechanics. It may well be asked, therefore, why we should undertake a lengthy discussion in terms of classical mechanics. Broadly speaking, the reasons are rooted in the close relationship between the classical and the quantum theories of molecular vibrations, in consequence of which an analysis on the basis of the former will give insight and information which will be very valuable with respect to the latter.

More particularly, our classical treatment will be based upon the assumption (in practice a very good one) that the vibrations are simple-harmonic. In this special case the advantages for later quantum-mechanical treatment are very considerable. We shall be led to introduce special kinds of coordinates in terms of which the classical vibrational problem assumes very simple forms; and when we come to the treatment on the basis of the Schrödinger wave-equation we shall find that the adoption of these same coordinates will effect a corresponding simplification. Moreover, the symmetry properties of the classical coordinates will have very important parallels in the symmetry properties of the wave-mechanical eigenfunctions, upon which the spectroscopic selection rules depend.

The simple-harmonic approximation leads to one especially noteworthy result concerning the actual values of the vibrational frequencies in any particular case. As observed in spectroscopy, a fundamental is the frequency of the radiation absorbed or emitted in consequence of a transition between quantized energy levels. We will designate it by ν_{qu}. On the other hand, the frequency-value calculated by classical theory on the basis

3

of a suitable model is a mechanical frequency. We will designate it by ν_{cl}. It transpires that $\nu_{qu} = \nu_{cl}$. This remarkable and conveniently uncomplicated result is peculiar to vibrations which are simple-harmonic.

1.2. The simple mechanical system to be studied

The remainder of this chapter will be devoted almost entirely to considering a simple mechanical system, without questioning whether such a system could serve usefully as a model of a molecule. The object of this exercise is to illustrate some features

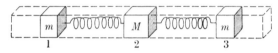

FIG. 1.1.   A simple mechanical system to illustrate the application
of classical mechanics.

of classical mechanics and, in particular, to familiarize the reader with some of the types of coordinates used. Suitable for our purposes is a system consisting of three masses joined in a straight line by two identical springs of negligible mass. We shall confine our attention to motions along the line of centres. In order to ensure the exclusion of motions in other directions, and also to exclude rotations of the individual masses, we may suppose each to consist of a uniform block of the same square cross-section, able to move in a piston-like manner inside a closely fitting and perfectly frictionless horizontal tube (see Fig. 1.1). For convenience the masses are numbered as shown. The masses 1 and 3 are equal, both having the value m. The central mass 2 has the different value M. When all three are at rest and at equilibrium, the configuration will obviously be such that the centres of masses 1 and 3 will be equidistant from the centre of mass 2.

1.3. Mechanical treatment in terms of simple displacement coordinates

We shall not be concerned with the actual positions of the mass-centres in their equilibrium configuration, but rather with

their individual displacements therefrom. We therefore introduce the respective displacement coordinates x_1, x_2, x_3. It
is essential that the reader should grasp clearly that these
coordinates are not simple positional coordinates measured
from some common origin, but that each is a displacement
measured from its own particular origin, namely, from the individual equilibrium position of the mass whose displacement
the coordinate describes. All three x-coordinates have the same
sense (e.g. from left to right in Fig. 1.1).

The potential energy of the system, calculated relative to the
value in the equilibrium configuration (which we shall take as
zero), will arise solely from the extension or compression of the
springs. We assume that these obey Hooke's law, i.e. that the
restoring force is directly proportional to the change in length.
This is a very good approximation in practice, especially when
(as in the present case) only very small changes are to be considered. If f is the force constant of each spring, this being the
restoring force per unit change of length, the potential energy
for a small extension or compression D will be given by

$$\int_0^D f\xi \, d\xi = \tfrac{1}{2}fD^2.$$

Now for the spring joining masses 1 and 2, which we shall call
spring 1, the change of length is

$$D_1 = x_2 - x_1,$$

and for the other spring, which we will call spring 2, the
corresponding change is

$$D_2 = x_3 - x_2.$$

Hence for the potential energy V of the system we may write

$$2V = f(x_2 - x_1)^2 + f(x_3 - x_2)^2. \tag{1.1}$$

The kinetic energy T may be expressed in the form

$$2T = m\dot{x}_1^2 + M\dot{x}_2^2 + m\dot{x}_3^2, \tag{1.2}$$

in which $\dot{x}_1$ is the time derivative of x_1, i.e. the velocity of mass 1,
and $\dot{x}_2$ and $\dot{x}_3$ are defined in an equivalent fashion.

5

The classical equation of motion for the ith mass has the form

$$\frac{d}{dt}\left(\frac{\partial T}{\partial \dot{x}_i}\right) + \frac{\partial V}{\partial x_i} = 0, \qquad (1.3)$$

in which t is the time. With a little thought those not already familiar with this equation will perceive that it merely expresses the fact that the rate of change of momentum (the first term) is equal to the force acting upon the mass (the negative of the second term). Using the value of V from eqn (1.1) and the value of T from eqn (1.2), we obtain the following equations of motion for masses 1, 2, and 3 respectively:

$$m\ddot{x}_1 + f(x_1 - x_2) = 0; \qquad (1.4)$$

$$M\ddot{x}_2 + f(2x_2 - x_1 - x_3) = 0; \qquad (1.5)$$

$$m\ddot{x}_3 + f(x_3 - x_2) = 0. \qquad (1.6)$$

Here $\ddot{x}_1$, $\ddot{x}_2$, $\ddot{x}_3$ are the second derivatives of the respective coordinates with respect to time, i.e. the accelerations.

In general the motion of the system will not be simple, but, as we shall see, it may always be regarded as a superposition of a certain number of simpler motions known as *normal modes*. In any one normal mode, considered alone, every mass performs a simple-harmonic motion with the same characteristic frequency ν. All the masses move in phase with one another, but generally with different amplitudes. The actual magnitudes of these amplitudes depend upon the initial conditions at $t = 0$, but in any particular normal mode their ratios to one another are determined by the nature of the system.

We therefore seek solutions of the type

$$x_1 = A_1^{(x)}\cos(\lambda^{\frac{1}{2}}t + \epsilon), \qquad (1.7)$$

$$x_2 = A_2^{(x)}\cos(\lambda^{\frac{1}{2}}t + \epsilon), \qquad (1.8)$$

$$x_3 = A_3^{(x)}\cos(\lambda^{\frac{1}{2}}t + \epsilon), \qquad (1.9)$$

in which $\lambda = 4\pi^2\nu^2$ and ϵ is a phase angle. The amplitudes of motion of the three masses are denoted by $A_1^{(x)}$, $A_2^{(x)}$, and $A_3^{(x)}$, the superscript indicating the type of coordinate to which they refer.

Considering mass 1, we substitute into the equation of motion, i.e. eqn (1.4), the value of x_1 in eqn (1.7) and of x_2 in eqn (1.8), noting from eqn (1.7) that $\ddot{x}_1 = -\lambda A_1^{(x)}\cos(\lambda^{\frac{1}{2}}t+\epsilon)$. We thus obtain

$$-m\lambda A_1^{(x)}\cos(\lambda^{\frac{1}{2}}t+\epsilon)+f(A_1^{(x)}-A_2^{(x)})\cos(\lambda^{\frac{1}{2}}t+\epsilon) = 0.$$

After dividing through by $\cos(\lambda^{\frac{1}{2}}t+\epsilon)$ this becomes

$$-m\lambda A_1^{(x)}+f(A_1^{(x)}-A_2^{(x)}) = 0,$$

which is the *amplitude equation* for mass 1. In the same way we may obtain the corresponding amplitude equations for masses 2 and 3. After slight rearrangements the three equations become respectively:

$$(f-m\lambda)A_1^{(x)} \qquad -fA_2^{(x)} \qquad\qquad = 0; \qquad (1.10)$$

$$-fA_1^{(x)}+(2f-M\lambda)A_2^{(x)} \qquad -fA_3^{(x)} = 0; \qquad (1.11)$$

$$-fA_2^{(x)}+(f-m\lambda)A_3^{(x)} = 0. \qquad (1.12)$$

The theory of equations tells us that the condition for these linear equations to hold simultaneously is that the determinant of the coefficients of the amplitudes must vanish, i.e. that

$$\begin{vmatrix} (f-m\lambda) & -f & 0 \\ -f & (2f-M\lambda) & -f \\ 0 & -f & (f-m\lambda) \end{vmatrix} = 0. \qquad (1.13)$$

Unless this condition is obeyed, the only possible solution is that in which $A_1^{(x)} = A_2^{(x)} = A_3^{(x)} = 0$. This, however, is trivial in that it represents no motion at all.

Eqn (1.13) is known as the *secular equation* of the motion and is the key to the whole classical treatment. By expanding the determinant we obtain an equation in λ (in our case it is a cubic) the solution of which gives us the roots λ and hence, since $\lambda = 4\pi^2\nu^2$, the frequencies of the normal modes. For our system there are just three roots and consequently just three normal modes, each with its own characteristic normal frequency. By expanding the secular determinant the reader may easily verify that the secular equation becomes

$$(f-m\lambda)(-2fm-fM+Mm\lambda)\lambda = 0.$$

We at once see that the roots, which we shall call λ_a, λ_b, and λ_n, are

$$\lambda_a = \frac{f}{m}, \tag{1.14}$$

$$\lambda_b = \frac{f(M+2m)}{Mm}, \tag{1.15}$$

$$\lambda_n = 0. \tag{1.16}$$

Thus two of the three normal modes have frequencies which are different from zero; they are genuine vibrations. The third mode, which has zero frequency, is said to be a *non-genuine vibration*. (This is the reason for the choice of the initial letter of 'non-genuine' to designate its root λ_n.) Its nature will become clear from a consideration of the relative amplitudes of motion of the three masses.

Let us deal first with the mode a of frequency λ_a, denoting the amplitudes by $A_{1a}^{(x)}$, $A_{2a}^{(x)}$, and $A_{3a}^{(x)}$. Substituting $\lambda_a = f/m$ into the amplitude equations (1.10), (1.11), and (1.12), we find at once from eqn (1.10) that

$$A_{2a}^{(x)} = 0. \tag{1.17}$$

From eqn (1.11), using eqn (1.17), we find that

$$A_{1a}^{(x)} = -A_{3a}^{(x)}. \tag{1.18}$$

Eqn (1.12) gives no further information. Thus in the normal mode a the central mass 2 remains throughout at rest, while the masses 1 and 3 oscillate in such a way that their displacements are always equal in magnitude and opposite in sign. The mode may be characterized by writing

$$A_{1a}^{(x)} : A_{2a}^{(a)} : A_{3a}^{(a)} = 1 : 0 : -1. \tag{1.19}$$

It should be noted that only the *ratios* of the amplitudes are uniquely defined. Their absolute magnitudes depend upon the values of x_1, x_3, $\dot{x}_1$, and $\dot{x}_3$ at $t = 0$. Since masses 1 and 3 are equal, it is clear that throughout the normal mode a the centre of mass of the system as a whole remains at rest.

The mode b can be similarly characterized by substitution of λ_b from eqn (1.15) into the amplitude equations (1.10), (1.11), and (1.12). This gives the result

$$A_{1b}^{(x)} : A_{2b}^{(x)} : A_{3b}^{(x)} = M : -2m : M. \tag{1.20}$$

Here the two outer masses move in the same sense while the central mass moves in the opposite sense. The displacement $\bar{x}$ of the centre of mass of the system is given by

$$(M+2m)\bar{x} = mx_1+Mx_2+mx_3. \tag{1.21}$$

By virtue of eqn (1.20) it follows that $\bar{x} = 0$. Thus in the normal mode b (as in mode a) the centre of mass of the system as a whole remains at rest.

The forms of the two modes are indicated in Fig. 1.2, in which the black circles represent the positions of the centres of the

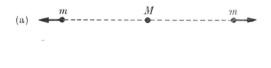

Fig. 1.2. Forms of the two normal vibrational modes of the system shown in Fig. 1.1.

individual masses at equilibrium and the arrows their displacements at some particular time.

Lastly we come to the mode which has zero frequency. By substituting $\lambda_n = 0$ into the amplitude equations (1.10), (1.11), and (1.12) we obtain

$$A_{1n}^{(x)} = A_{2n}^{(x)} = A_{3n}^{(x)}.$$

In this non-genuine vibrational mode the displacements of all three masses are equal in magnitude and in sign. The motion can thus be identified as a translation of the system as a whole, without any change in the lengths of the springs (i.e. without any contribution to the potential energy V). Obviously this mode differs from both modes a and b in that the centre of gravity of the system as a whole does not remain at rest.

Any general motion of the system must consist of a superposition of the three normal modes, each with its own set of amplitudes and its own phase. The motions of the individual masses will not, in general, be simple.

9

1.4. Use of mass-weighted displacement coordinates

In the preceding section we chose as coordinates the actual displacements of the individual masses from their equilibrium positions. This, however, is by no means the only possible choice. For example, it is possible to convert the expression for the kinetic energy into a simpler form by using the so-called mass-weighted displacements q, defined as

$$q_1 = m^{\frac{1}{2}}x_1,$$
$$q_2 = M^{\frac{1}{2}}x_2,$$
$$q_3 = m^{\frac{1}{2}}x_3.$$

In place of eqn (1.2) we obtain

$$2T = \dot{q}_1^2 + \dot{q}_2^2 + \dot{q}_3^2. \tag{1.22}$$

As an expression for the potential energy, substitution of $x_1 = m^{-\frac{1}{2}}q_1$ etc. into eqn (1.1) gives

$$2V = f\left(\frac{q_1^2}{m} + \frac{2q_2^2}{M} + \frac{q_3^2}{m} - \frac{2q_1q_2}{\sqrt{(Mm)}} - \frac{2q_2q_3}{\sqrt{(Mm)}}\right). \tag{1.23}$$

We see that the introduction of mass-weighted coordinates does indeed result in a simplification of the expression for the kinetic energy, but only at the expense of some complication in the expression for the potential energy.

We now use the basic classical equation in the form

$$\frac{\mathrm{d}}{\mathrm{d}t}\left(\frac{\partial T}{\partial \dot{q}_i}\right) + \frac{\partial V}{\partial q_i} = 0,$$

which is to be compared with eqn (1.3), and seek solutions of the form

$$q_1 = A_1^{(q)}\cos(\lambda^{\frac{1}{2}}t + \epsilon), \tag{1.24}$$
$$q_2 = A_2^{(q)}\cos(\lambda^{\frac{1}{2}}t + \epsilon), \tag{1.25}$$
$$q_3 = A_3^{(q)}\cos(\lambda^{\frac{1}{2}}t + \epsilon). \tag{1.26}$$

Proceeding in exactly the same way as when working with x-coordinates, we easily deduce the following amplitude equations:

$$\left(\frac{f}{m} - \lambda\right)A_1^{(q)} - \frac{f}{(Mm)^{\frac{1}{2}}}A_2^{(q)} = 0; \tag{1.27}$$

$$\frac{-f}{(Mm)^{\frac{1}{2}}}A_1^{(q)} + \left(\frac{2f}{M} - \lambda\right)A_2^{(q)} - \frac{f}{(Mm)^{\frac{1}{2}}}A_3^{(q)} = 0; \tag{1.28}$$

$$-\frac{f}{(Mm)^{\frac{1}{2}}}A_2^{(q)} + \left(\frac{f}{m} - \lambda\right)A_3^{(q)} = 0. \tag{1.29}$$

These should be compared with eqns (1.10), (1.11), and (1.12). The secular equation in determinantal form follows at once:

$$\begin{vmatrix} \left(\dfrac{f}{m}-\lambda\right) & \dfrac{-f}{\sqrt{(Mm)}} & 0 \\[2mm] \dfrac{-f}{\sqrt{(Mm)}} & \left(\dfrac{2f}{M}-\lambda\right) & \dfrac{-f}{\sqrt{(Mm)}} \\[2mm] 0 & \dfrac{-f}{\sqrt{(Mm)}} & \left(\dfrac{f}{m}-\lambda\right) \end{vmatrix} = 0.$$

Expansion gives a cubic equation with exactly the same roots as we found when using x-coordinates, namely

$$\lambda_a = \frac{f}{m},$$

$$\lambda_b = \frac{f(M+2m)}{Mm},$$

$$\lambda_n = 0.$$

Successive substitution of these λ-values into the amplitude equations (1.27), (1.28), and (1.29) leads to the result

$$A_{1a}^{(q)} : A_{2a}^{(q)} : A_{3a}^{(q)} = 1 : 0 : -1, \tag{1.30}$$

$$A_{1b}^{(q)} : A_{2b}^{(q)} : A_{3b}^{(q)} = M^{\frac{1}{2}} : -2m^{\frac{1}{2}} : M^{\frac{1}{2}}, \tag{1.31}$$

$$A_{1n}^{(q)} : A_{2n}^{(q)} : A_{3n}^{(q)} = m^{\frac{1}{2}} : M^{\frac{1}{2}} : m^{\frac{1}{2}}. \tag{1.32}$$

As we have previously noted, the absolute values of the amplitudes of motion in any mode depend upon the particular initial conditions, and only their ratios are uniquely determined. For reasons that will become apparent later, it is of great advantage to introduce for each normal mode certain pure numbers l which are in the same ratios as the amplitudes, but whose values are independent of the actual magnitudes of these amplitudes. For the mode a, for example, we define three such quantities as

$$l_{1a} = \frac{A_{1a}^{(q)}}{\{(A_{1a}^{(q)})^2+(A_{2a}^{(q)})^2+(A_{3a}^{(q)})^2\}^{\frac{1}{2}}}, \tag{1.33}$$

$$l_{2a} = \frac{A_{2a}^{(q)}}{\{(A_{1a}^{(q)})^2+(A_{2a}^{(q)})^2+(A_{3a}^{(q)})^2\}^{\frac{1}{2}}}, \tag{1.34}$$

$$l_{3a} = \frac{A_{3a}^{(q)}}{\{(A_{1a}^{(q)})^2+(A_{2a}^{(q)})^2+(A_{3a}^{(q)})^2\}^{\frac{1}{2}}}. \tag{1.35}$$

We note first that
$$l_{1a}:l_{2a}:l_{3a} = A_{1a}^{(q)}:A_{2a}^{(q)}:A_{3a}^{(q)}.$$
Secondly we note that the values of l_{1a}, l_{2a}, and l_{3a} are independent of the magnitudes of the particular set of amplitudes $A_{1a}^{(q)}$, $A_{2a}^{(q)}$, and $A_{3a}^{(q)}$ used to define them. This is at once made clear by considering the use, in place of some chosen set, of another set each member of which is k times greater. The factor k obviously cancels out from the expressions defining l_{1a}, l_{2a}, and l_{3a}. Thirdly, it is important to note that the form of the definitions is specially chosen so as to ensure that
$$l_{1a}^2+l_{2a}^2+l_{3a}^2 = 1. \qquad (1.36)$$
In view of this property, l_{1a}, l_{2a}, and l_{3a} are said to be *normalized* or, more precisely, normalized to unity. The full implications of this will be explained at a later stage. For mode b we introduce the corresponding quantities l_{1b}, l_{2b}, and l_{3b}, and for the non-genuine vibrational mode the quantities l_{1n}, l_{2n}, and l_{3n}, their definitions being analogous to those in eqns (1.33), (1.34), and (1.35). Their actual values are obtained by making use of eqns (1.30), (1.31), and (1.32) which give the ratios of the amplitudes of motion. For example, combining eqn (1.30) with eqn (1.33) we find
$$l_{1a} = \frac{A_{1a}^{(q)}}{\{(A_{1a}^{(q)})^2+0+(A_{1a}^{(q)})^2\}^{\frac{1}{2}}} = \frac{1}{2^{\frac{1}{2}}}.$$
In the same way the values of the other l-numbers are found to be
$$l_{2a} = 0,$$
$$l_{3a} = -\frac{1}{2^{\frac{1}{2}}},$$
$$l_{1b} = l_{3b} = \left\{\frac{M}{2(M+2m)}\right\}^{\frac{1}{2}},$$
$$l_{2b} = -\left(\frac{2m}{M+2m}\right)^{\frac{1}{2}},$$
$$l_{1n} = l_{3n} = \left(\frac{m}{M+2m}\right)^{\frac{1}{2}},$$
$$l_{2n} = \left(\frac{M}{M+2m}\right)^{\frac{1}{2}}.$$

1.5. Definition and properties of normal coordinates

We now make use of the l-numbers in order to generate a set of new and very important coordinates, Q_a, Q_b, Q_n, called the *normal coordinates* of the system. They are defined by the three equations

$$q_1 = l_{1a} Q_a + l_{1b} Q_b + l_{1n} Q_n, \qquad (1.37)$$

$$q_2 = l_{2a} Q_a + l_{2b} Q_b + l_{2n} Q_n, \qquad (1.38)$$

$$q_3 = l_{3a} Q_a + l_{3b} Q_b + l_{3n} Q_n. \qquad (1.39)$$

This may well be regarded as a rather indirect way of defining the new coordinates. However, the three equations may be solved to give expressions for Q_a, Q_b, and Q_n as linear combinations of q_1, q_2, and q_3. These expressions will be deduced later in this section.

In passing, we may call attention to a convenient way of expressing the content of eqns (1.37), (1.38), and (1.39). The l-numbers are written as an array:

$$\begin{bmatrix} l_{1a} & l_{1b} & l_{1n} \\ l_{2a} & l_{2b} & l_{2n} \\ l_{3a} & l_{3b} & l_{3n} \end{bmatrix}. \qquad (1.40)$$

This array is referred to as the *matrix* which effects the linear transformation of the set of Q-coordinates into the set of q-coordinates. Matrices are conventionally enclosed in square brackets, as shown in (1.40), in order to distinguish them from determinants. The relevant properties of matrices will be explained in Chapter 3. We shall later make extensive use of them in discussing the general classical treatment of molecular vibrations.

The very great importance of normal coordinates lies in the fact that the expressions for the kinetic and potential energies assume specially simple forms when written in terms of them. Consider first the kinetic energy of our system of three masses, the expression for which, in terms of the q-coordinates, was given in eqn (1.22). Transforming to normal coordinates we have

$$2T = \dot{q}_1^2 + \dot{q}_2^2 + \dot{q}_3^2$$
$$= (l_{1a} \dot{Q}_a + l_{1b} \dot{Q}_b + l_{1n} \dot{Q}_n)^2 + (l_{2a} \dot{Q}_a + l_{2b} \dot{Q}_b + l_{2n} \dot{Q}_n)^2 +$$
$$+ (l_{3a} \dot{Q}_a + l_{3b} \dot{Q}_b + l_{3n} \dot{Q}_n)^2.$$

As the reader may easily verify, insertion of the l-values given at the end of the preceding section causes this expression for $2T$ to reduce to

$$2T = \dot{Q}_a^2 + \dot{Q}_b^2 + \dot{Q}_n^2. \tag{1.41}$$

Thus the simplicity which followed from the introduction of q-coordinates is fully retained in terms of normal coordinates. As we shall see, this is a general property of normal coordinates, so that in any mechanical problem we may write

$$2T = \sum_i \dot{Q}_i^2. \tag{1.42}$$

Considering now the potential energy, we similarly transform from q-coordinates to normal coordinates by substituting from eqns (1.37), (1.38), and (1.39) into eqn (1.23). After insertion of the actual l-values for our problem, the expression for the potential energy reduces to

$$2V = \left(\frac{f}{m}\right)Q_a^2 + \left\{\frac{f(M+2m)}{Mm}\right\}Q_b^2.$$

Remembering the values of the roots given in eqns (1.14) and (1.15), we see that

$$2V = \lambda_a Q_a^2 + \lambda_b Q_b^2.$$

Moreover, because $\lambda_n = 0$, we may go further and write

$$2V = \lambda_a Q_a^2 + \lambda_b Q_b^2 + \lambda_n Q_n^2. \tag{1.43}$$

This is a particular example of another general property of normal coordinates. In fact, as we shall see later, in any mechanical problem we may write

$$2V = \sum_i \lambda_i Q_i^2. \tag{1.44}$$

Thus both the kinetic energy and the potential energy of a system assume extremely simple forms when expressed in terms of normal coordinates.

This simplicity is naturally passed on to the equations of motion. In view of eqns (1.41) and (1.43), the general equation in the form

$$\frac{\mathrm{d}}{\mathrm{d}t}\left(\frac{\partial T}{\partial \dot{Q}_i}\right) - \frac{\partial V}{\partial Q_i} = 0$$

at once reduces to

$$\ddot{Q}_a - \lambda_a Q_a = 0, \tag{1.45}$$

$$\ddot{Q}_b - \lambda_b Q_b = 0, \tag{1.46}$$

$$\ddot{Q}_n - \lambda_n Q_n = \ddot{Q}_n = 0, \tag{1.47}$$

the solutions of which are

$$Q_a = A_a^{(Q)}\cos(\lambda_a^{\frac{1}{2}}t + \epsilon_a),$$
$$Q_b = A_b^{(Q)}\cos(\lambda_b^{\frac{1}{2}}t + \epsilon_b),$$
$$Q_n = kt.$$

Here $A_a^{(Q)}$ and $A_b^{(Q)}$ are the amplitudes of the genuine vibrational modes a and b respectively, and k is a constant. Since $\dot{Q}_n = k$, by setting $k = 0$ we annihilate the contribution of the translation to the kinetic energy of the system. In any case its contribution to the potential energy is zero. Thus by setting $k = 0$ we can effectively exclude the translation from the problem.

Let us now examine the appearance of the normal coordinates when they are expressed as linear combinations of the mass-weighted coordinates q. To this end we substitute the l-values, as given at the end of Section 1.4, into eqns (1.37), (1.38), and (1.39), and then solve these equations for Q_a, Q_b, and Q_n. As the reader may easily confirm, the solutions are

$$Q_a = \frac{1}{2^{\frac{1}{2}}}q_1 - \frac{1}{2^{\frac{1}{2}}}q_3, \tag{1.48}$$

$$Q_b = \left\{\frac{M}{2(M+2m)}\right\}^{\frac{1}{2}}q_1 - \left(\frac{2m}{M+2m}\right)^{\frac{1}{2}}q_2 + \left\{\frac{M}{2(M+2m)}\right\}^{\frac{1}{2}}q_3, \tag{1.49}$$

$$Q_n = \left(\frac{m}{M+2m}\right)^{\frac{1}{2}}q_1 + \left(\frac{M}{M+2m}\right)^{\frac{1}{2}}q_2 + \left(\frac{m}{M+2m}\right)^{\frac{1}{2}}q_3. \tag{1.50}$$

Remembering the values of the l-numbers, we perceive that these solutions may be written as

$$Q_a = l_{1a}q_1 + l_{2a}q_2 + l_{3a}q_3, \tag{1.51}$$

$$Q_b = l_{1b}q_1 + l_{2b}q_2 + l_{3b}q_3, \tag{1.52}$$

$$Q_n = l_{1n}q_1 + l_{2n}q_2 + l_{3n}q_3. \tag{1.53}$$

In other words, we have discovered that the matrix effecting the linear transformation of the q-coordinates into the normal coordinates, i.e.

$$\begin{bmatrix} l_{1a} & l_{2a} & l_{3a} \\ l_{1b} & l_{2b} & l_{3b} \\ l_{1n} & l_{2n} & l_{3n} \end{bmatrix}, \tag{1.54}$$

15

is identical with the matrix which effects the inverse transformation, except that the rows and columns of the latter have been transposed. This remarkable and in practice very convenient result is by no means fortuitous. It is in fact an example of a general result which we shall presently establish. Here we merely note that it is not obtained unless the l-numbers for each mode are normalized to unity, i.e. obey a condition like that of eqn (1.36).

We have seen that Q_a has the frequency corresponding to the root λ_a. We also know from eqn (1.30) that the mode a is characterized by $q_1 = -q_3$ and $q_2 = 0$. If we now substitute these relative values of the q-coordinates into the eqns (1.51), (1.52), and (1.53), which define the normal coordinates, we find that Q_b and Q_n both vanish and only Q_a survives. Similarly, if we take from eqn (1.31) the relative q-values which characterize the mode b and substitute them into the equations defining the normal coordinates, we find that Q_a and Q_n both vanish and only Q_b survives. The relative q-values for the translational mode are obtainable from eqn (1.32). When they are substituted into the equations defining the normal coordinates, only Q_n survives.

1.6. Use of internal coordinates

In terms either of a set of three x-coordinates or of a set of three q-coordinates, the treatment of the motions of our simple system necessarily had to deal with three normal modes. Solution of the cubic secular equation accordingly gave one zero root, indicating that one of the modes was not a genuine vibration. As far as the genuine vibrations of the system are concerned, this mode (actually the translation of the system as a whole) had to be discarded at the end of the treatment. Now the existence of this translation as one of the normal modes could have been foreseen, and by the use of so-called internal coordinates it could have been excluded right from the start. This sort of procedure is indeed adopted in most treatments of molecular vibrational problems.

For the simple system which we have been considering, we

must choose two internal coordinates which will suffice to express, not only the potential energy, but also the purely internal (i.e. vibrational) kinetic energy. Using these two coordinates, we shall need only to deal with the two genuine vibrational modes a and b; in other words we shall have isolated the purely vibrational problem.

An appropriate choice is the changes D_1 and D_2 of the lengths of the two springs. These internal coordinates are given by

$$D_1 = x_2 - x_1, \tag{1.55}$$

$$D_2 = x_3 - x_2. \tag{1.56}$$

The potential energy is very simply expressible as

$$2V = f(D_1^2 + D_2^2).$$

The treatment of the kinetic energy is less straightforward. The total value, including the translational kinetic energy, is given by

$$2T = m\dot{x}_1^2 + M\dot{x}_2^2 + m\dot{x}_3^2. \tag{1.57}$$

We have to exclude from this the contribution from the translational mode. This can be done by introducing a condition which makes this contribution zero, i.e. which ensures that there is no linear momentum of the system as a whole. The condition is

$$m\dot{x}_1 + M\dot{x}_2 + m\dot{x}_3 = 0. \tag{1.58}$$

As to the x-coordinates themselves, it follows from eqn (1.58) that

$$mx_1 + Mx_2 + mx_3 = \text{constant},$$

i.e. that the centre of gravity of the system remains stationary at a constant distance from some arbitrary zero position. We are not in the least interested in such a displacement, and so we may set it equal to zero, and write

$$mx_1 + Mx_2 + mx_3 = 0. \tag{1.59}$$

This condition in no way affects the vibrational problem; for, as we have seen, the centre of gravity of the system does indeed remain at rest throughout both genuine vibrational modes.

We are now in a position to deal with the task of expressing the purely vibrational kinetic energy in terms of the internal coordinates D_1 and D_2. By differentiating eqns (1.55) and (1.56) with respect to time, we obtain $\dot{D}_1 = \dot{x}_2 - \dot{x}_1$ and $\dot{D}_2 = \dot{x}_3 - \dot{x}_2$, which, with eqn (1.58), provide three equations solvable for $\dot{x}_1$, $\dot{x}_2$, and $\dot{x}_3$. We thus find

$$\dot{x}_1 = -\left(\frac{M+m}{M+2m}\right)\dot{D}_1 - \left(\frac{m}{M+2m}\right)\dot{D}_2, \qquad (1.60)$$

$$\dot{x}_2 = \left(\frac{m}{M+2m}\right)\dot{D}_1 - \left(\frac{m}{M+2m}\right)\dot{D}_2, \qquad (1.61)$$

$$\dot{x}_3 = \left(\frac{m}{M+2m}\right)\dot{D}_1 + \left(\frac{M+m}{M+2m}\right)\dot{D}_2. \qquad (1.62)$$

We now substitute these values of $\dot{x}_1$, $\dot{x}_2$, and $\dot{x}_3$ into eqn (1.57) and so obtain the purely vibrational kinetic energy in terms of $\dot{D}_1$ and $\dot{D}_2$, having excluded the translational kinetic energy by virtue of the condition expressed in eqn (1.58). We shall continue to make use of the same symbol T to denote the vibrational energy as we did to denote the total kinetic energy. The meaning will always be clear from the context, and especially from the type of coordinates (D or x) in terms of which T is expressed. For the system we are considering we find

$$2T = \left\{\frac{m(M+m)}{M+2m}\right\}\dot{D}_1^2 + \left(\frac{2m^2}{M+2m}\right)\dot{D}_1\dot{D}_2 + \left\{\frac{m(M+m)}{M+2m}\right\}\dot{D}_2^2.$$
$$(1.63)$$

Having now expressed both the potential energy and the vibrational kinetic energy in terms of our two internal coordinates, we may proceed to write down and solve the equations of motion. Using the basic classical equation in the form

$$\frac{\mathrm{d}}{\mathrm{d}t}\left(\frac{\partial T}{\partial \dot{D}_i}\right) + \frac{\partial V}{\partial D_i} = 0,$$

and assuming solutions of the form

$$D_1 = A_1^{(D)}\cos(\lambda^{\frac{1}{2}}t + \epsilon),$$

$$D_2 = A_2^{(D)}\cos(\lambda^{\frac{1}{2}}t + \epsilon),$$

we obtain the following two amplitude equations:

$$\left\{f - \frac{m(M+m)\lambda}{M+2m}\right\}A_1^{(D)} - \left(\frac{m^2\lambda}{M+2m}\right)A_2^{(D)} = 0; \quad (1.64)$$

$$-\left(\frac{m^2\lambda}{M+2m}\right)A_1^{(D)} + \left\{f - \frac{m(M+m)\lambda}{M+2m}\right\}A_2^{(D)} = 0. \quad (1.65)$$

The secular equation in determinantal form is therefore

$$\begin{vmatrix} \left\{f - \dfrac{m(M+m)\lambda}{M+2m}\right\} & -\dfrac{m^2\lambda}{M+2m} \\[2ex] -\dfrac{m^2\lambda}{M+2m} & \left\{f - \dfrac{m(M+m)\lambda}{M+2m}\right\} \end{vmatrix} = 0.$$

On expansion it gives a quadratic equation for λ, the roots of which are easily seen to be

$$\lambda_a = \frac{f}{m},$$

$$\lambda_b = f\left(\frac{M+2m}{Mm}\right).$$

These are, of course, exactly the same roots as we obtained when using simple displacement coordinates x or mass-weighted displacement coordinates q. The advantage of the use of internal coordinates D is that it reduces the degree of the secular equation. In our simple case this advantage is not great, the reduction being only from a cubic to a quadratic. In the general three-dimensional problem, however, the use of internal coordinates brings a larger (and consequently very welcome) simplification.

Finally, as in the earlier treatments, we may successively substitute the roots λ_a and λ_b into the amplitude equations (1.64) and (1.65), and so obtain the ratio of the amplitudes in each of the two vibrational modes. For mode a we find that

$$A_{1a}^{(D)} = A_{2a}^{(D)}.$$

The corresponding result for mode b is

$$A_{1b}^{(D)} = -A_{2b}^{(D)}.$$

Thus in mode a both springs are at all times extended (or both at all times compressed) to the same extent; whereas in mode b, when one is extended the other is compressed. These descriptions of the vibrational modes are, of course, in complete agreement with our previous findings as represented in Fig. 1.2.

1.7. The shape of things to come

Before we can apply the methods of the present chapter to the vibrations of molecules, we must first consider the questions that arise in choosing an acceptable molecular model. Also, if we are to use internal coordinates, we must consider certain difficulties which are involved in separating off the non-genuine vibrations in order to isolate the genuine vibrational problem. These matters will be discussed in Chapter 2. There will then follow, in Chapters 3 and 4, an account of matrix algebra and vector algebra, in terms of which the classical treatment of vibrations assumes its most elegant and convenient form. Our attention will be confined, however, to those aspects which are directly relevant to the vibrational problem. After these preparations, Chapters 5 and 6 will explain and exemplify the details of the classical treatment in terms of internal coordinates. The elements of symmetry theory will then be expounded (Chapters 7–9), but only to the extent required for a full understanding of its great importance for our particular purposes. The ensuing general discussion of the applications of symmetry considerations to molecular vibrations (Chapters 10–16) will be illustrated by particular examples and will bring to an end our account of the subject in terms of classical mechanics.

The remaining eight chapters of the book will be concerned with the quantum-mechanical theory of molecular vibrational spectroscopy.

2 Approximations in the treatment of molecular vibrations using mechanical models

2.1. Choice of molecular model

THE treatment of a simple mechanical system in the preceding chapter will have illustrated some of the methods of classical mechanics. We may now inquire whether the adoption of systems of this general type as models will permit the use of similar methods for the treatment of the vibrations of molecules.

Doubtless the reader will be familiar with the general idea of a molecule as consisting of a number of very small and massive atomic nuclei held in a fairly rigid configuration by directed chemical bonds which originate from the presence of electrons of relatively little mass. A complete mechanical treatment would, of course, cover the motions of the electrons as well as those of the nuclei. However, because of the very great difference of mass between the two kinds of particle, it is permissible, as a very good approximation, to regard their respective motions as mechanically separable. This is usually referred to as the Born–Oppenheimer approximation, and it underlies most theoretical treatments of nuclear motions, including molecular vibrations. In practice it means that we need not include the electrons explicitly in the mechanical equations, although their presence does in fact make itself felt in the dependence of the potential energy of the system upon the nuclear displacement coordinates.

We thus choose an idealized mechanical model of a molecule, in which the nuclei are represented by point masses with a certain equilibrium configuration, but subject to a certain intramolecular force-field, in that deformations (such as alterations of internuclear distances etc.) bring into play restoring

21

forces and so give rise to vibrational motions. In order to be able to carry through our mathematical treatment, we shall make the assumption that the force field is a quadratic one, i.e. that the vibrational motions are representable as superpositions of simple-harmonic normal modes. In fact, molecular vibrational amplitudes are small, and consequently the simple-harmonic approximation is generally a fairly good one.

2.2. The exclusion of translational motions

In Chapter 1 we discussed a simple linear system constrained so as to be capable of motion in one direction only, and we saw how the translation of the system as a whole could be excluded from the mechanical problem. Considering now a 3-dimensional system of point masses capable of motion in any direction, we can similarly exclude all translational possibilities by introducing three conditions which annihilate the linear momenta of the system as a whole in all three mutually orthogonal directions x, y, and z of space. These conditions are analogous to the one expressed in eqn (1.58), and may be written as

$$\sum_i m_i \dot{x}_i = \sum_i m_i \dot{y}_i = \sum_i m_i \dot{z}_i = 0, \qquad (2.1)$$

where $\dot{x}_i$, $\dot{y}_i$, and $\dot{z}_i$ are the time derivatives of the Cartesian displacements of a particular mass m_i and the summation extends over all the masses. As in eqn (1.59), we may ensure that the conditions of eqns (2.1) are obeyed by writing

$$\sum_i m_i x_i = \sum_i m_i y_i = \sum_i m_i z_i = 0. \qquad (2.2)$$

Translational motion is said to be 'separable', in the sense that it can always be excluded from consideration in the above manner without at all affecting the validity of the remaining equations for other kinds of motion. The underlying reason for this is that there is no interaction between translational motion and the other kinds of motion. In particular, a system will vibrate in exactly the same way whether or not it is simultaneously translating uniformly through space in any direction.

2.3. Rotation–vibration interactions

Translation, however, is not the only kind of 'external' motion which our model is able to perform in addition to its internal, genuine vibrational modes. It can also rotate as a whole. Indeed free molecules, in whose vibrations we are interested, will certainly be rotating as well as vibrating. By introducing suitable conditions (see Section 2.4) it is possible to isolate the purely vibrational problem which would apply to a non-rotating (as well as non-translating) molecule; but, since all free molecules must in reality be rotating, the question then arises as to whether the resulting mathematical formulation is appropriate to describe the vibrational motions of such molecules. The answer is that it is not completely appropriate. This is owing to the fact that (as we shall see below) rotational motion necessarily interacts with vibrational motion, with the consequence that the vibrations of a molecule are not the same when it is rotating as when it is not. However, the rotation–vibration interactions are generally quite weak, and therefore the purely vibrational problem (in which rotational effects are assumed to be absent) represents a correspondingly good approximation for free molecules.

Some insight into the general nature of rotation–vibration interactions and of the special forces involved, can be obtained from the following simple considerations and examples. The kinetic energy of a rotating system in its equilibrium configuration (i.e. non-vibrating) depends upon the moment of inertia of the system about the axis of rotation, and hence upon the distances of all the point masses from this axis. When the system is vibrating, these distances are in general not constant, and the vibrational displacement coordinates accordingly become involved. Nor, in general, is the mean value of the moment of inertia, averaged over a complete vibrational period, exactly equal to the value in the equilibrium configuration; instead it is dependent upon the nature and amplitude of the vibrational mode. Even in the absence of vibration, the purely rotational problem of a non-rigid model is more complicated than it would be if the model were rigid. This is due to the *centrifugal forces* which have their origin in the rotation of the model as a whole. We know from Newton's law that a mass which is moving in a certain direction will continue to move with constant linear momentum unless it is acted upon by a force. It follows that a mass which is moving in a circular path with constant angular velocity must be constrained to do so by a force which is constantly altering its linear momentum. Because the angular momentum is constant, this force must be directed towards the centre of the circle. Consider a molecular model of the linear m-M-m type treated in Chapter 1, but now able to rotate as a whole. Suppose that it is indeed rotating with uniform angular velocity about an axis through the central point mass and at right angles to the line of the masses. The central mass will remain stationary, but the two outer ones will move in a circular path. They can only do so provided that each has acting upon it a force directed towards the rotation axis. Such forces can only be provided by the springs (or, in a molecule, the chemical bonds), and these must therefore be stretched accordingly. An observer who is rooted in the model so as to share (and so be unaware of) its rotation, will see merely that the model has taken up a new 'equilibrium configuration' in which the springs have become extended, and he will attribute this to the action of forces acting *outwards* upon the two outer masses. These are the *centrifugal forces*. The higher the angular velocity of the system, the greater the centrifugal forces and hence the extension of the springs. The consequent dependence of moment of inertia upon rotational velocity complicates the purely rotational problem. The effects of centrifugal forces also bring additional complications to the vibrational problem of a rotating molecule. As we shall see in detail later, normal modes and frequencies depend upon the geometry of the equilibrium configuration of a molecule, as well as upon the force field and the masses. Although it is not strictly permissible to consider centrifugal effects alone, it will be easily appreciated that the distortions to which they give rise will in general alter the geometry of the nuclear configuration and so lead to interactions between rotation and vibration. As a very simple example, consider a molecular model of the type *m-M-m* which is not linear, i.e. in which the equilibrium inter-bond angle α (see Fig. 2.1(a)) is less than 180°. When the model rotates uniformly about the axis shown in the figure by the vertical line, centrifugal forces

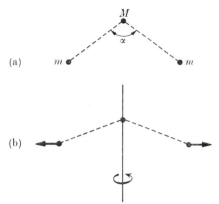

FIG. 2.1. (a) A non-linear *m-M-m* system and (b) the effect of centrifugal forces due to the rotation indicated by the curved arrow.

acting on the two masses *m* will not only stretch the bonds but will also increase the bond angle—see Fig. 2.1(b). In the simple-harmonic approximation the force field will be unaffected, since the force constants are independent of the magnitude of the displacements; however, the normal frequencies will depend upon the actual value of the interbond angle in the non-vibrating configuration. Rotation–vibration interactions of the kinds so far mentioned are inevitable, but fortunately they are very weak, so that their effects upon the vibrational problem may, to a very good approximation, be neglected.

We shall now consider briefly the nature of forces of a different kind which are brought into play by rotation. These are the *Coriolis forces*. Although perhaps generally less familiar than centrifugal forces, they usually give rise to larger rotation–vibration interactions. Coriolis forces resemble centrifugal forces in that their origin lies in mechanical inertia, but they differ essentially in the following important respect. Centrifugal forces occur in a rotating system even when the constituent masses are at rest relative to a frame of reference which is rooted

25

in the system so as to rotate with it: Coriolis forces occur only when (as for example in a genuine vibration) a mass is moving within this frame of reference, and more particularly when its distance from the axis of rotation is changing with time.

Consider a mass m which is part of a system rotating about an axis through the point A and normal to the plane of the figure, as indicated by the curved arrow in Fig. 2.2. Suppose

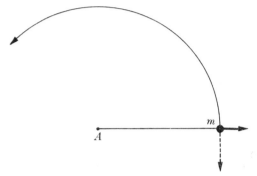

FIG. 2.2. Diagram to illustrate Coriolis force.

that, relative to a frame of reference fixed in the system and rotating along with it, the mass is moving with a velocity represented in the figure by the solid arrow directed away from the axis. This motion tends to cause the angular momentum of the mass to become greater. Because of inertia, however, its angular momentum tends to remain constant. The result is that the mass tends to lag behind in the rotational motion of the system. An observer rooted in the system (and so unaware of its rotation) will perceive this effect, and will attribute it to a force acting at right angles both to the radial velocity of the mass and to the axis of rotation of the system as a whole. This is the so-called Coriolis force. In magnitude it is proportional both to the radial component of the velocity of the mass in the system-fixed frame of reference and also to the angular velocity of overall rotation. As an example of the influence of Coriolis forces upon vibrations, consider again the linear m-M-m model capable of motions in three dimensions. In addition to force constants for the stretching of the two springs (or bonds), we introduce a new one concerned with forces tending to restore the linear configuration when the inter-bond angle is made less than 180°. There will be two vibrational modes of bond stretching *a* and *b*, exactly like those depicted in Fig. 1.2, and in addition the model will have a new genuine vibrational bending mode *c* in which all the masses move at right angles to the line which passes through them in their equilibrium configuration. The form of this mode is shown in Fig. 2.3. Now suppose that

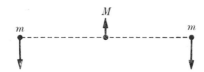

Fig. 2.3. Form of normal 'bending' mode
of a linear *m*-*M*-*m* system.

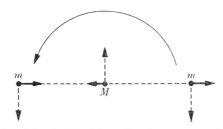

Fig. 2.4. Coriolis interaction between rotation
and vibration for a linear *m*-*M*-*m* system.

the whole model is rotating anticlockwise about an axis through its centre of mass and normal to the line passing through the masses in their equilibrium configuration. In the absence of any vibration the only effect of the uniform rotation is slightly to stretch the bonds. Now suppose that the model is vibrating in the unsymmetrical bond-stretching normal mode—as in Fig. 1.2(b). Coriolis forces will come into play, and consequently the vibrational motion will not be exactly the same as it would have been if the model had not been rotating. Fig. 2.4 shows the directions of these forces (dotted arrows) in relation to the vibrational velocities of the masses (solid arrows) at some particular stage of the mode—let us say the stage at which each mass is passing through its equilibrium configuration and

27

thus has its maximum vibrational velocity. (It is important to grasp clearly that the solid arrows in Fig. 2.4 represent *velocities*, and that their significance is thus quite different from that of the arrows in Figs. 1.2 and 2.3, which represent displacements.) We see that in this particular case the Coriolis forces are such as to tend to bend the model. Because they are proportional to the vibrational velocities, the forces will oscillate synchronously with the vibration, and will therefore tend to force the model to perform the motions of mode *c* with the frequency of mode *b*. When occurring freely, however, the mode *c* will have its own normal frequency, which will certainly not be equal to that of mode *b*. Without going into the complicated details, we can readily appreciate that Coriolis forces inevitably give rise to interactions between rotational and vibrational motions, and in certain cases to consequent interactions between one vibrational mode and another of a rotating molecule.

Fortunately Coriolis forces (like centrifugal forces) are sufficiently small that in most cases their effects upon the vibrational problem may be ignored without serious error. This means that, having excluded both translation and rotation, we are left with a purely vibrational problem which, although strictly only appropriate to a non-translating and non-rotating molecule, nevertheless is a good approximation for a molecule which is freely rotating.

2.4. The exclusion of rotational motions

If a molecule is not subject to any external forces, the non-genuine vibrational modes of zero frequency (i.e. the translations and rotations) make no contribution to its potential energy, which is thus purely vibrational in origin and expressible in terms of internal coordinates only. It is in the expression for the total kinetic energy (which along with the potential energy forms the starting point for the application of classical mechanics) that the translational and rotational motions make their appearance in the problem. As we shall see in Section 5.3, the total kinetic energy is in fact the sum of a number of essentially different parts. The purely vibrational part, representing the kinetic energy which the vibrating molecule would continue to possess if translations and rotations were completely absent, is in fact the only part which we shall retain in formulating the (approximate) vibrational problem. There is also another part representing the purely translational and purely rotational kinetic energies which the molecule would continue to possess if it were not vibrating. In pursuing our interest in vibrational motions, we are perfectly entitled to exclude this external part from our consideration. As far as the translational kinetic energy is concerned, we have already seen how it may be excluded (without affecting the vibrational problem) by the introduction of the three conditions expressed in eqns (2.2). Even if we also disregard the purely rotational kinetic energy, we are still not left merely with the purely vibrational part, for there remains in addition a part representing rotation–vibration interactions. Only with the adoption of an approximation in which these interactions are regarded as negligible does it become possible to formulate the vibrational problem without reference to rotational motions.

For a non-linear molecule containing N nuclei the procedure adopted is to introduce, in addition to the three relationships of eqns (2.2), three further relationships between the $3N$ nuclear displacement coordinates. In this book we are concerned, not so much with the details of the way in which rotations are thus approximately excluded, but rather with the treatment of the purely vibrational problem that results. The relevant relationships will therefore be quoted and their significance briefly indicated, but we shall not discuss them further. The non-linear molecule possesses three mutually orthogonal principal axes of inertia passing through its centre of mass. We choose these as the axes of a system of Cartesian coordinates (X, Y, Z) which is fixed in the molecule and rotates along with it. Let the time-independent quantities X_{0i}, Y_{0i}, Z_{0i} be the corresponding coordinates of the equilibrium position of the ith nucleus. Also let the vibrational displacement coordinates x_i, y_i, z_i of this nucleus be related to a system whose origin is at the equilibrium position and whose axes are parallel respectively to the axes of X, Y, and Z. Then the three relationships in question are

$$
\left.
\begin{aligned}
\sum_i m_i(Y_{0i}z_i - Z_{0i}y_i) &= 0 \\
\sum_i m_i(Z_{0i}x_i - X_{0i}z_i) &= 0 \\
\sum_i m_i(X_{0i}y_i - Y_{0i}x_i) &= 0
\end{aligned}
\right\}. \tag{2.3}
$$

It is not difficult to see that the time derivative of the left-hand side of the first of these equations is an approximation to the angular momentum of the molecule about the X-axis, in which the vibrational displacements y_i and z_i have been regarded as negligibly small relative to Y_{0i} and Z_{0i} respectively. The second of the equations (2.3) is similarly related to the angular momentum about the Y-axis, and the third to the angular momentum about the Z-axis. Thus in the stated approximation the three conditions (2.3) ensure zero angular momentum with respect to the molecule-fixed system.

Along with eqns (2.2), these conditions represent six linear relationships connecting the $3N$ displacement coordinates of a molecule containing N nuclei. The number of independent, linear equations with which to describe the purely internal problem is thus $3N-6$. The molecule will possess this number of vibrational normal modes with non-zero frequencies. It may well be convenient to carry through the classical treatment on the basis of a set of $3N-6$ appropriately chosen internal coordinates D, in terms of which the potential energy expression may be easily formulated. The $3N-6$ linear equations which define these coordinates D in terms of the Cartesian displacement coordinates, together with the 6 relations of eqns (2.2) and (2.3), give a total of $3N$, by the solution of which we can express each of the $3N$ Cartesian displacement coordinates in terms of the $3N-6$ internal coordinates D. Differentiation with respect to time gives us all the $\dot{x}_i, \dot{y}_i$, and $\dot{z}_i$ in terms of the time derivatives of the D-coordinates. By substitution in the expression

$$
2T = \sum_i m_i(\dot{x}_i^2 + \dot{y}_i^2 + \dot{z}_i^2)
$$

we obtain the kinetic energy in terms of the internal coordinates only. This is the value required for the purely vibrational problem, which is now amenable to treatment by a generalization of the methods used in treating a simple system in Chapter 1.

The reader is advised at this point to refer back to Chapter 1 and especially to Section 1.6, where an example of the use of a set of just two internal coordinates is carried through. The use of the two eqns (1.55) and (1.56) which define the D-coordinates, along with the condition (1.59) which excludes the single non-genuine vibration, gives each of the three displacement co-ordinates x, and hence their time derivatives $\dot{x}$ for substitution into the general expression (1.57) for the kinetic energy. This is seen to be just a special instance of the procedure, outlined in the present section, whereby the purely vibrational kinetic energy may be expressed in terms of any conveniently chosen set of internal coordinates. In Chapter 5 we shall discuss further the general formulation of the total kinetic energy and the approximation involved in using the purely internal part for the treatment of the vibrations of free molecules.

2.5. Normal modes of linear molecules

For any molecule containing N nuclei, three relations between the $3N$ nuclear displacement coordinates are required to account for translational motions. For a non-linear molecule, i.e. for one with a non-vanishing moment of inertia about each of its three principal axes, a further three relationships are required to account for rotational motions. Linear molecules, however, present a special case. In general the kinetic energy of a system rotating with angular velocity ω is given by $\frac{1}{2}I\omega^2$, where I is the moment of inertia with respect to the axis of rotation. Now for a linear molecule, the line of the nuclei is a principal axis and the moment of inertia of the molecule with respect to it is zero. It follows that rotation about this axis cannot contribute anything to the kinetic energy or, in other words, such rotation is automatically excluded. This being so, it is not necessary to introduce any condition to ensure its exclusion. Instead of the three relationships of eqn (2.3) we need only two; in fact, if we choose the line of nuclei as the Z-axis, it is the third which becomes redundant.

It follows that the number of independent coordinates for the description of the internal motions of the molecule is now $3N-5$, as compared with $3N-6$ for a non-linear molecule. This means that a linear molecule will have one more normal vibrational coordinate than if it were non-linear. We may see how this comes about by comparing a linear m-M-m model with the corresponding model which has an inter-bond angle of less

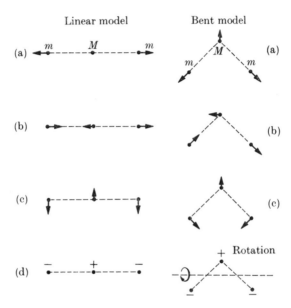

Fig. 2.5. Forms of normal vibrational modes of linear and bent m-M-m models.

than $180°$. Here the value of $3N$ is 9 and so the linear model must have $9-5 = 4$ normal vibrational modes. We have already become acquainted with three of them, the symmetrical bond-stretching mode a, the unsymmetrical bond-stretching mode b, and the bending mode c. The forms of the first two are shown in Fig. 1.2; that of mode c is shown in Fig. 2.3. The nature of the single remaining vibrational mode is easily appreciated. If we regard mode c (see Fig. 2.3) as taking place in the plane of the paper, we perceive the possibility of a fourth, entirely independent mode d, which is exactly similar to mode c

except that it takes place in a plane at right angles to the plane of the paper. It will clearly be possible to stimulate either mode c or mode d without in either case involving the other. It is also obvious that both must necessarily have precisely the same frequency. (In fact they provide an example of what is called a mutually degenerate pair.) For the purposes of comparison, the forms of all the four vibrational modes of the linear m-M-m model are shown on the left-hand side of Fig. 2.5. In the case of mode d, the signs $+$ and $-$ signify respectively movements towards and away from the reader and normal to the plane of the figure.

The corresponding modes for the bent model are shown on the right-hand side of Fig. 2.5. The first three are respectively analogous to the modes a, b, and c of the linear model, and all involve only motions in the plane of the three point-masses. They are all genuine vibrations. On the other hand, the fourth mode (although otherwise quite analogous to mode d of the linear model) is seen to amount simply to a rotation of the model as a whole about the principal axis which lies in the plane of the model and is normal to the line bisecting the bond angle. From this simple example we see how it is that the linear model has one more genuine vibrational mode than does the corresponding bent model—in fact $3N-5$ instead of $3N-6$. In this book we shall proceed in general as though the molecules under consideration were non-linear, and so shall use the number $3N-6$. *Mutatis mutandis*, however, the arguments and conclusions will apply equally well to linear molecules.

3 Matrices and their relevant properties

3.1. Introduction

THE classical treatment of molecular vibrations assumes its most elegant and concise form when expressed with the help of matrices. In this chapter, therefore, some aspects of matrix algebra will be briefly reviewed, though only so far as is relevant to the particular applications that we shall later have in hand. Those readers who are already quite familiar with the subject may accordingly omit this chapter and proceed directly to the next. Even there we shall not directly tackle the vibrational problem, but shall discuss another mathematical theme, that of vectors and their uses, which will also be of great value to us. Thereafter, in Chapter 5, we shall be in a position to undertake the general classical treatment of the mechanics of molecular vibrations.

3.2. Matrices and matrix multiplication

In Chapter 1 we encountered an array (1.40) of coefficients which, as was then remarked, forms a matrix. We are here generally considering a set of quantities and the formation from them of a new set, each member of which is a linear combination of all the members of the original set. As a simple example, take the four quantities x_1, x_2, x_3, and x_4, and from them form the new quantities x_1', x_2', x_3', and x_4', according to

$$\left. \begin{aligned} x_1' &= A_{11}x_1 + A_{12}x_2 + A_{13}x_3 + A_{14}x_4 \\ x_2' &= A_{21}x_1 + A_{22}x_2 + A_{23}x_3 + A_{24}x_4 \\ x_3' &= A_{31}x_1 + A_{32}x_2 + A_{33}x_3 + A_{34}x_4 \\ x_4' &= A_{41}x_1 + A_{42}x_2 + A_{43}x_3 + A_{44}x_4 \end{aligned} \right\}. \tag{3.1}$$

The matrix which effects the linear transformation of the set of xs into the set of x's is then obtained by writing the coefficients of (3.1) in an array as

$$\begin{bmatrix} A_{11} & A_{12} & A_{13} & A_{14} \\ A_{21} & A_{22} & A_{23} & A_{24} \\ A_{31} & A_{32} & A_{33} & A_{34} \\ A_{41} & A_{42} & A_{43} & A_{44} \end{bmatrix}. \tag{3.2}$$

Matrices are conventionally enclosed in square brackets, as indicated, in order to distinguish them from determinants. The matrix (3.2) happens to have four rows and four columns, and so is called a square matrix. Not all matrices are square, however; they may have any number of rows and any number of columns.

The individual members of the array forming the matrix are referred to as the *elements*. In accordance with the usage in (3.2), each matrix element is given two subscripts, the first to designate the row in which the element occurs and the second to designate the column. In this book we shall use non-italic capital letters in bold type to denote matrices. Thus for the matrix (3.2) we shall use the symbol **A**.

Now suppose that, from the set of quantities x'_1, x'_2, x'_3, x'_4 we form yet another set x''_1, x''_2, x''_3, x''_4 by means of a different linear transformation with the matrix **B**. Clearly, starting with x_1, x_2, x_3, x_4, transformation to x''_1, x''_2, x''_3, x''_4 could be effected by a single linear transformation. Let the matrix of this be **C**. Then **C**, whose effect is identical with the effect of **A** followed by **B**, is called the *matrix product* **BA**. Note particularly that the order of writing the two factor matrices in this product is such that their order of application is found by reading from *right to left*. The word 'product' is used by analogy with ordinary algebra. We must not expect, however, that the products of arrays will have the same properties as the products of ordinary numbers.

We shall now establish the rule for forming the product **BA** of two matrices. Let **A** be effective in transforming a set of quantities x_j into a new set x'_i. Then if A_{ij} is the element of **A**

in the ith row and the jth column, we have by the definition of $\mathbf{A}$,
$$x'_i = \sum_j A_{ij} x_j.$$

Thus for the effect of $\mathbf{B}$ we have
$$x''_k = \sum_i B_{ki} x'_i$$
$$= \sum_i B_{ki} \sum_j A_{ij} x_j = \sum_j \left(\sum_i B_{ki} A_{ij} \right) x_j. \tag{3.3}$$

But if the matrix product $\mathbf{BA}$ is $\mathbf{C}$, then
$$x''_k = \sum_j C_{kj} x_j. \tag{3.4}$$

Comparing eqn (3.3) with eqn (3.4), we find
$$C_{kj} = \sum_i B_{ki} A_{ij}. \tag{3.5}$$

This is the desired rule for the formation of the matrix product $\mathbf{BA} = \mathbf{C}$. Expressed in words, it is as follows. In order to find the element of the product matrix in the kth row and the jth column, first select the kth row of $\mathbf{B}$ and the jth column of $\mathbf{A}$. Then multiply each element of the selected row by the corresponding element of the selected column, and add together the results so obtained.

Although at first sight the procedure may appear a little complicated, it is soon found in practice to be very easy and straightforward. The novice is advised to work through some simple numerical examples. Thus consider the product
$$\begin{bmatrix} 1 & 2 \\ 3 & 4 \end{bmatrix} \begin{bmatrix} 5 & 6 \\ 7 & 8 \end{bmatrix} = \begin{bmatrix} C_{11} & C_{12} \\ C_{21} & C_{22} \end{bmatrix}.$$

Following the procedure of eqn (3.5), we have:
$$C_{11} = (1 \times 5) + (2 \times 7) = 19;$$
$$C_{12} = (1 \times 6) + (2 \times 8) = 22;$$
$$C_{21} = (3 \times 5) + (4 \times 7) = 43;$$
and
$$C_{22} = (3 \times 6) + (4 \times 8) = 50.$$

The desired product matrix is thus
$$\begin{bmatrix} 19 & 22 \\ 43 & 50 \end{bmatrix}.$$

While we have this simple example before us, we may use it to illustrate an important general property of matrix products. The reader may easily confirm that, if we interchange the factor matrices in the product expression, we obtain

$$\begin{bmatrix} 5 & 6 \\ 7 & 8 \end{bmatrix} \begin{bmatrix} 1 & 2 \\ 3 & 4 \end{bmatrix} = \begin{bmatrix} 23 & 34 \\ 31 & 56 \end{bmatrix}.$$

The product thus depends upon the order in which the factors are written or, in other words, matrix multiplication is in general *non-commutative*. For this reason it is necessary to adopt some terminology to distinguish between, say, the product **BA** and the product **AB**. In **BA**, the matrix **A** is conventionally said to be *pre-multiplied* by **B**, or alternatively **B** is said to be *post-multiplied* by **A**. In the product **AB**, however, the convention states that **B** is pre-multiplied by **A** or that **A** is post-multiplied by **B**.

Although matrix multiplication is non-commutative, we may easily verify that it is *associative*, i.e. if **A**, **B**, and **C** be any suitable matrices then

$$\mathbf{A(BC)} = \mathbf{(AB)C}.$$

Thus although the order of writing the matrices in a product is of great importance, the order of carrying out the multiplications of the adjacent pairs of factors is immaterial.

Behind all matrix multiplication lies the limitation that the process is impossible unless the matrices are suitably matched, i.e. for the product **BA** to be meaningful, the matrix **B** must have a number of columns equal to the number of rows possessed by **A**. If this condition is satisfied, the number of elements B_{ij} in the ith row of **B** will be equal to the number of elements A_{jk} in the kth column of **A**, and it will be possible to form the product element in accordance with eqn (3.5). The two matrices are then said to be *conformable* for the formation of the product **BA**.

3.3. Non-square matrices, including single-column and single-row matrices

So far we have been discussing only square matrices, which have equal numbers of rows and columns. In general, however, a matrix can be rectangular with more columns than rows, or vice versa. In the limit a matrix may consist of a single column or a single row. In all cases the rule of multiplication is the same as has already been given in eqn (3.5). Of course it is essential that two matrices be conformable for the multiplication, otherwise the product is not defined. For example, the matrices

$$\mathbf{B} = \begin{bmatrix} B_{11} & B_{12} \\ B_{21} & B_{22} \end{bmatrix} \quad \text{and} \quad \mathbf{A} = \begin{bmatrix} A_{11} \\ A_{21} \end{bmatrix}$$

are conformable only for the product **BA**, the value of which is easily seen to be the single-column matrix

$$\begin{bmatrix} B_{11} A_{11} + B_{12} A_{21} \\ B_{21} A_{11} + B_{22} A_{21} \end{bmatrix}.$$

It is easy to see that, in general, the product **CD** of two conformable matrices is a matrix with the same number of rows as **C** and the same number of columns as **D**. Furthermore, two matrices that are conformable for the product **CD** are not necessarily conformable for the product **DC**.

We now return to the initial example of Section 3.2 and write the original set of quantities x_1, x_2, x_3, and x_4 as a single-column matrix **X** and similarly the transformed set x_1', x_2', x_3', and x_4' as another single-column matrix **X'**. We then see that the content of eqns (3.1) can be expressed as a matrix product formation,

$$\begin{bmatrix} x_1' \\ x_2' \\ x_3' \\ x_4' \end{bmatrix} = \begin{bmatrix} A_{11} & A_{12} & A_{13} & A_{14} \\ A_{21} & A_{22} & A_{23} & A_{24} \\ A_{31} & A_{32} & A_{33} & A_{34} \\ A_{41} & A_{42} & A_{43} & A_{44} \end{bmatrix} \begin{bmatrix} x_1 \\ x_2 \\ x_3 \\ x_4 \end{bmatrix}, \tag{3.6}$$

or simply

$$\mathbf{X}' = \mathbf{A}\mathbf{X}.$$

In the same way we may write the set x_1'', x_2'', x_3'', and x_4'' as the single-column matrix **X''** and so arrive at the elegantly concise statement

$$\mathbf{X}'' = \mathbf{B}\mathbf{X}' = \mathbf{B}\mathbf{A}\mathbf{X}.$$

In our later treatment of molecular vibrations we shall have frequent occasion to employ matrix algebraic expressions of this kind in dealing with linear transformations.

3.4. Addition of matrices. Partitioned matrices

The sum of two matrices only has a meaning when they are conformable for addition, i.e. when each has the same number of rows and columns as the other. The situation is then extremely simple. The sum is a matrix, each of whose elements is simply the sum of the corresponding elements of the two summed matrices.

From this and the rule of multiplication it is easy to show, for example, that

$$[A+B]C = AC+BC$$

and that $\quad [A+B][C+D] = AC+AD+BC+BD.$

Thus the *distributive* law is obeyed. There are evidently some notable similarities, as well as some notable differences, between matrix algebra and ordinary algebra.

Consider now two matrices like those shown in (3.7), which are conformable for product formation in that the first has the same number of columns, i.e. $n+m$, as the second has rows.

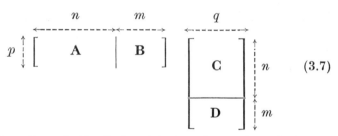

$$(3.7)$$

The number of rows in the first matrix is p; the number of columns in the second is q. By means of the line drawn across it, the first matrix is said to be *partitioned* into the two *sub-matrices* **A** and **B**, the former having p rows and n columns and the latter having p rows and m columns. The other matrix is likewise partitioned into the sub-matrices **C** and **D**, the situation being so arranged that **A** and **C** are conformable for the formation of the product **AC**, and likewise **B** and **D** are conformable for the formation of **BD**. From the nature of matrix multiplication and addition, the reader will easily perceive that the value of the product is **AC+BD**.

We shall also encounter the case of the product of two matrices which are partitioned in a way different from the above, as in

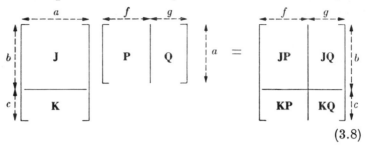

$$(3.8)$$

Here the product matrix is partitioned into four sub-matrices, and there is no question of matrix addition arising.

A special form of matrix which will later play a very important part in the theory of vibrations consists of a square matrix in which the only non-zero elements occur in square blocks situated (like beads on a string) along the diagonal which runs from the top left-hand corner to the bottom right-hand corner. Such a matrix is said to have the *diagonal-block form*. The individual blocks may involve any number of rows (and of course the same number of columns). We speak of a 1×1 block, a 2×2 block, etc. A conventional method of representation is illustrated by

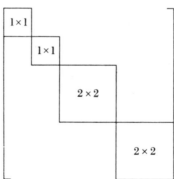

which shows a 6×6 matrix with the diagonal-block form consisting of two 1×1 blocks and two 2×2 blocks. The drawing of blocks in this way implies by convention that all elements outside the blocks are zero-elements. Suppose now that we have two different matrices both having the same diagonal-block form, i.e. the same numbers of blocks of the same sizes, arranged in the same order. Then, as may be verified without difficulty, their product is another matrix which also has the same diagonal-block form, each of its blocks being simply the product of the corresponding blocks of the original matrices. For example

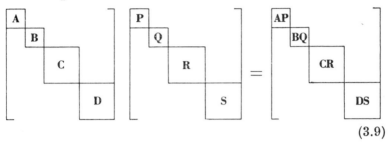

$$(3.9)$$

3.5. The transpose of a matrix and of a matrix product

To transpose a matrix is to interchange rows and columns. In this way we obtain from a matrix **A** the so-called *transpose* of **A**, which we shall designate by $\mathbf{A}^\dagger$. By convention this symbol is often used for the conjugate complex transpose of **A**, i.e. the matrix obtained from **A** by first transposing it and then replacing each of the elements by its conjugate complex. However, since we shall be concerned exclusively with *real* matrices, no possibility of confusion will arise.

If the element of **A** in the ith row and the jth column is A_{ij}, it follows that the corresponding element of $\mathbf{A}^\dagger$ (i.e. the one in the ith row and the jth column) is A_{ji}. Consider now the product **AB** with another matrix **B** whose typical element is B_{ij}. The typical element of the product is $\sum_k A_{ik} B_{kj}$. To obtain the typical element of the transpose $[\mathbf{AB}]^\dagger$ of the product, we simply have to interchange i and j, giving $\sum_k A_{jk} B_{ki}$. But it is easy to see that this is the typical element of the product $\mathbf{B}^\dagger\mathbf{A}^\dagger$, for which the general rule of multiplication gives $\sum_k B_{ki} A_{jk}$. We have thus proved that

$$[\mathbf{AB}]^\dagger = \mathbf{B}^\dagger\mathbf{A}^\dagger. \qquad (3.10)$$

In words, the transpose of a product of two matrices is the product of their respective transposes in reverse order. We may note that, when the matrices in question are not square, the reversal of order is necessary so that they shall remain conformable for product formation after having been transposed.

The rule for the formation of the transpose of a product may be extended to any number of factors. For instance,

$$[\mathbf{ABCD}]^\dagger = [\mathbf{CD}]^\dagger[\mathbf{AB}]^\dagger = \mathbf{D}^\dagger\mathbf{C}^\dagger\mathbf{B}^\dagger\mathbf{A}^\dagger.$$

The transpose of a single-column matrix is clearly a single-row matrix, and the linear transformation in eqn (3.6) can be re-written in the completely equivalent transposed form

$$[x_1' \quad x_2' \quad x_3' \quad x_4'] = [x_1 \quad x_2 \quad x_3 \quad x_4] \begin{bmatrix} A_{11} & A_{21} & A_{31} & A_{41} \\ A_{12} & A_{22} & A_{32} & A_{42} \\ A_{13} & A_{23} & A_{33} & A_{43} \\ A_{14} & A_{24} & A_{34} & A_{44} \end{bmatrix}.$$

It follows that there must be some ambiguity in a statement which merely refers to 'the matrix which effects a certain linear transformation'; for the matrix is not completely defined unless we are told whether it effects the transformation by pre-multiplying a single column or post-multiplying a single row. The conventions we shall adopt will be made clear in due course.

3.6. Symmetric and diagonal matrices. Null and unit matrices

A matrix $\mathbf{A}$ is said to be *symmetric* when $A_{ij} = A_{ji}$, i.e. when the element in any chosen position is equal to the element in the position obtained by reflection in the diagonal from the top left-hand to the bottom right-hand corner of the array. It is obvious that all symmetric matrices must be square. A particular case is the so-called *diagonal matrix*, for which $A_{ij} = 0$ unless $i = j$, so that all elements are zero except those on the diagonal. Obviously for any symmetric matrix (and this includes all diagonal ones) the transpose must be identical with the matrix itself.

Consider a matrix product of the form $\mathbf{DAD}^\dagger$ in which $\mathbf{A}$ is a square matrix and $\mathbf{D}$ is any matrix which is conformable for

the multiplications. The transpose of the product is given by
$$[\mathbf{DAD}^\dagger]^\dagger = \mathbf{DA}^\dagger\mathbf{D}^\dagger.$$
If now $\mathbf{A}$ is a symmetric matrix, so that $\mathbf{A}^\dagger = \mathbf{A}$, we see that the transpose of the product $\mathbf{DAD}^\dagger$ must be identical with the product itself; in other words the product is a symmetric matrix. This holds in particular when $\mathbf{A}$ is a diagonal matrix. We shall encounter cases of this sort in our application of matrix algebra to the molecular vibrational problem. For example, we shall have to deal with a matrix $\mathbf{G}$ which is defined as equal to $\mathbf{BM}^{-1}\mathbf{B}^\dagger$, where $\mathbf{M}^{-1}$ is a diagonal matrix of reciprocal masses. Without knowing anything at all about the matrix $\mathbf{B}$, we can state that $\mathbf{G}$ must be symmetric. We shall also have to deal with another matrix $\mathbf{G}$ which is equal to $\mathbf{UGU}^\dagger$. Knowing only that $\mathbf{G}$ is symmetric, we can be sure that $\mathbf{G}$ must be symmetric also.

We shall later encounter matrices of two specially interesting types: the null and the unit matrices. A *null matrix* is one for which every element is zero. Such a matrix is often represented by the symbol $\mathbf{O}$. It is obvious that multiplication of any matrix by a null matrix gives a product which is also a null matrix, as in
$$\mathbf{AO} = \mathbf{OA} = \mathbf{O}.$$
However, if the product $\mathbf{AB}$ of two matrices is a null matrix, it does not follow that either $\mathbf{A}$ or $\mathbf{B}$ must be a null matrix. A simple example will serve to illustrate this. Using
$$\mathbf{A} = \begin{bmatrix} 1 & -2 \\ -2 & 4 \end{bmatrix} \quad \text{and} \quad \mathbf{B} = \begin{bmatrix} 2 & 4 \\ 1 & 2 \end{bmatrix}$$
the reader will be able easily to confirm that the product $\mathbf{AB}$ is a null matrix. (Incidentally he will find that the product $\mathbf{BA}$ is not a null matrix.)

A *unit matrix* is defined as a diagonal matrix each of whose diagonal elements is unity. The corresponding conventional symbol is $\mathbf{E}$ (from the German *Einheit*, meaning unity). In certain circumstances it is useful to append a subscript denoting the dimension of the unit matrix, i.e. the number of diagonal entries it possesses. Another way of defining a unit matrix is to say that the element in the ith row and the jth column is equal to δ_{ij}. Here δ_{ij} is the so-called '*Kronecker*' *delta*, which is defined as being equal to unity when $i = j$, but zero otherwise. It is a quantity which we shall find useful in a number of different contexts.

It is easily seen that a matrix remains unaffected when it is premultiplied or post-multiplied by a unit matrix, as in

$$\mathbf{EA} = \mathbf{AE} = \mathbf{A}.$$

3.7. The inverse of a matrix

If the product $\mathbf{AB}$ of two square $n \times n$ matrices is equal to the unit matrix $\mathbf{E}_n$, then $\mathbf{B}$ is said to be the *inverse* of $\mathbf{A}$ and is written $\mathbf{A}^{-1}$. Thus by definition

$$\mathbf{AA}^{-1} = \mathbf{E}. \tag{3.11}$$

Pre-multiplication of both sides of eqn (3.11) by $\mathbf{A}^{-1}$ gives

$$\mathbf{A}^{-1}\mathbf{AA}^{-1} = \mathbf{A}^{-1}\mathbf{E} = \mathbf{A}^{-1} = \mathbf{EA}^{-1}.$$

Thus $$[\mathbf{A}^{-1}\mathbf{A}]\mathbf{A}^{-1} = \mathbf{EA}^{-1}$$

and therefore $$\mathbf{A}^{-1}\mathbf{A} = \mathbf{E}. \tag{3.12}$$

Eqns (3.11) and (3.12) show that a matrix and its inverse commute.

Not all matrices possess inverses. In order for an inverse to exist, a matrix must be square and must consist of elements whose determinant does not vanish. If the second condition stated is not satisfied, the matrix is said to be *singular*. Singular matrices have no inverses.

Provided that a matrix is non-singular, it is always possible to find its inverse by the straightforward method which we will now use to derive, for example, the inverse of the matrix

$$\begin{bmatrix} 1 & 3 & 5 \\ 2 & 1 & 3 \\ 5 & 4 & 3 \end{bmatrix}.$$

Writing

$$\begin{bmatrix} 1 & 3 & 5 \\ 2 & 1 & 3 \\ 5 & 4 & 3 \end{bmatrix} \begin{bmatrix} a & b & c \\ d & e & f \\ g & h & i \end{bmatrix} = \begin{bmatrix} 1 & 0 & 0 \\ 0 & 1 & 0 \\ 0 & 0 & 1 \end{bmatrix},$$

our task is to evaluate the nine elements $a, b, ..., i$. The ordinary rules of matrix multiplication give us the necessary nine equations. For the successive elements of the first row on the right-hand side we find

$$a + 3d + 5g = 1,$$
$$2a + d + 3g = 0,$$
$$5a + 4d + 3g = 0.$$

Hence, on solution, $a = -\frac{3}{11}$, $d = \frac{3}{11}$, and $g = \frac{1}{11}$. Similarly for the second and third rows,

$$b + 3e + 5h = 0,$$
$$2b + e + 3h = 1,$$
$$5b + 4e + 3h = 0,$$

whence $b = \frac{1}{3}$, $e = -\frac{2}{3}$, and $h = \frac{1}{3}$, and

$$c + 3f + 5i = 0,$$
$$2c + f + 3i = 0,$$
$$5c + 4f + 3i = 1,$$

whence $c = \frac{4}{33}$, $f = \frac{7}{33}$, and $i = -\frac{5}{33}$. The required inverse matrix is thus found to be

$$\begin{bmatrix} -\frac{3}{11} & \frac{1}{3} & \frac{4}{33} \\ \frac{3}{11} & -\frac{2}{3} & \frac{7}{33} \\ \frac{1}{11} & \frac{1}{3} & -\frac{5}{33} \end{bmatrix}.$$

This method, though always applicable, becomes rather cumbersome for large matrices. A neater method is available, but we need not discuss it here.

As a very simple illustration of a singular matrix we may take

$$\begin{bmatrix} 1 & 3 \\ 2 & 6 \end{bmatrix}.$$

We note that the determinant of its elements does indeed vanish. Let us attempt to find its inverse by the method used above. We accordingly write

$$\begin{bmatrix} 1 & 3 \\ 2 & 6 \end{bmatrix} \begin{bmatrix} a & b \\ c & d \end{bmatrix} = \begin{bmatrix} 1 & 0 \\ 0 & 1 \end{bmatrix}$$

and seek to evaluate a, b, c, and d. The equations we obtain for a and c are

$$a+3c = 1,$$
$$2a+6c = 0.$$

They are obviously incompatible with any finite values of a and c. Similarly for b and d we obtain

$$b+3d = 0,$$
$$2b+6d = 1,$$

which are likewise incompatible with any finite values of b and d. The matrix does not possess an inverse.

One type of inverse which will be of interest to us later is that of a matrix of diagonal-block form, it being assumed that none of the blocks is in itself singular. Here, as the reader will easily be able to convince himself, exactly the same diagonal-block form is preserved in the inverse, but each block is simply replaced by the inverse of the corresponding one in the original matrix.

Consider now the inverse of the product **AB** of two non-singular matrices. Pre-multiplying by $\mathbf{B^{-1}A^{-1}}$ we obtain

$$\mathbf{B^{-1}A^{-1}AB = B^{-1}[A^{-1}A]B = B^{-1}EB}$$
$$= \mathbf{B^{-1}B = E.}$$

This proves that $\mathbf{B^{-1}A^{-1}}$ is the inverse of **AB**, i.e.

$$[\mathbf{AB}]^{-1} = \mathbf{B^{-1}A^{-1}}. \tag{3.13}$$

The form of eqn (3.13) for the inverse of a product should be compared with the form of eqn (3.10) for the transpose of a product.

We conclude this section with a proof of the very interesting and important fact that the transpose of the inverse of a matrix is equal to the inverse of the transpose; or, in other words, that the operations of transposing and inverse-formation commute with one another. If **A** is any matrix possessing an inverse, we can write

$$[\mathbf{A^{-1}A}]^\dagger = \mathbf{E}^\dagger = \mathbf{E.}$$

Therefore
$$\mathbf{A}^\dagger[\mathbf{A^{-1}}]^\dagger = \mathbf{E.}$$

But by definition
$$\mathbf{A}^\dagger[\mathbf{A}^\dagger]^{-1} = \mathbf{E.}$$

Comparison of the last two equations shows that

$$[\mathbf{A}^{-1}]^{\dagger} = [\mathbf{A}^{\dagger}]^{-1}, \tag{3.14}$$

which completes the proof.

3.8. Orthogonal matrices

A matrix $\mathbf{A}$ is an *orthogonal matrix* if its inverse is identical with its transpose, i.e.

$$\mathbf{A}^{-1} = \mathbf{A}^{\dagger}. \tag{3.15}$$

We have already encountered one example in the array (1.40) which effected the transformation of normal coordinates Q into mass-weighted coordinates q according to eqns (1.37), (1.38), and (1.39); for we found that the array (1.54), which effected the inverse transformation, was merely the transpose of (1.40). When we come to generalize the methods of Chapter 1 we shall find that orthogonal matrices play a very important part in the treatment of molecular vibrations.

Every orthogonal matrix must of course be square, otherwise it could not possess an inverse. The types of relationship which must exist between the elements can be illustrated by considering, for simplicity, an orthogonal matrix with 3 rows and 3 columns. By virtue of (3.15) we may write

$$\begin{bmatrix} A_{11} & A_{12} & A_{13} \\ A_{21} & A_{22} & A_{23} \\ A_{31} & A_{32} & A_{33} \end{bmatrix} \begin{bmatrix} A_{11} & A_{21} & A_{31} \\ A_{12} & A_{22} & A_{32} \\ A_{13} & A_{23} & A_{33} \end{bmatrix} = \begin{bmatrix} 1 & 0 & 0 \\ 0 & 1 & 0 \\ 0 & 0 & 1 \end{bmatrix}. \tag{3.16}$$

Hence, by carrying out the multiplication we obtain

$$A_{11}^2 + A_{12}^2 + A_{13}^2 = 1,$$
$$A_{21}^2 + A_{22}^2 + A_{23}^2 = 1,$$
$$A_{31}^2 + A_{32}^2 + A_{33}^2 = 1,$$
$$A_{11}A_{21} + A_{12}A_{22} + A_{13}A_{23} = 0,$$
$$A_{11}A_{31} + A_{12}A_{32} + A_{13}A_{33} = 0, \text{ etc.}$$

It becomes clear that the elements of any orthogonal matrix must obey relations of the following kinds: (*a*) the sum of the squares of the elements of any row must be equal to unity; (*b*) for any two rows the sum of the products of corresponding elements must be equal to zero. These conclusions follow from eqn (3.16) which is of the form $AA^\dagger = E$. However, because the matrix A is orthogonal and because a matrix and its inverse commute, we may equally well write $A^\dagger A = E$. Exactly similar considerations then show that the elements of the columns of A must obey relations of exactly the same kind as (*a*) and (*b*) above for the rows.

Consider now the situation when an orthogonal matrix A transforms a set of quantities x (written as a single-column matrix X) into a set of new quantities x' (the single-column matrix X'). We may write this in the form

$$X' = AX.$$

Pre-multiplying both sides by A^{-1}, we obtain

$$X = A^{-1}X' \qquad (3.17)$$

and hence
$$X^\dagger = [X']^\dagger [A^{-1}]^\dagger = [X']^\dagger A. \qquad (3.18)$$

The product $X^\dagger X$, being the product of a single row and a single column, is clearly a 1×1 matrix, i.e. a simple number. Its value is $\sum_i x_i^2$, where x_i is an element of the matrix X (or $X^\dagger$) and the summation extends over all the elements. Using the values of X and $X^\dagger$ in eqns (3.17) and (3.18), we have

$$X^\dagger X = [X']^\dagger A\, A^{-1} X' = [X']^\dagger X'.$$

We have thus proved that the transformation effected by the orthogonal matrix is such that it leaves unchanged the sum of the squares of the elements; thus

$$\sum_i x_i^2 = \sum_i (x_i')^2.$$

A transformation of this kind is called an *orthogonal transformation*. (The reason for the choice of this name will become clear in the following chapter when we come to discuss vectors.) It also follows that, if a transformation is such that it leaves unaltered the sum of the squares of the elements of a single-column matrix, then the matrix which effects the transformation must be an orthogonal one.

3.9. Quadratic forms expressed by means of matrices

The kinetic energy of a molecular model, and its potential energy in the case of simple-harmonic vibrations, are quadratic forms, i.e. they are sums of terms of the second degree involving either a set of time-derivatives of displacement coordinates or a set of the coordinates themselves. For our purposes it is therefore of interest that such forms are expressible with the aid of suitable matrices.

First consider two different sets of quantities, the set $x_1, x_2, ..., x_i, ..., x_n$ and the set $y_1, y_2, ..., y_j, ..., y_n$. The general second-degree form (called a bilinear form) is $\sum_{i,j} A_{ij} x_i x_j$, where the double summation is extended over both sets. If we form the sets respectively into the single-column matrices $\mathbf{X}$ and $\mathbf{Y}$, it is easily seen that the general bilinear form can be expressed as $\mathbf{X}^\dagger \mathbf{A} \mathbf{Y}$, where $\mathbf{A}$ is a square $(n \times n)$ matrix with the element A_{ij} in the ith row and the jth column.

A quadratic form arises when the two sets become identical. Consider the product $\mathbf{X}^\dagger \mathbf{A} \mathbf{X}$, using the same matrix A as before. The coefficient of the term in $x_i x_j$ becomes $A_{ij} + A_{ji}$ when $i \neq j$, whereas when $i = j$ it is merely A_{ii}. Thus although a general quadratic form $\sum_{i,j} B_{ij} x_i x_j$ is expressible as a matrix product of the form $\mathbf{X}^\dagger \mathbf{B} \mathbf{X}$, the off-diagonal elements of $\mathbf{B}$ are not uniquely determined. For example the quadratic form $5x_1^2 + 6x_1 x_2 + 2x_2^2$ is expressible as

$$[x_1 \quad x_2] \begin{bmatrix} 5 & 3 \\ 3 & 2 \end{bmatrix} \begin{bmatrix} x_1 \\ x_2 \end{bmatrix}. \tag{3.19}$$

But it is equally well expressible by using

$$\begin{bmatrix} 5 & 4 \\ 2 & 2 \end{bmatrix} \quad \text{or} \quad \begin{bmatrix} 5 & 0 \\ 6 & 2 \end{bmatrix}$$

as the central square matrix. Indeed, in this simple case, any 2×2 matrix will serve, provided it has the diagonal elements 5 and 2, and that the sum of the off-diagonal elements is 6. The particular one chosen for use in (3.19) is unique in being symmetric. In order to express vibrational potential energies we shall use symmetric matrices whose elements are force constants.

In terms of the time derivatives $\dot{x}$ of displacement coordinates, the kinetic energy is the sum of terms of the type $\frac{1}{2}m_i\dot{x}_i^2$, where m_i is a nuclear mass. Because of the absence of 'cross terms' involving products of the type $\dot{x}_i\dot{x}_j$, the matrix **M** in the expression

$$2T = \dot{\mathbf{X}}^\dagger\mathbf{M}\dot{\mathbf{X}}$$

is uniquely determined as a *diagonal* matrix of the nuclear masses involved.

4   Vectors and matrices

4.1. Vectors in physical space

A DISPLACEMENT in physical space (for example, the displacement of a nucleus from its equilibrium position in a molecule) has both magnitude and direction. It is therefore a *vector*, and may be represented by a directed line (i.e. an arrow) drawn from a fixed origin O (representing zero displacement) to a point P in physical space. The magnitude of the vector is represented by the length $OP = p$ of the line, and the direction of the vector by the direction of the line. The meaning of 'direction' here includes the sense, the arrowhead showing that the line is drawn from O to P, rather than from P to O. In this book we shall use lower-case bold italic type to denote vectors: the one at present question in is thus p.

Multiplication of p by a simple numerical factor k gives the product kp, representable by a line from O drawn as before, but of length kp. If k is positive, the sense is the same as for p; but if k is negative, the sense is reversed. The sum of two vectors p and q is represented by the diagonal OR of the parallelogram $OPQR$, two of whose sides are formed by the lines OP and OQ which represent the two vectors p and q, as in Fig. 4.1. The magnitude $OR = r$ of the sum will of course depend, not only upon the magnitudes of p and q, but also upon the angle between them. From the geometry of Fig. 4.1 it is clear that in vector algebra, as in ordinary algebra,

$$p+q = q+p,$$

and that, if k is a scalar factor,

$$k(p+q) = kp+kq. \tag{4.1}$$

A little thought about the geometry of vector summation will also show that, for three vectors f, g, and h,

$$f+g+h = (f+g)+h = f+(g+h) = (f+h)+g.$$

51

We see that any vector in physical space may be expressed in a variety of ways as a linear combination (vector sum) of a set of three vectors. The only restriction on the choice of the three reference vectors is that they must not be coplanar. Thus suppose that we first choose arbitrarily a vector

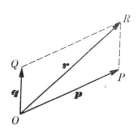

FIG. 4.1.   Sum of two vectors.

a. By multiplying this by suitable scalar factors we can generate from it all vectors which lie along the line of **a.** Now let us arbitrarily choose a second vector **b**, which is not collinear with **a.** Linear combinations of the type $k_a \mathbf{a} + k_b \mathbf{b}$ will generate all vectors lying in the plane containing **a** and **b.** Now in order to be able to generate all vectors lying in the three-dimensional space, it is sufficient to choose a third vector **c** which is not coplanar with **a** and **b.** Any general vector **p** can then be written as a linear combination of the form

$$\mathbf{p} = k_a \mathbf{a} + k_b \mathbf{b} + k_c \mathbf{c}, \qquad (4.2)$$

where the ks are appropriate scalar factors. This is expressed by saying that the three chosen vectors form a *basis* for the vector space, or alternatively that they *span* the space.

The directions (including the senses) of the three basis vectors may be used as axes for an associated coordinate system. The numbers k_a, k_b, k_c in eqn (4.2) thus become the coordinates of the end-point P of the line OP which represents the general vector **p.** They may also be regarded as the components of **p** in the particular chosen basis, or (more correctly) as the magnitudes of these components. We shall refer to them indiscriminately as the coordinates or the components of the vector in the relevant basis.

The choice of the three basis vectors is clearly by no means unique, being subject only to the restriction that they must not be coplanar. This is equivalent to saying that no one of them must be expressible as a linear combination of the other two, i.e. that they must be linearly independent. The basis vectors may be inclined at any non-zero angle to one another, in which case

52

the axes of the associated coordinate system will be similarly inclined. Also they may have any non-zero magnitudes; these merely determine the scales in which the coordinates or the components of a vector are expressed.

A very useful choice of basis (and the kind with which we shall be almost exclusively concerned) is that in which the basis vectors are all at right angles (i.e. orthogonal) to one another, and all of the same unit length. A set of vectors of this kind is said to constitute an *orthonormal basis*. The associated coordinate system is then of the familiar Cartesian type, with the same scale for all three axes. Such a system may, of course, be chosen in an unlimited number of ways, according to its spatial orientation. It is customary to represent mutually orthogonal unit vectors by the symbol e with suitable subscripts. For an ordinary Cartesian system in physical space we should therefore have the basis vectors e_x, e_y, and e_z, corresponding to the coordinates x, y, and z. However, for subsequent extension of the concept of vector space, it is more convenient to call the Cartesian co-ordinates x_1, x_2, and x_3 (rather than x, y, and z). Accordingly the orthonormal basis of ordinary space is made up of the corre-sponding vectors e_1, e_2, e_3, and a general vector p is expressible in the form

$$p = x_1 e_1 + x_2 e_2 + x_3 e_3. \qquad (4.3)$$

We shall adopt this notation from now onwards.

Once a basis has been chosen, and the coordinates of a general vector p defined in relation to it, we may use these coordinates as elements of a single-column matrix $\mathbf{P}$. This matrix will then represent the vector in the chosen basis. Alternatively we could have formed a corresponding single-row matrix to represent the vector. In order to avoid confusion we shall write such a row as $\mathbf{P}^{\dagger}$. We thus lay down the convention that a matrix of the coordinates of a vector shall always be understood to be a *column matrix*, and that if at any time we wish to refer to the corresponding row matrix, we shall designate it as the transpose of the column.

Let us now see, with this convention, how vector algebraic expressions can be re-written in matrix form. Consider first the addition of two vectors, which is normally written as the simple equation

$$p+q = r. \tag{4.4}$$

If the coordinates of p are x_{1p}, x_{2p}, x_{3p} and those of q are x_{1q}, x_{2q}, x_{3q}, it is clear that the coordinates of their vector sum r will be given by

$$x_{1r} = x_{1p}+x_{1q}, \quad x_{2r} = x_{2p}+x_{2q}, \quad x_{3r} = x_{3p}+x_{3q}.$$

Thus, using column matrices, we can write

$$\mathbf{P}+\mathbf{Q} = \mathbf{R}. \tag{4.5}$$

We note that this matrix form is exactly analogous to the form (4.4) in vector algebra.

The vector kp, obtained by multiplying p by the scalar factor k, is represented by a column matrix, each element of which is k times greater than the corresponding element of the matrix $\mathbf{P}$. We may write this as $k\mathbf{P}$, and consequently extend the analogies between vector and matrix algebraic expressions. For example, if p and q be two vectors, we found above (see eqn (4.1)) that

$$k(p+q) = kp+kq.$$

In matrix form this becomes

$$k(\mathbf{P}+\mathbf{Q}) = k\mathbf{P}+k\mathbf{Q}.$$

4.2. The scalar product of two vectors and its expression in matrix form

In the preceding section we have dealt briefly with the vector analogues of addition and of multiplication by a scalar factor— operations which are familiar when applied to ordinary numbers. We now complete the picture by defining for two vectors the analogue of the operation of product formation, and we do this in such a way as to ensure that the product (like the product of two numbers) is a scalar quantity. The so-called *scalar product* of two vectors p and q in ordinary space is conventionally written as $p \cdot q$, and is defined as

$$p \cdot q = pq \cos \theta, \tag{4.6}$$

where p and q are the respective magnitudes of the vectors and θ is the angle between them. Because of the convention of writing the scalar product by placing a dot between the two factors, it is often known as their *dot product*. We shall adopt this convenient name. Generally speaking, some such special designation is required to distinguish the dot product of two vectors from another quite different quantity known as their vector product. In this book, however, we shall not have any occasion to make use of vector products, and so need not consider their nature.

Let us now examine how the dot product fits into the algebra of vectors. First of all, from the definition in eqn (4.6), it is clear that $p.q = q.p$, i.e. that dot product formation is commutative. Consider the dot product of a vector p and another vector which is the sum of the two vectors r and s. It is equal to the ordinary product of the magnitude of p and the magnitude of the projection of $(r+s)$ upon p. Now from the geometry of vector addition it follows that the magnitude of this projection is simply the sum of the magnitudes of the projections of r and s individually. Therefore

$$p.(r+s) = (p.r)+(p.s).$$

By simple extension we find that

$$(p+q).(r+s) = (p.r)+(p.s)+(q.r)+(q.s)$$

and that, in general, the algebraic rules for dot product formation are quite similar to those of ordinary algebra.

In connection with dot products, there are two special cases that deserve careful note. First, if the two vectors are at right angles (i.e. orthogonal) to one another, then $\cos \theta = \cos 90° = 0$. The dot product therefore vanishes. Secondly, in the dot product of a vector with itself $\theta = 0$ and $\cos \theta = 1$. In this case, therefore, the dot product is simply the square of the magnitude of the vector, as in

$$p.p = p^2.$$

Consider now a Cartesian system whose basis consists of the three unit vectors e_1, e_2, e_3. Because they are mutually orthogonal, we have

$$e_1.e_2 = e_2.e_3 = e_3.e_1 = 0. \tag{4.7}$$

The dot product of any one of them with itself must be unity, so

$$e_1.e_1 = e_2.e_2 = e_3.e_3 = 1. \tag{4.8}$$

The contents of eqns (4.7) and (4.8) may be conveniently summarized by writing

$$e_i.e_j = \delta_{ij}, \tag{4.9}$$

where the quantity on the right-hand side is the Kronecker delta which we previously used in Section 3.6. In this basis let any two vectors be expressible as follows:

$$\boldsymbol{p} = x_{1p}\,\boldsymbol{e}_1 + x_{2p}\,\boldsymbol{e}_2 + x_{3p}\,\boldsymbol{e}_3,$$
$$\boldsymbol{q} = x_{1q}\,\boldsymbol{e}_1 + x_{2q}\,\boldsymbol{e}_2 + x_{3q}\,\boldsymbol{e}_3.$$

Their dot product is then

$$\boldsymbol{p}\cdot\boldsymbol{q} = (x_{1p}\,\boldsymbol{e}_1 + x_{2p}\,\boldsymbol{e}_2 + x_{3p}\,\boldsymbol{e}_3)\cdot(x_{1q}\,\boldsymbol{e}_1 + x_{2q}\,\boldsymbol{e}_2 + x_{3q}\,\boldsymbol{e}_3)$$
$$= x_{1p}\,x_{1q}(\boldsymbol{e}_1\cdot\boldsymbol{e}_1) + x_{2p}\,x_{2q}(\boldsymbol{e}_2\cdot\boldsymbol{e}_2) + x_{3p}\,x_{3q}(\boldsymbol{e}_3\cdot\boldsymbol{e}_3) +$$
$$+ (x_{1p}\,x_{2q} + x_{1q}\,x_{2p})(\boldsymbol{e}_1\cdot\boldsymbol{e}_2) + (x_{2p}\,x_{3q} + x_{2q}\,x_{3p})(\boldsymbol{e}_2\cdot\boldsymbol{e}_3) +$$
$$+ (x_{3p}\,x_{1q} + x_{3q}\,x_{1p})(\boldsymbol{e}_3\cdot\boldsymbol{e}_1).$$

By virtue of eqn (4.9) the last three terms vanish and the expression for the dot product reduces to

$$\boldsymbol{p}\cdot\boldsymbol{q} = x_{1p}\,x_{1q} + x_{2p}\,x_{2q} + x_{3p}\,x_{3q}. \tag{4.10}$$

We can now write down the condition under which two vectors $\boldsymbol{p}$ and $\boldsymbol{q}$ shall be mutually orthogonal. It is that their dot product must vanish, i.e. that

$$x_{1p}\,x_{1q} + x_{2p}\,x_{2q} + x_{3p}\,x_{3q} = 0. \tag{4.11}$$

By making $\boldsymbol{q}$ identical with $\boldsymbol{p}$, we find from eqn (4.10) the dot product of $\boldsymbol{p}$ with itself:

$$\boldsymbol{p}\cdot\boldsymbol{p} = x_{1p}^2 + x_{2p}^2 + x_{3p}^2. \tag{4.12}$$

This result is obviously in agreement with our earlier conclusion that the dot product of a vector with itself is simply the square of its magnitude.

It remains to discover how dot product formation can be expressed in matrix form. In this connection we recall our convention that the matrix representing a vector should always be a single *column* of its coordinates, and not a single row. Adhering to this convention, we see that the dot product $\boldsymbol{p}\cdot\boldsymbol{q}$ can be written in the matrix form $\mathbf{P}^\dagger\mathbf{Q}$. For

$$\mathbf{P}^\dagger\mathbf{Q} = \begin{bmatrix} x_{1p} & x_{2p} & x_{3p} \end{bmatrix}\begin{bmatrix} x_{1q} \\ x_{2q} \\ x_{3q} \end{bmatrix}$$
$$= x_{1p}\,x_{1q} + x_{2p}\,x_{2q} + x_{3p}\,x_{3q}$$
$$= \boldsymbol{p}\cdot\boldsymbol{q}. \tag{4.13}$$

Despite the fact that we thus use a matrix product as the equivalent of a dot product, we encounter no difficulty from the fact that the latter is commutative. Thus

$$q \cdot p = Q^\dagger P = \begin{bmatrix} x_{1q} & x_{2q} & x_{3q} \end{bmatrix} \begin{bmatrix} x_{1p} \\ x_{2p} \\ x_{3p} \end{bmatrix}$$

$$= x_{1q} x_{1p} + x_{2q} x_{2p} + x_{3q} x_{3p}.$$

In short,
$$P^\dagger Q = Q^\dagger P = [P^\dagger Q]^\dagger.$$

This is so because the matrix product is in each case a 1×1 matrix, i.e. a simple number.

The similarity between vector algebra and matrix algebra goes even further than this. Consider, for example, a dot product of the type

$$(p+q) \cdot (r+s) = (p \cdot r) + (p \cdot s) + (q \cdot r) + (q \cdot s)$$
$$= (r \cdot p) + (s \cdot p) + (r \cdot q) + (s \cdot q)$$
$$= (r+s) \cdot (p+q).$$

In matrix form this becomes

$$[P^\dagger + Q^\dagger][R+S] = P^\dagger R + P^\dagger S + Q^\dagger R + Q^\dagger S$$
$$= R^\dagger P + S^\dagger P + R^\dagger Q + S^\dagger Q$$
$$= [R^\dagger + S^\dagger][P+Q].$$

The continued correspondence between the two kinds of algebra is evident.

4.3. Vectors in configuration space

In the two preceding sections our consideration of vectors has been deliberately confined to the familiar three-dimensional space of our physical experience. Accordingly we were able throughout to visualise the vectors clearly and could, if required, have constructed three-dimensional models to represent them. We now proceed to consider, not merely the spatial displacement of one nucleus, but the simultaneous displacements of the N nuclei which a molecule contains. This means taking into account a total of $3N$ Cartesian displacement coordinates, say x_1, x_2, and x_3 for the first nucleus, x_4, x_5, and x_6 for the second, and so on. In order to do this, we regard the whole set of co-ordinates as defining a vector in a $3N$-dimensional space which we call *configuration space*.

Obviously this configuration space and the concept of a vector in it are pure mathematical abstractions. Our minds have no experience and no powers to enable us to form any mental picture of them; our hands, no matter how skilful, can never construct any physical model to represent them. If the abstractions are to be practically useful, we must endow them with mathematically formulated properties; naturally we must do this in such a way that our familiar physical space and the familiar vectors in it are seen as special cases of the generalized concepts.

The three displacement coordinates of any one nucleus are related to its own individual basis, which consists of three mutually orthogonal unit vectors. In thinking of the molecular configuration as a whole, we take all the $3N$ unit vectors (three for each of the N nuclei) and form them conceptually into a single orthonormal basis for configuration space. Clearly the term 'orthonormal' is here being used in a new and generalized sense, and in order to clarify it we must decide what we mean, first by the magnitude of a vector in configuration space, and secondly by the orthogonality of two such vectors. Both meanings follow from the adoption of a satisfactory definition of a dot product. Let x_{ip} (where i may have any value from 1 to $3N$) be the co-ordinates of a general vector $\boldsymbol{p}$. We may utilize them as the elements of a single-column matrix $\mathbf{P}$. If $\mathbf{Q}$ be the corresponding single-column matrix for another vector $\boldsymbol{q}$ with the coordinates x_{iq}, we may satisfactorily define the dot product of $\boldsymbol{p}$ and $\boldsymbol{q}$ as

$$\boldsymbol{p} \cdot \boldsymbol{q} = \mathbf{P}^\dagger \mathbf{Q} = \sum_i x_{ip} x_{iq}.$$

The analogy with the three-dimensional case is obvious—see eqn (4.13).

We are now in a position to generalize the meaning of ortho-gonality by stating (compare with eqn (4.11)) that $\boldsymbol{p}$ and $\boldsymbol{q}$ are mutually orthogonal if

$$\boldsymbol{p} \cdot \boldsymbol{q} = \sum_i x_{ip} x_{iq} = 0. \tag{4.14}$$

We can also define the magnitude p of a vector $\boldsymbol{p}$ by the equation

$$p^2 = \boldsymbol{p} \cdot \boldsymbol{p} = \sum_i x_{ip}^2. \tag{4.15}$$

The meaning of the description 'orthonormal', as applied to our basis of configuration space, is now clear. It is fully expressed in the concise equation

$$\boldsymbol{e}_i \cdot \boldsymbol{e}_j = \delta_{ij}, \tag{4.16}$$

where either i or j may have any value from 1 to $3N$. This general equation clearly includes eqn (4.9) as a special case.

Once a basis has been chosen, configuration space is such that a single point P in it represents the whole configuration of a molecule (or, more precisely, the change of configuration relative to that at equilibrium). The 'line' drawn from the origin to P represents the configuration vector $\boldsymbol{p}$ which is characterized by a set of $3N$ coordinates or components. The values of these for a certain configuration will of course depend upon the particular chosen basis. The space and the vectors in it are mathematical abstractions; but the algebra of the vectors, and likewise that of the corresponding matrices, follow the same lines as for ordinary vectors (see Sections 4.1 and 4.2).

4.4. Basis vectors as matrix elements

We shall soon wish to discuss the important effects of changes in the choice of basis for configuration space. Such changes are, of course, changes of sets of basis vectors, and it will be convenient to be able to express them in matrix form. In all our discussions of matrices up to this point, we have tacitly assumed that the elements composing them were scalar quantities, but there is nothing improper in using non-scalar quantities as matrix elements, provided always that they are quantities of such a kind that it is meaningful to subject them to the operations of the matrix algebra in which they become involved.

The transformation of a set $x_1, x_2, ..., x_n$ of scalar quantities into a set $x_1', x_2', ..., x_n'$ of linear combinations can, as we have seen, be expressed in the matrix form

$$\mathbf{X}' = \mathbf{AX}.$$

Here **A** is a suitable square matrix, and the form becomes possible by using the sets of scalars $x_1, x_2, ..., x_n$, and $x'_1, x'_2, ..., x'_n$ respectively to form the two single-column matrices **X** and **X′**. A change from the set $b_1, b_2, ..., b_n$ of basis vectors to the set $b'_1, b'_2, ..., b'_n$ of linear combinations (i.e. vector sums), in the change to a new basis, can likewise be expressed in matrix form, if we use the original set of basis vectors as the elements of a single-column or single-row matrix **B** and similarly the new set as the elements of a corresponding matrix **B′**.

The reader will recall that for sets of scalar quantities we adopted the convention that the matrix **X** formed from them should always be a single *column*, and that the corresponding row should always be written as $\mathbf{X}^\dagger$. Now, for sets of basis vectors, we shall adopt just the opposite convention. We accordingly lay down that, when a set of basis vectors is used to form a matrix **B**, the matrix shall always be understood to be a single *row*, and that the corresponding column shall always be written as $\mathbf{B}^\dagger$.

This decision, though at first sight it may well seem capricious, is in fact made in the interest of consistency between the linear transformation of the basis vectors considered individually and the same transformation of them as a set. This is perhaps best seen from a simple illustration. (The situation is exactly similar in the general case, but for those less mathematically experienced its essential features tend perhaps to be obscured by the extra complexity of generalized notation.) Consider, then, the vectors e_1, e_2, e_3 forming an orthonormal basis of three-dimensional space. Their coordinates are respectively $(1, 0, 0)$, $(0, 1, 0)$, and $(0, 0, 1)$. In each case the scalar coordinates can be used (in conformity with our convention) to form a single-column matrix, which then represents the base vector in question. Let **A** be the 3×3 matrix which effects the linear transformation in which we are interested. In the case of e_1, for example, we then have

$$\begin{bmatrix} A_{11} & A_{12} & A_{13} \\ A_{21} & A_{22} & A_{23} \\ A_{31} & A_{32} & A_{33} \end{bmatrix} \begin{bmatrix} 1 \\ 0 \\ 0 \end{bmatrix} = \begin{bmatrix} A_{11} \\ A_{21} \\ A_{31} \end{bmatrix}.$$

Thus the transformed basis vector e'_1 has the coordinates A_{11}, A_{21},

A_{31} and may be written

$$e'_1 = A_{11} e_1 + A_{21} e_2 + A_{31} e_3. \qquad (4.17)$$

Similarly, for the other two basis vectors we obtain

$$e'_2 = A_{12} e_1 + A_{22} e_2 + A_{32} e_3,$$

and

$$e'_3 = A_{13} e_1 + A_{23} e_2 + A_{33} e_3. \qquad (4.18)$$

To be consistent, we must obtain the same expressions for e'_1, e'_2, e'_3 when we consider the basis vectors as a set and use them as the elements of a matrix. The desired result is indeed achieved if the matrix so formed is a single *row*. Thus the reader will easily see that the equation

$$[e'_1 \quad e'_2 \quad e'_3] = [e_1 \quad e_2 \quad e_3]\begin{bmatrix} A_{11} & A_{12} & A_{13} \\ A_{21} & A_{22} & A_{23} \\ A_{31} & A_{32} & A_{33} \end{bmatrix} \qquad (4.19)$$

is in agreement with (4.17) and (4.18), whereas the equation

$$\begin{bmatrix} e'_1 \\ e'_2 \\ e'_3 \end{bmatrix} = \begin{bmatrix} A_{11} & A_{12} & A_{13} \\ A_{21} & A_{22} & A_{23} \\ A_{31} & A_{32} & A_{33} \end{bmatrix}\begin{bmatrix} e_1 \\ e_2 \\ e_3 \end{bmatrix}$$

is not.

4.5. Change of basis and its effect on the coordinates of a fixed vector

In dealing with the problem of the vibrations of a molecule we shall from time to time find it advantageous to change from one basis to another or, otherwise expressed, from one set of co-ordinates to another. We wish now to examine how the concomitant changes of basis and coordinates are related to one another. Consider the change from basis I to basis II in n-dimensional space, the sets of basis vectors forming the elements of the respective single-row matrices $\mathbf{B}^{\mathrm{I}}$ and $\mathbf{B}^{\mathrm{II}}$. If $\mathbf{A}$ is the square matrix concerned, we may represent the change diagrammatically in the form

$$\begin{matrix} \xleftarrow{\quad n \quad} & \xleftarrow{\quad n \quad} & \xleftarrow{\quad n \quad} \\ [\quad \mathbf{B}^{\mathrm{II}} \quad] = [\quad \mathbf{B}^{\mathrm{I}} \quad]\begin{bmatrix} & & \\ & \mathbf{A} & \\ & & \end{bmatrix} n. & & \end{matrix} \qquad (4.20)$$

In basis I let the coordinates of a certain fixed point P, i.e. of a certain fixed vector $\boldsymbol{p}$, form the elements of the single-column matrix $\mathbf{P^I}$. After the change to basis II, let the consequently changed coordinates of the same fixed point or vector form the changed single-column matrix $\mathbf{P^{II}}$. Then we may write

$$\boldsymbol{p} = \begin{bmatrix} \mathbf{B^I} \end{bmatrix} \begin{bmatrix} \mathbf{P^I} \end{bmatrix} n, \qquad (4.21)$$

and likewise

$$\boldsymbol{p} = \begin{bmatrix} \mathbf{B^{II}} \end{bmatrix} \begin{bmatrix} \mathbf{P^{II}} \end{bmatrix} n. \qquad (4.22)$$

Substituting the expression for $\mathbf{B^{II}}$ from eqn (4.20) into eqn (4.22) we obtain

$$\boldsymbol{p} = \begin{bmatrix} \mathbf{B^I} \end{bmatrix} \begin{bmatrix} \mathbf{A} \end{bmatrix} \begin{bmatrix} \mathbf{P^{II}} \end{bmatrix}. \qquad (4.23)$$

Comparison of eqn (4.23) with eqn (4.21) shows that

$$\begin{bmatrix} \mathbf{P^I} \end{bmatrix} = \begin{bmatrix} \mathbf{A} \end{bmatrix} \begin{bmatrix} \mathbf{P^{II}} \end{bmatrix},$$

whence, premultiplying both sides by $\mathbf{A^{-1}}$, we obtain the desired relationship between the coordinates in basis II and those in basis I:

$$\begin{bmatrix} \mathbf{P^{II}} \end{bmatrix} = \begin{bmatrix} \mathbf{A^{-1}} \end{bmatrix} \begin{bmatrix} \mathbf{P^I} \end{bmatrix}. \qquad (4.24)$$

In order to make possible a direct comparison with the corresponding relationship between basis II and basis I, we transpose both sides of eqn (4.24), thus obtaining

$$[\quad (\mathbf{P^{II}})^\dagger \quad] = [\quad (\mathbf{P^I})^\dagger \quad] \begin{bmatrix} & & \\ & (\mathbf{A}^{-1})^\dagger & \\ & & \end{bmatrix}. \qquad (4.25)$$

This now has the same form as eqn (4.20), which expresses the basis change.

We see that the relation between the associated changes of basis and of coordinates is in general a complicated one. However, there is one special case in which it is extremely simple: when the matrix $\mathbf{A}$ is orthogonal, then $\mathbf{A}^{-1} = \mathbf{A}^\dagger$ and so $[\mathbf{A}^{-1}]^\dagger = \mathbf{A}$. We thus prove that when the matrix effecting a change of basis is an orthogonal one, the coordinates of a fixed vector are changed in precisely the same way as are the basis vectors, i.e. if some chosen original basis vector is transformed into a certain linear combination of all the original basis vectors, then the corresponding original coordinate of any fixed vector is consequently transformed into precisely the same linear combination of all the original coordinates of that vector. We shall have occasion later to invoke this result.

4.6. Matrices and symmetry operations

Both in classical mechanics and in quantum mechanics, the treatment of molecular vibrations is greatly simplified by taking molecular symmetry into account, and in due course we shall have a great deal to say about so-called symmetry operations. By their nature as reflections in planes and rotations around axes, etc., they are all operations which, when performed upon a vector, change its coordinates while leaving its magnitude unaffected. Clearly, therefore, symmetry operations can be represented by matrices, and indeed by *orthogonal* matrices.

The effect of performing the symmetry operation R upon the vector $\boldsymbol{p}$ to produce the new vector $\boldsymbol{p}'$ may be conventionally expressed either in the form

$$\boldsymbol{p} \xrightarrow{\ R\ } \boldsymbol{p}',$$

or in the form

$$\mathscr{R}\boldsymbol{p} = \boldsymbol{p}',$$

where $\mathscr{R}$ is the *operator* of the operation R. For the reader who is not already familiar with operators, we may quote $\sqrt{}$ and $\partial/\partial x$ as examples of operators applicable to ordinary scalars. They correspond, of course, to the respective operations of taking the square root and the partial derivative with respect to x. In themselves they have no values; but when applied, for instance, to the quantity $x^2 y^2$, they produce the results $\pm xy$ and $2xy^2$ respectively. So it is with all operators. In themselves they are not *quantities* of any kind, but when written on the left of a quantity they signify the performance of a certain operation upon it.

If $\mathscr{R}$ is a symmetry operator and k is a scalar factor, it is easily seen that

$$\mathscr{R}(k\boldsymbol{p}) = k\mathscr{R}\boldsymbol{p} = k\boldsymbol{p}'. \qquad (4.26)$$

Also if $\boldsymbol{r}$ is the sum of two vectors $\boldsymbol{p}$ and $\boldsymbol{q}$, we see that

$$\mathscr{R}\boldsymbol{r} = \mathscr{R}(\boldsymbol{p}+\boldsymbol{q}) = \mathscr{R}\boldsymbol{p}+\mathscr{R}\boldsymbol{q}. \qquad (4.27)$$

By virtue of eqns (4.26) and (4.27) the operator $\mathscr{R}$ is called a *linear operator*.

Let us now consider in detail the performance of a certain symmetry operation R upon a vector $\boldsymbol{p}$ in ordinary physical space. We choose as basis a set of three mutually orthogonal unit vectors $\boldsymbol{e}_1$, $\boldsymbol{e}_2$, and $\boldsymbol{e}_3$. In this basis let x_1, x_2, x_3 be the coordinates of the vector $\boldsymbol{p}$. (The coordinate system is of course of the familiar Cartesian type, the three coordinates corresponding to the customary x, y, and z.) Under the symmetry operation R, the vector $\boldsymbol{p}$ is transformed into another vector $\mathscr{R}\boldsymbol{p} = \boldsymbol{p}'$, whose coordinates (in the same basis as for $\boldsymbol{p}$) we shall write as x_1', x_2', and x_3'. Then, since $\mathscr{R}$ is a linear operator,

$$\boldsymbol{p}' = \mathscr{R}\boldsymbol{p} = \mathscr{R}(x_1\boldsymbol{e}_1+x_2\boldsymbol{e}_2+x_3\boldsymbol{e}_3)$$

$$= x_1\mathscr{R}\boldsymbol{e}_1+x_2\mathscr{R}\boldsymbol{e}_2+x_3\mathscr{R}\boldsymbol{e}_3. \qquad (4.28)$$

Now using the coordinates of $\boldsymbol{p}$ and $\boldsymbol{p}'$ respectively as elements for the single-column matrices $\mathbf{P}$ and $\mathbf{P}'$ we can express the effect of the operation in the matrix form $\mathbf{P}' = \mathbf{RP}$, where $\mathbf{R}$

is the matrix representing R. Written out in full, this becomes

$$\begin{bmatrix} x'_1 \\ x'_2 \\ x'_3 \end{bmatrix} = \begin{bmatrix} R_{11} & R_{12} & R_{13} \\ R_{21} & R_{22} & R_{23} \\ R_{31} & R_{32} & R_{33} \end{bmatrix} \begin{bmatrix} x_1 \\ x_2 \\ x_3 \end{bmatrix}.$$

Hence

$$\boldsymbol{p}' = x'_1 \boldsymbol{e}_1 + x'_2 \boldsymbol{e}_2 + x'_3 \boldsymbol{e}_3$$
$$= (R_{11} x_1 + R_{12} x_2 + R_{13} x_3) \boldsymbol{e}_1 + (R_{21} x_1 + R_{22} x_2 + R_{23} x_3) \boldsymbol{e}_2 +$$
$$+ (R_{31} x_1 + R_{32} x_2 + R_{33} x_3) \boldsymbol{e}_3.$$

On rearranging terms this gives

$$\boldsymbol{p}' = x_1(R_{11} \boldsymbol{e}_1 + R_{21} \boldsymbol{e}_2 + R_{31} \boldsymbol{e}_3) + x_2(R_{12} \boldsymbol{e}_1 + R_{22} \boldsymbol{e}_2 + R_{32} \boldsymbol{e}_3) +$$
$$+ x_3(R_{13} \boldsymbol{e}_1 + R_{23} \boldsymbol{e}_2 + R_{33} \boldsymbol{e}_3). \qquad (4.29)$$

Comparing with eqn (4.28) and equating coefficients of x_1, x_2, and x_3, we obtain

$$\left. \begin{array}{l} \mathscr{R}\boldsymbol{e}_1 = R_{11} \boldsymbol{e}_1 + R_{21} \boldsymbol{e}_2 + R_{31} \boldsymbol{e}_3 \\ \mathscr{R}\boldsymbol{e}_2 = R_{12} \boldsymbol{e}_1 + R_{22} \boldsymbol{e}_2 + R_{32} \boldsymbol{e}_3 \\ \mathscr{R}\boldsymbol{e}_3 = R_{13} \boldsymbol{e}_1 + R_{23} \boldsymbol{e}_2 + R_{33} \boldsymbol{e}_3 \end{array} \right\}. \qquad (4.30)$$

The array of the coefficients of the basis vectors on the right-hand sides of eqns (4.30) is seen to be the transpose of the matrix **R** which we have taken to represent the symmetry operation. If, therefore, we use the set of unit vectors as the elements of a *single-row* matrix **B**, the effect of **R** upon them will be expressible as **BR**. This is to be contrasted with the form **RP** which expresses the effect of R upon the vector $\boldsymbol{p}$, when this is written as a *single-column* matrix **P**.

4.7. An example of an operation and a matrix representing it

As a simple illustration, take the case when R is a rotation through the angle θ about the x_3 axis. Under certain circumstances and with appropriate values of θ, this may correspond to a molecular symmetry operation. We now choose an orthonormal basis and seek the matrix **R** which represents R. It is at once obvious that rotation of a general vector around the x_3 axis leaves its x_3 coordinate unaffected, but changes its coordinates

x_1 and x_2 to different values x_1' and x_2'. Fig. 4.2 shows a projection of the vector upon the $x_1 x_2$ plane. The x_3 axis is normal to the plane of the figure and directed towards the reader. The lines OP and OP' of length l represent the projections of the vector

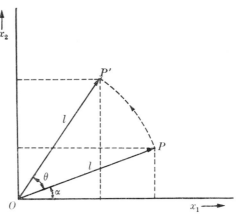

FIG. 4.2. Rotation of a vector.

before and after the operation respectively. The angle between them is the angle of rotation θ. Let α be the angle between OP and the x_1 axis. Then we have

$$x_1 = l \cos \alpha \Big\}$$
and
$$x_2 = l \sin \alpha \Big\}. \qquad (4.31)$$

Under the operation R,

$$x_1 \xrightarrow{R} x_1' = l \cos(\alpha+\theta) = l \cos \alpha \cos \theta - l \sin \alpha \sin \theta \qquad (4.32)$$

and

$$x_2 \xrightarrow{R} x_2' = l \sin(\alpha+\theta) = l \sin \alpha \cos \theta + l \cos \alpha \sin \theta. \qquad (4.33)$$

Substituting from eqn (4.31) into eqns (4.32) and (4.33), we obtain
$$x_1' = \cos \theta \, x_1 - \sin \theta \, x_2,$$
$$x_2' = \sin \theta \, x_1 + \cos \theta \, x_2.$$

Remembering that $x_3' = x_3$, we may now write the effect of the operation in the matrix form

$$\begin{bmatrix} x_1' \\ x_2' \\ x_3' \end{bmatrix} = \begin{bmatrix} \cos \theta & -\sin \theta & 0 \\ \sin \theta & \cos \theta & 0 \\ 0 & 0 & 1 \end{bmatrix} \begin{bmatrix} x_1 \\ x_2 \\ x_3 \end{bmatrix}. \qquad (4.34)$$

The square matrix on the right-hand side of the equation is the desired matrix **R** which represents the operation of anticlockwise rotation through θ about the x_3 axis. The reader will easily be able to verify that it is indeed an orthogonal matrix, satisfying the criteria set out in Section 3.8.

For the sake of completeness we shall also derive this matrix by considering the effect of the operation upon the unit base vectors, e_1, e_2, and e_3. From Fig. 4.3, which shows the $x_1 x_2$

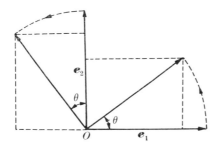

FIG. 4.3.   Rotation of unit basis vectors.

plane, it is seen that R changes e_1 into a vector whose resolved parts in the x_1 and x_2 directions are respectively $\cos\theta$ and $\sin\theta$. Thus

$$e_1 \xrightarrow{\ R\ } e_1' = \cos\theta\, e_1 + \sin\theta\, e_2. \tag{4.35}$$

Similarly

$$e_2 \xrightarrow{\ R\ } e_2' = -\sin\theta\, e_1 + \cos\theta\, e_2. \tag{4.36}$$

The basis vector e_3 is unaffected, i.e.

$$e_3 \xrightarrow{\ R\ } e_3' = e_3. \tag{4.37}$$

We saw in the preceding section that the matrix **R** which represents the operation is the matrix which transforms the set of basis vectors when these are written as a single-row matrix. The result of eqns (4.35), (4.36), and (4.37) can now be expressed in the form

$$\begin{bmatrix} e_1' & e_2' & e_3' \end{bmatrix} = \begin{bmatrix} e_1 & e_2 & e_3 \end{bmatrix} \begin{bmatrix} \cos\theta & -\sin\theta & 0 \\ \sin\theta & \cos\theta & 0 \\ 0 & 0 & 1 \end{bmatrix}.$$

The square matrix is, of course, identical with that found in eqn (4.34).

4.8. The effect of a change of basis upon a matrix which represents a certain operation

Just as the coordinates of a fixed vector depend upon the choice of the basis of the vector space, so also do the elements of a matrix which represents a particular operation. We shall now discuss the important question of the nature of this dependence. Let $\mathbf{A}$ be the matrix which effects the change from basis I to basis II, i.e. which transforms the row matrix of the basis vectors of basis I into the corresponding row matrix of basis II. Let $\mathbf{P}^{I}$ and $\mathbf{P}^{II}$ be the column matrices of one and the same vector in the bases I and II respectively. Then, according to eqn (4.24), we have $\mathbf{P}^{II} = \mathbf{A}^{-1}\mathbf{P}^{I}$. Now consider the operation R which changes the vector into another whose coordinates in the two bases form the columns $(\mathbf{P}')^{I}$ and $(\mathbf{P}')^{II}$ respectively. In basis I let the matrix representing R be $\mathbf{R}^{I}$. The change of basis from I to II will mean that the elements of the matrix representing R must change. Nevertheless the changed matrix $\mathbf{R}^{II}$ will still be essentially equivalent to $\mathbf{R}^{I}$, since they both represent one and the same operation.

In the basis I we have

$$(\mathbf{P}')^{I} = \mathbf{R}^{I}\mathbf{P}^{I}. \tag{4.38}$$

Now from $\mathbf{P}^{II} = \mathbf{A}^{-1}\mathbf{P}^{I}$ we have

$$\mathbf{P}^{I} = \mathbf{A}\mathbf{P}^{II}, \tag{4.39}$$

and similarly for the transformed vector we have

$$(\mathbf{P}')^{I} = \mathbf{A}(\mathbf{P}')^{II}. \tag{4.40}$$

Substituting eqns (4.39) and (4.40) into eqn (4.38), we obtain

$$\mathbf{A}(\mathbf{P}')^{II} = \mathbf{R}^{I}\mathbf{A}\mathbf{P}^{II},$$

whence $\quad\quad (\mathbf{P}')^{II} = (\mathbf{A}^{-1}\mathbf{R}^{I}\mathbf{A})\mathbf{P}^{II}.$

But by definition $\quad (\mathbf{P}')^{II} = \mathbf{R}^{II}\mathbf{P}^{II}$

and so we have proved that

$$\mathbf{R}^{II} = \mathbf{A}^{-1}\mathbf{R}^{I}\mathbf{A}. \tag{4.41}$$

Because they both represent the same operation, $\mathbf{R}^{II}$ and $\mathbf{R}^{I}$ are said to be *equivalent matrices*, and in consequence of the change of basis $\mathbf{R}^{I}$ is said to have been subjected to a *similarity transformation*. It is obvious that, for any given square matrix, the number of equivalent matrices is unlimited, and all of them are related to one another by similarity transformations of the type seen in eqn (4.41) or of the entirely equivalent type

$$\mathbf{R}^{II} = \mathbf{CR}^{I}\mathbf{C}^{-1},$$

which is obtained by writing $\mathbf{A}^{-1} = \mathbf{C}$.

4.9. The character of a square matrix and its invariance under similarity transformations

A very important quantity associated with a square matrix is the sum of its diagonal elements. In matrix theory this quantity is usually referred to as the *trace* or, from the equivalent German expression, the *spur*. In symmetry theory, where the matrix is regarded as representing a symmetry operation, it is called the *character*. We shall adopt this name and the symbol χ. Thus for a square matrix $\mathbf{R}$, whose general element is R_{ij}, the character is defined by the equation

$$\chi_R = \sum_i R_{ii}. \tag{4.42}$$

The importance of the character in symmetry theory depends upon the fact (which we are about to prove) that all the unlimited number of equivalent matrices of a given size have one thing in common: they all have the same character. In other words, the character of a matrix is invariant under all similarity transformations. We shall make repeated use of this fact in our treatment of molecular vibrations.

Probably the simplest way of presenting the proof is to start with any two square matrices $\mathbf{C}$ and $\mathbf{D}$ of the size in question, and to establish the fact that the products $\mathbf{CD}$ and $\mathbf{DC}$ both have the same character. This may be done as follows. Remembering that the ith diagonal element of $\mathbf{CD}$ is $\sum_k C_{ik} D_{ki}$, and that the kth diagonal element of $\mathbf{DC}$ is $\sum\limits_i D_{ki} C_{ik}$, we have

$$\chi_{(CD)} = \sum_i \sum_k C_{ik} D_{ki} = \sum_k \sum_i D_{ki} C_{ik} = \chi_{(DC)}.$$

Now in the matrix product $\mathbf{A^{-1}RA}$, let us write $\mathbf{A^{-1}} = \mathbf{C}$ and $\mathbf{RA} = \mathbf{D}$. We thus find

$$\chi_{(A^{-1}RA)} = \chi_{(CD)} = \chi_{(DC)} = \chi_{(RAA^{-1})} = \chi_R, \qquad (4.43)$$

which proves that the character of the matrix $\mathbf{R}$ is invariant under the similarity transformation $\mathbf{A^{-1}RA}$.

4.10. Epilogue for Chapters 3 and 4

Our review of the immediately relevant aspects of matrices and vectors is now complete, and we are ready to begin a general treatment of the molecular vibrational problem by classical mechanics. Only one important matter concerning matrices remains to be presented, namely, that of eigenvalues and eigenvectors. We shall not have occasion to invoke these, however, until considerably later in this book. It will be advantageous to postpone discussion of them until then.

5 Classical treatment of the molecular vibrational problem using internal coordinates

5.1. Cartesian displacement coordinates and the total kinetic energy

In this chapter we shall consider the classical mechanical treatment of the vibrations of a molecule containing N nuclei, and shall make extensive use of matrices, the relevant aspects of which have been presented in Chapters 3 and 4. Before introducing the internal coordinates in terms of which the vibrational problem will be treated, we first choose a Cartesian coordinate system in terms of which the displacement of each of the N nuclei can be expressed. Thus for the first nucleus we have three displacement coordinates x_1, x_2, and x_3; for the second we have three more, which we call x_4, x_5, and x_6; and so on up to the Nth nucleus with the displacement coordinates x_{3N-2}, x_{3N-1}, and x_{3N}. As has already been explained in Section 4.3, the complete set of $3N$ Cartesian displacement coordinates may be considered as defining a vector in $3N$-dimensional configuration space, and this vector may be represented by a single-column matrix $\mathbf{X}$ whose elements are the coordinates.

The total kinetic energy of the system, including translational and rotational contributions as well as the vibrational kinetic energy, is easily expressible in terms of the time derivatives $\dot{x}_i$ of the Cartesian displacement coordinates. We shall use the symbol $\overline{T}$ for the total kinetic energy, in order to distinguish it from the purely vibrational kinetic energy T; only the latter will interest us in the treatment of the vibrational problem. If m_j is the mass of the jth nucleus we have

$$2\overline{T} = (m_1\dot{x}_1^2 + m_1\dot{x}_2^2 + m_1\dot{x}_3^2) + (m_2\dot{x}_4^2 + m_2\dot{x}_5^2 + m_2\dot{x}_6^2) + \ldots +$$
$$+ (m_N\dot{x}_{3N-2}^2 + m_N\dot{x}_{3N-1}^2 + m_N\dot{x}_{3N}^2). \quad (5.1)$$

As explained in Section 3.9, this quadratic form may be expressed as a matrix product

$$2\bar{T} = \dot{\mathbf{X}}^{\dagger}\mathbf{M}\dot{\mathbf{X}}, \qquad (5.2)$$

where $\dot{\mathbf{X}}$ is a single-column matrix whose elements are the time-derivatives of the $3N$ x-coordinates, and $\mathbf{M}$ is a diagonal matrix of the nuclear masses, each of which appears three times. Thus reading from the top left-hand element of $\mathbf{M}$ to the bottom right-hand one, the diagonal elements are

$$m_1,\ m_1,\ m_1,\ m_2,\ m_2,\ m_2,\ ...,\ m_N,\ m_N,\ m_N.$$

Our particular choice of Cartesian coordinates is not, of course, a unique one. Instead of the $3N$ x-coordinates, we could choose any other basis which spans the configuration space. The new coordinates for a given configuration would then be linear combinations of the previous ones, and the new single-column matrix $\mathbf{X}'$ would be related to $\mathbf{X}$ by an equation of the type $\mathbf{X}' = \mathbf{AX}$, in which $\mathbf{A}$ is a square matrix with $3N$ rows and $3N$ columns.

5.2. Introduction of internal coordinates

In the treatment of the purely vibrational problem it is advantageous to introduce a set of internal coordinates which suffice to describe the *relative* positions of the nuclei, without being concerned either with the position of the molecule as a whole in physical space or with its overall orientation. For a non-linear molecule the number of internal coordinates will be $3N-6$, i.e. the total number of displacement coordinates less the three required to define the position of the centre of mass of the system and the three required to define its overall orientation in space. The particular case of linear molecules has been considered in Section 2.5, and it has been shown that the number of internal coordinates required is $3N-5$. In what follows we shall assume that the molecule under consideration is non-linear. *Mutatis mutandis*, the arguments for linear molecules are exactly similar.

Let $\mathbf{B}$ be the matrix which generates the internal coordinates D from the Cartesian coordinates x. Clearly $\mathbf{B}$ will have $3N-6$

rows and $3N$ columns. By attaching 6 more rows we may construct a square matrix. The attached rows form a $6 \times 3N$ matrix which we shall call $\mathbf{B_0}$. The square matrix which we have constructed is thus partitioned into $\mathbf{B}$ and $\mathbf{B_0}$. When used to premultiply $\mathbf{X}$, it will therefore give a column which is similarly partitioned:

$$
\begin{array}{c} 3N-6 \\ \\ 6 \end{array}
\begin{bmatrix} \mathbf{D} \\ \hline \mathbf{D_0} \end{bmatrix}
=
\begin{array}{c} 3N-6 \\ \\ 6 \end{array}
\begin{bmatrix} \mathbf{B} \\ \hline \mathbf{B_0} \end{bmatrix}
\begin{bmatrix} \mathbf{X} \end{bmatrix} . \tag{5.3}
$$

The upper portion $\mathbf{D}$ of the product column is given by

$$\mathbf{D} = \mathbf{BX}. \tag{5.4}$$

Its elements are the $3N-6$ internal coordinates. The lower portion, which we will call $\mathbf{D_0}$ by analogy with $\mathbf{B_0}$, contains the remaining six coordinates which are concerned with overall translational and rotational displacements of the molecule. We may refer to them as *external* coordinates. As we have seen, they correspond to non-genuine vibrations with zero frequency. It is for this reason that we have elected to distinguish them by zero subscripts. The part $\mathbf{D_0}$ is given by

$$\mathbf{D_0} = \mathbf{B_0}\,\mathbf{X} \tag{5.5}$$

which is, of course, exactly analogous to eqn (5.4).

The matrix partitioned into $\mathbf{B}$ and $\mathbf{B_0}$ is square, and so it will in general have an inverse, the use of which will enable us to express the Cartesian coordinates in terms of the internal and external coordinates. This is written in the form

$$
\begin{bmatrix} \mathbf{X} \end{bmatrix}
=
\overset{\overset{\displaystyle 3N-6 \qquad 6}{\longleftrightarrow\ \longleftrightarrow}}{\begin{bmatrix} \mathbf{A} & \mathbf{A_0} \end{bmatrix}}
\begin{bmatrix} \mathbf{D} \\ \hline \mathbf{D_0} \end{bmatrix}
\begin{array}{l} 3N-6 \\ \\ 6 \end{array}
\tag{5.6}
$$

in which we have partitioned the square matrix into $\mathbf{A}$ and $\mathbf{A_0}$ so as to make $\mathbf{A}$ conformable for multiplication with $\mathbf{D}$, and $\mathbf{A_0}$ conformable for multiplication with $\mathbf{D_0}$.

5.3. The kinetic energy in terms of internal and external coordinates

We now substitute the expression for $\mathbf{X}$ from eqn (5.6) into the matrix equation (5.2) for the total kinetic energy. This gives

$$2\overline{T} = [\dot{\mathbf{D}}^\dagger \mid \dot{\mathbf{D}}_0^\dagger] \left[\begin{array}{c} \mathbf{A}^\dagger \\ \hline \mathbf{A}_0^\dagger \end{array} \right] \left[\begin{array}{c} \mathbf{M} \end{array} \right] \left[\begin{array}{c|c} \mathbf{A} & \mathbf{A}_0 \end{array} \right] \left[\begin{array}{c} \dot{\mathbf{D}} \\ \hline \dot{\mathbf{D}}_0 \end{array} \right]. \quad (5.7)$$

On multiplying out the three central matrices in eqn (5.7), the following result is obtained:

$$2\overline{T} = [\dot{\mathbf{D}}^\dagger \mid \dot{\mathbf{D}}_0^\dagger] \left[\begin{array}{c|c} \mathbf{A}^\dagger\mathbf{M}\mathbf{A} & \mathbf{A}^\dagger\mathbf{M}\mathbf{A}_0 \\ \hline \mathbf{A}_0^\dagger\mathbf{M}\mathbf{A} & \mathbf{A}_0^\dagger\mathbf{M}\mathbf{A}_0 \end{array} \right] \left[\begin{array}{c} \dot{\mathbf{D}} \\ \hline \dot{\mathbf{D}}_0 \end{array} \right].$$

Finally, completing the matrix multiplications, we find that

$$2\overline{T} = \dot{\mathbf{D}}^\dagger\mathbf{A}^\dagger\mathbf{M}\dot{\mathbf{D}} + \dot{\mathbf{D}}^\dagger\mathbf{A}^\dagger\mathbf{M}\mathbf{A}_0\dot{\mathbf{D}}_0 + \dot{\mathbf{D}}_0^\dagger\mathbf{A}_0^\dagger\mathbf{M}\mathbf{A}\dot{\mathbf{D}} + \dot{\mathbf{D}}_0^\dagger\mathbf{A}_0^\dagger\mathbf{M}\mathbf{A}_0\dot{\mathbf{D}}_0. \quad (5.8)$$

For a discussion of matrix addition and the nature of the products of partitioned matrices, the reader may refer back to Section 3.4.

From eqn (5.8) we see that the total kinetic energy of the molecule is the sum of four parts. The first part is concerned solely with the internal coordinates D, and thus represents the kinetic energy of the genuine vibrations of the molecule in the absence of translational and rotational motions. In contrast, the fourth part is not at all concerned with the internal coordinates, and thus represents the kinetic energy of translation and rotation of the molecule in the absence of vibrational motions. The two remaining parts are concerned simultaneously with both internal and external coordinates. They thus represent the contribution to the total kinetic energy from interactions of vibrational with other kinds of motion, and in particular with rotational motion (see Section 2.3).

For the treatment of the vibrational problem we shall use only the first term of the expression for the total kinetic energy given in eqn (5.8). Thus for the vibrational kinetic energy T we shall write

$$2T = \dot{\mathbf{D}}^\dagger \mathbf{A}^\dagger \mathbf{M} \mathbf{A} \dot{\mathbf{D}}. \tag{5.9}$$

Clearly we are fully justified in omitting the last term of eqn (5.8), for this is independent of the internal coordinates D. If the molecule were in a rotationless state we should also be justified in omitting the second and third terms. In reality, of course, the free molecule which we are considering will necessarily be rotating, and consequently there will necessarily be a contribution to the kinetic energy from rotation–vibration interactions. Our adoption of eqn (5.9) for the kinetic energy thus involves an approximation in assuming that the second and third terms of eqn (5.8) are so small that they may be set equal to zero. This is equivalent to the assumption that

$$\left.\begin{array}{l} \mathbf{A}^\dagger \mathbf{M} \mathbf{A}_0 = \mathbf{O} \\ \mathbf{A}_0^\dagger \mathbf{M} \mathbf{A} = \mathbf{O} \end{array}\right\}. \tag{5.10}$$

and

In fact it is easily seen that the two equations (5.10) are equivalent to one another. This follows at once from the fact that, since $\mathbf{M}$ is a diagonal matrix, $\mathbf{A}_0^\dagger \mathbf{M} \mathbf{A}$ is simply the transpose of $\mathbf{A}^\dagger \mathbf{M} \mathbf{A}_0$. If $\mathbf{A}^\dagger \mathbf{M} \mathbf{A}_0$ is a null matrix, then so also must be its transpose, for obviously the transpose of a null matrix is also a null matrix.

We have already pointed out in Chapter 2 that the approximation involved in the neglect of rotation–vibration interactions is quite a good one. In practice it may be effected by constructing the matrix $\mathbf{B}_0$ of eqn (5.3) in such a way that it generates from $\mathbf{X}$ the six linear combinations which, when set equal to zero, give the conditions that the translational momentum is zero and that (in the approximation under discussion) the rotational momentum is also zero. The linear combinations in question are those given, with slightly different notation, in eqns (2.2) and (2.3). Their application here means setting $\mathbf{D}_0 = \mathbf{O}$ and likewise $\dot{\mathbf{D}}_0 = \mathbf{O}$, with the consequence that only the first term on the right-hand side of eqn (5.8) survives. We thus arrive at eqn (5.9) for the kinetic energy to be used in the treatment of the vibrational problem. It should not be forgotten, however, that the neglect of rotation–vibration interactions always involves an approximation, albeit in general quite a good one. We may note here that no such considerations arise in the case of the potential energy, which of course depends solely upon the internal coordinates.

5.4. The Wilson G matrix

In terms of internal coordinates, the purely vibrational kinetic energy T may be very simply expressed by the use of the matrix G first introduced by E. Bright Wilson Jr. and defined by

$$G = BM^{-1}B^{\dagger}. \tag{5.11}$$

Here B is the matrix, already considered in the preceding section, which generates the $3N-6$ internal coordinates from the $3N$ Cartesian coordinates in accordance with eqn (5.4), and M^{-1} is the inverse of the diagonal matrix M of eqn (5.2), and is thus a diagonal matrix each of whose non-zero elements is the inverse of the corresponding element of M.

In order to understand the significance of G, we need first to derive some further relations between the matrices B, B_0, A, and A_0. These follow from the fact that the square matrix made up of the parts A and A_0 (see eqn (5.6)) is the inverse of the square matrix made up of the parts B and B_0 (see eqn (5.3)). When multiplied together the two square matrices must therefore give a unit matrix with $3N$ rows and $3N$ columns. This we will write as E_{3N}. Consider first the following product

$$\left[\begin{array}{c} B \\ \hline B_0 \end{array}\right] \left[\begin{array}{c|c} A & A_0 \end{array}\right] = \left[\begin{array}{c|c} BA & BA_0 \\ \hline B_0 A & B_0 A_0 \end{array}\right].$$

Because this must equal E_{3N}, we have

$$BA = E_{3N-6}, \tag{5.12}$$
$$B_0 A_0 = E_6, \tag{5.13}$$
$$BA_0 = O, \tag{5.14}$$
$$B_0 A = O. \tag{5.15}$$

It should be noted that, although $\mathbf{BA} = \mathbf{E}$, it would be improper to say that $\mathbf{B}$ and $\mathbf{A}$ are the inverses of one another; for neither $\mathbf{B}$ nor $\mathbf{A}$ is square, and only for square matrices is the inverse defined. The same is true for $\mathbf{B_0}$ and $\mathbf{A_0}$.

Secondly, consider the reverse product

$$\left[\begin{array}{c|c} \mathbf{A} & \mathbf{A_0} \end{array} \right] \left[\begin{array}{c} \mathbf{B} \\ \hline \mathbf{B_0} \end{array} \right] = \left[\begin{array}{c} \mathbf{AB} \end{array} \right] + \left[\begin{array}{c} \mathbf{A_0 B_0} \end{array} \right].$$

Here all we can deduce is that

$$\mathbf{AB} + \mathbf{A_0 B_0} = \mathbf{E}_{3N}. \tag{5.16}$$

In what follows we shall require the value of $\mathbf{B}^\dagger \mathbf{A}^\dagger$ obtained from eqn (5.16). This is given by

$$\mathbf{B}^\dagger \mathbf{A}^\dagger = [\mathbf{AB}]^\dagger = [\mathbf{E}_{3N} - \mathbf{A_0 B_0}]^\dagger = \mathbf{E}_{3N} - \mathbf{B}_0^\dagger \mathbf{A}_0^\dagger. \tag{5.17}$$

We are now in a position to investigate the significance of the matrix $\mathbf{G}$ defined in eqn (5.11). To this end consider the result of post-multiplying $\mathbf{G}$ by $\mathbf{A}^\dagger \mathbf{MA}$, i.e.

$$\mathbf{GA}^\dagger \mathbf{MA} = \mathbf{BM}^{-1} \mathbf{B}^\dagger \mathbf{A}^\dagger \mathbf{MA}.$$

Substituting from eqn (5.17) the value of $\mathbf{B}^\dagger \mathbf{A}^\dagger$, we find

$$\mathbf{GA}^\dagger \mathbf{MA} = \mathbf{BM}^{-1}[\mathbf{E}_{3N} - \mathbf{B}_0^\dagger \mathbf{A}_0^\dagger]\mathbf{MA}$$
$$= \mathbf{BM}^{-1}\mathbf{MA} - \mathbf{BM}^{-1}\mathbf{B}_0^\dagger \mathbf{A}_0^\dagger \mathbf{MA}. \tag{5.18}$$

According to eqns (5.10), the approximation we are using for the kinetic energy involves assuming $\mathbf{A}_0^\dagger \mathbf{MA} = \mathbf{O}$. The second term on the right-hand side of eqn (5.18) therefore vanishes. The first term is simply equal to $\mathbf{BA}$. But according to eqn (5.12) $\mathbf{BA} = \mathbf{E}_{3N-6}$. We have therefore proved that

$$\mathbf{GA}^\dagger \mathbf{MA} = \mathbf{E}_{3N-6},$$

or

$$\mathbf{G}^{-1} = \mathbf{A}^\dagger \mathbf{MA}. \tag{5.19}$$

The significance of $\mathbf{G}^{-1}$ becomes fully apparent when we recall that the vibrational kinetic energy T has the value given by eqn (5.9), namely,

$$2T = \dot{\mathbf{D}}^\dagger \mathbf{A}^\dagger \mathbf{MA} \dot{\mathbf{D}}.$$

Accordingly we may write

$$2T = \dot{\mathbf{D}}^\dagger \mathbf{G}^{-1} \dot{\mathbf{D}}. \tag{5.20}$$

This important relation plays a leading part in the setting up of the secular equation of the vibrational problem.

The reader should note the analogy between the expressions for the total kinetic energy and the purely vibrational energy, as given respectively in eqn (5.2) and eqn (5.20). In the expression for the vibrational kinetic energy in terms of the internal coordinates the matrix $\mathbf{G}^{-1}$ is seen to play a role analogous to that of the matrix $\mathbf{M}$ in the expression for the total kinetic energy in terms of the Cartesian coordinates. The matrix $\mathbf{G}$ (or $\mathbf{G}^{-1}$) is square and has $3N-6$ rows and $3N-6$ columns. Unlike $\mathbf{M}$ (or $\mathbf{M}^{-1}$), $\mathbf{G}$ is not a diagonal matrix. It is, however, symmetric; this was proved in Section 3.6 to be a consequence of the definition $\mathbf{G} = \mathbf{B}\mathbf{M}^{-1}\mathbf{B}^{\dagger}$. By eqn (5.19), $\mathbf{G}^{-1} = \mathbf{A}^{\dagger}\mathbf{M}\mathbf{A}$. It can therefore be shown, using the same process of deduction, that $\mathbf{G}^{-1}$ is also symmetric.

Before any use can be made of the $\mathbf{G}$ matrix in a particular problem, it must of course be constructed. The explanation of how this can be done will be found in Chapter 6.

5.5. The potential energy and the F matrix

The potential energy V is a function only of the internal coordinates D. Since the nuclear displacements involved are small, it may be expanded as a Taylor series:

$$V(D_1, D_2, D_3,...) = V_0 + \left(\frac{\partial V}{\partial D_1}\right)_0 D_1 + \left(\frac{\partial V}{\partial D_2}\right)_0 D_2 + ... +$$
$$+ \frac{1}{2}\left(\frac{\partial^2 V}{\partial D_1^2}\right)_0 D_1^2 + \frac{1}{2}\left(\frac{\partial^2 V}{\partial D_2^2}\right)_0 D_2^2 + ... +$$
$$+ \left(\frac{\partial^2 V}{\partial D_1 \partial D_2}\right)_0 D_1 D_2 + \left(\frac{\partial^2 V}{\partial D_2 \partial D_3}\right)_0 D_2 D_3 + ...$$

The higher terms may be regarded as negligible. The first term V_0 is the value of the potential energy at the equilibrium configuration (i.e. when $D_1 = D_2 = D_3 = ... = 0$). This may be taken as the zero from which vibrational potential energies are measured. The zero subscripts on the differential coefficients indicate that their values are those at the equilibrium configuration. At this configuration the value of V is a minimum, and therefore all the first differential coefficients, $(\partial V/\partial D_1)_0$, $(\partial V/\partial D_2)_0$, etc., must vanish. We are thus left with only the quadratic terms, i.e.

$$V = \frac{1}{2}\left(\frac{\partial^2 V}{\partial D_1^2}\right)_0 D_1^2 + \frac{1}{2}\left(\frac{\partial^2 V}{\partial D_2^2}\right)_0 D_2^2 + \ldots +$$

$$+ \left(\frac{\partial^2 V}{\partial D_1 \partial D_2}\right)_0 D_1 D_2 + \left(\frac{\partial^2 V}{\partial D_2 \partial D_3}\right)_0 D_2 D_3 + \ldots.$$

The so-called *force constants* are defined as

$$\left. \begin{array}{ll} F_{11} = \left(\dfrac{\partial^2 V}{\partial D_1^2}\right)_0, & F_{22} = \left(\dfrac{\partial^2 V}{\partial D_2^2}\right)_0, \quad \text{etc.} \\[3mm] F_{12} = \left(\dfrac{\partial^2 V}{\partial D_1 \partial D_2}\right)_0, & F_{23} = \left(\dfrac{\partial^2 V}{\partial D_2 \partial D_3}\right)_0, \quad \text{etc.} \end{array} \right\} \quad (5.21)$$

The expression for the potential energy may accordingly be written as

$$2V = F_{11} D_1^2 + F_{22} D_2^2 + \ldots + 2F_{12} D_1 D_2 + 2F_{23} D_2 D_3 + \ldots. \tag{5.22}$$

This quadratic form can be expressed in matrix notation as

$$2V = \mathbf{D}^\dagger \mathbf{F} \mathbf{D} \tag{5.23}$$

where $\mathbf{F}$ is a square matrix whose elements are the force constants F_{ij}. Like $\mathbf{G}$, $\mathbf{F}$ has $3N-6$ rows and $3N-6$ columns and, since $F_{ij} = \left(\dfrac{\partial^2 V}{\partial D_i \partial D_j}\right)_0 = F_{ji}$, it is a symmetric matrix. Taken together, eqn (5.20) for the kinetic energy and eqn (5.23) for the potential energy form the basis of the classical treatment of the molecular vibrational problem.

5.6. The secular equation of the vibrational problem

The classical treatment proceeds along the lines which were exemplified in Chapter 1 for a simple mechanical system. We begin with the classical equations of motion, one for each internal coordinate D_i, in the form

$$\frac{\mathrm{d}}{\mathrm{d}t}\frac{\partial T}{\partial \dot{D}_i} + \frac{\partial V}{\partial D_i} = 0, \tag{5.24}$$

and make use of the values of T and V expressed in eqns (5.20) and (5.23). For newcomers to the subject who are not particularly gifted mathematically or who are not accustomed to reading mathematical shorthand, the unavoidable complexity of the notation in the general case is likely to obscure the essential content of the argument. For the sake of clarity, therefore, we shall present in full and simple form the successive steps of the treatment in the particular case where the number of internal coordinates is three. (This corresponds to a non-linear triatomic molecule, for which $3N - 6 = 3$.) The reader will appreciate that the treatment of the general case will follow exactly the same lines and consequently will give a result of exactly the same form.

Writing G_{11}^{-1}, G_{12}^{-1}, etc. for the elements of the $\mathbf{G}^{-1}$ matrix, the expanded form of eqn (5.20) for the kinetic energy is

$$2T = G_{11}^{-1}\dot{D}_1^2 + G_{22}^{-1}\dot{D}_2^2 + G_{33}^{-1}\dot{D}_3^2 +$$
$$+ 2G_{12}^{-1}\dot{D}_1\dot{D}_2 + 2G_{23}^{-1}\dot{D}_2\dot{D}_3 + 2G_{13}^{-1}\dot{D}_1\dot{D}_3,$$

whence
$$\frac{\partial T}{\partial \dot{D}_1} = G_{11}^{-1}\dot{D}_1 + G_{12}^{-1}\dot{D}_2 + G_{13}^{-1}\dot{D}_3,$$

and
$$\frac{\mathrm{d}}{\mathrm{d}t}\frac{\partial T}{\partial \dot{D}_1} = G_{11}^{-1}\ddot{D}_1 + G_{12}^{-1}\ddot{D}_2 + G_{13}^{-1}\ddot{D}_3. \tag{5.25}$$

Similarly, writing F_{11}, F_{12}, etc. for the elements of the $\mathbf{F}$ matrix (i.e. the force constants), the expanded form of eqn (5.23) for the potential energy is

$$2V = F_{11}D_1^2 + F_{22}D_2^2 + F_{33}D_3^2 +$$
$$+ 2F_{12}D_1D_2 + 2F_{23}D_2D_3 + 2F_{13}D_1D_3,$$

whence
$$\frac{\partial V}{\partial D_1} = F_{11}D_1 + F_{12}D_2 + F_{13}D_3. \tag{5.26}$$

We now seek solutions of the type

$$\left.\begin{array}{l} D_1 = A_1\cos(\lambda^{\frac{1}{2}}t + \epsilon) \\ D_2 = A_2\cos(\lambda^{\frac{1}{2}}t + \epsilon) \\ D_3 = A_3\cos(\lambda^{\frac{1}{2}}t + \epsilon) \end{array}\right\}, \tag{5.27}$$

in which $\lambda = 4\pi^2\nu^2$ and ϵ is a phase angle. The common frequency with which all the nuclei vibrate in any particular normal mode is ν, and their amplitudes are respectively A_1, A_2, and A_3. This form of solution requires that

$$\left.\begin{array}{l} \ddot{D}_1 = -\lambda D_1 \\ \ddot{D}_2 = -\lambda D_2 \\ \ddot{D}_3 = -\lambda D_3 \end{array}\right\}. \tag{5.28}$$

Substitution of eqns (5.28) into eqn (5.25) gives

$$\frac{\mathrm{d}}{\mathrm{d}t}\frac{\partial T}{\partial \dot{D}_1} = -\lambda G_{11}^{-1}D_1 - \lambda G_{12}^{-1}D_2 - \lambda G_{13}^{-1}D_3. \tag{5.29}$$

We now substitute eqns (5.29) and (5.26) into the general equation of motion (5.24) for D_1. The result is

$$(F_{11}-\lambda G_{11}^{-1})D_1 + (F_{12}-\lambda G_{12}^{-1})D_2 + (F_{13}-\lambda G_{13}^{-1})D_3 = 0.$$

Inserting the values of the internal coordinates from eqn (5.27) and dividing through by $\cos(\lambda^{\frac{1}{2}}t + \epsilon)$, we obtain the amplitude equation

$$(F_{11}-\lambda G_{11}^{-1})A_1 + (F_{12}-\lambda G_{12}^{-1})A_2 + (F_{13}-\lambda G_{13}^{-1})A_3 = 0. \tag{5.30}$$

In precisely the same way, by considering in turn the equations of motion for the other two internal coordinates D_2 and D_3, we obtain two further amplitude equations. They are entirely analogous to eqn (5.30), as

$$(F_{21}-\lambda G_{21}^{-1})A_1 + (F_{22}-\lambda G_{22}^{-1})A_2 + (F_{23}-\lambda G_{23}^{-1})A_3 = 0, \tag{5.31}$$

$$(F_{31}-\lambda G_{31}^{-1})A_1 + (F_{32}-\lambda G_{32}^{-1})A_2 + (F_{33}-\lambda G_{33}^{-1})A_3 = 0. \tag{5.32}$$

The reader will note that in the above treatment account has been taken of the symmetric nature of both $\mathbf{G}^{-1}$ and $\mathbf{F}$, in consequence of which $G_{13}^{-1} = G_{31}^{-1}$, $F_{23} = F_{32}$, etc.

The theory of equations tells us that the condition under which all three amplitude equations shall hold simultaneously is that the determinant of the coefficients shall vanish, i.e.

$$\begin{vmatrix} F_{11}-\lambda G_{11}^{-1} & F_{12}-\lambda G_{12}^{-1} & F_{13}-\lambda G_{13}^{-1} \\ F_{21}-\lambda G_{21}^{-1} & F_{22}-\lambda G_{22}^{-1} & F_{23}-\lambda G_{23}^{-1} \\ F_{31}-\lambda G_{31}^{-1} & F_{32}-\lambda G_{32}^{-1} & F_{33}-\lambda G_{33}^{-1} \end{vmatrix} = 0.$$

This is a form of the secular equation of the vibrational problem. It is conveniently written in terms of $\mathbf{F}$ and $\mathbf{G}$ as

$$|\mathbf{F} - \lambda \mathbf{G}^{-1}| = 0 \qquad (5.33)$$

the vertical lines indicating that the array in question is to be regarded as a determinant, and not as a matrix. In our particular case, expansion of the determinant gives the secular equation as a cubic in λ. Solution of this cubic gives the three roots and hence, since $\lambda_i = 4\pi^2 \nu_i^2$, the three normal vibrational frequencies of the molecule under consideration.

The general case of a molecule with any number N of nuclei may be treated in exactly the same way. Increase in N merely increases the dimensions of the matrices $\mathbf{F}$ and $\mathbf{G}^{-1}$, and tends to make the algebra more voluminous. The result, however, is of exactly the same form. Thus, although we have deduced the secular equation (5.33) only for the particular case of three internal coordinates, it remains valid in the general case. For a non-linear molecule with N nuclei, expansion of the secular determinant gives an equation of order $3N - 6$, solution of which gives this number of normal vibrational frequencies. For a linear molecule, the corresponding number is $3N - 5$.

The secular equation (5.33) can be converted into two other forms which, though entirely equivalent, are more convenient for practical use. Since the rules for the multiplication of determinants are the same as those for the multiplication of matrices, we may pre-multiply the left-hand side of eqn (5.33) by $|\mathbf{G}|$ and so obtain

$$|\mathbf{GF} - \lambda \mathbf{GG}^{-1}| = 0,$$

i.e. $\qquad\qquad\qquad |\mathbf{GF} - \lambda \mathbf{E}| = 0. \qquad (5.34)$

Alternatively we may post-multiply by $|\mathbf{G}|$. This gives the equivalent form

$$|\mathbf{FG} - \lambda \mathbf{E}| = 0. \qquad (5.35)$$

It should be noted that the secular equation can be set up, and by its solution the normal frequencies can be obtained, without at any stage introducing the normal coordinates. In fact, the normal coordinates corresponding to the normal frequencies can be deduced from the secular equation. We shall postpone consideration of how this can be done until we have learned how the secular equation can itself be given a more amenable form by taking molecular symmetry into account.

First of all, however, we shall make good the omission in the foregoing treatment (see the end of Section 5.4), by considering how the **G** matrix can be constructed. This is the theme of the next chapter, where the general method is explained and its application illustrated by the case of the PCl_3 molecule.

6 Setting up the G matrix

6.1. Setting up G by direct evaluation of $\mathbf{BM^{-1}B^\dagger}$

IN order to obtain the secular equation for any particular molecule in the form $|\mathbf{FG}-\lambda\mathbf{E}| = 0$ it is obviously necessary to begin by choosing suitable internal coordinates, and then to set up the corresponding matrices $\mathbf{F}$ and $\mathbf{G}$. In general the $\mathbf{F}$ matrix is unlikely to present serious difficulties, for the internal coordinates will have been chosen in such a way that the force constants (i.e. the elements of $\mathbf{F}$) are simply related to them. On the other hand, the $\mathbf{G}$ matrix (which of course must be based upon the same internal coordinates as the $\mathbf{F}$ matrix), is not so straightforward.

By definition $$\mathbf{G} = \mathbf{BM^{-1}B^\dagger}, \qquad (6.1)$$

where $\mathbf{B}$ (see Section 5.2) is the matrix which generates the $3N-6$ internal coordinates D from the $3N$ Cartesian displacement coordinates according to the matrix equation

$$\mathbf{D} = \mathbf{BX}. \qquad (6.2)$$

If, therefore, we can set up the $\mathbf{B}$ matrix, expressing each of our chosen internal coordinates as a linear combination of Cartesian nuclear displacement coordinates, we shall be able to obtain $\mathbf{G}$ by substitution into eqn (6.1). For reasonably simple types of internal coordinate this can be done without great difficulty. For example, suppose that a certain internal coordinate D is the change in the distance between two nuclei which we shall designate by the numbers 1 and 2. (If the two atoms in question are joined by a chemical bond, D will be an internal coordinate of bond stretching—a fairly common type.) Let the actual Cartesian positional coordinates of the two nuclei be respectively ξ_1, η_1, ζ_1 and ξ_2, η_2, ζ_2. Then the distance r between them (i.e. the bond length if they are bonded together) will be

$$r = \{(\xi_1-\xi_2)^2+(\eta_1-\eta_2)^2+(\zeta_1-\zeta_2)^2\}^{\frac{1}{2}}. \qquad (6.3)$$

The internal displacement coordinate D is thus equal to Δr, where Δ indicates a small increase. Hence

$$D = \frac{\partial r}{\partial \xi_1}\Delta\xi_1 + \frac{\partial r}{\partial \eta_1}\Delta\eta_1 + \frac{\partial r}{\partial \zeta_1}\Delta\zeta_1 +$$

$$+ \frac{\partial r}{\partial \xi_2}\Delta\xi_2 + \frac{\partial r}{\partial \eta_2}\Delta\eta_2 + \frac{\partial r}{\partial \zeta_2}\Delta\zeta_2. \quad (6.4)$$

Now the quantities $\Delta\xi_1$, $\Delta\eta_1$, $\Delta\zeta_1$, $\Delta\xi_2$, $\Delta\eta_2$, and $\Delta\zeta_2$ are simply the Cartesian *displacement* coordinates of the two nuclei. We may designate them as $x_1, x_2, ..., x_6$. We have thus succeeded in expressing D as a linear combination of the x-coordinates. In other words we have derived the relevant row of the matrix **B**. Only the first six elements of it are different from zero, because only nuclei 1 and 2 are concerned. In fact we find at once from eqn (6.3) that the first element, i.e. $\partial r/\partial\xi_1$, has the value

$$\frac{\xi_1 - \xi_2}{r} = \cos\theta_x,$$

where θ_x is the angle between the axis of ξ (or of x) and the line joining masses 1 and 2. Similarly we find $\partial r/\partial\eta_1 = \cos\theta_y$ and $\partial r/\partial\zeta_1 = \cos\theta_z$. The partial differential coefficients with respect to the positional coordinates of nucleus 2 are equal in magnitude but opposite in sign, as compared to those for nucleus 1. The first six elements of the row of **B** in question are thus $\cos\theta_x$, $\cos\theta_y$, $\cos\theta_z$, $-\cos\theta_x$, $-\cos\theta_y$, and $-\cos\theta_z$, and all subsequent elements are zero. In similar ways the remaining rows of **B** could be derived, and the complete matrix used in eqn (6.1) in order to generate **G**. Although this sort of procedure is perhaps the most straightforward, it is less elegant and less convenient than the vector method, first introduced by Wilson, which we shall now proceed to examine.

6.2. The Wilson vector method for setting up G

We necessarily start from the definitions of **G** and **B** given in eqns (6.1) and (6.2). The elements of the column matrix **X** in the latter are $3N$ in number, and may be considered to be made up of N sets of three, each set being associated with a particular nucleus. Thus the first three elements (reading from the top of the column downwards) are the Cartesian displacements of the first nucleus, the next three are the corresponding displacements of the second nucleus, and so on. The three elements associated with the tth nucleus may be regarded as the components of the displacement vector of this nucleus. We shall call this vector $x^{(t)}$. Pre-multiplication of **X** by the matrix **B** gives the column **D** whose $3N-6$ elements are the internal coordinates. The particular internal coordinate D_i is produced by the pre-multiplication of the column **X** by the ith row of **B**. As in the case of **X**, we regard the $3N$ elements of this row as consisting of N sets of three, so that in the generation of D_i the first set of three multiplies with the first set of three in **X**, the second set in the ith row of **B** multiplies with the second set in **X**, and so on. We now regard the three elements of the tth set in the **B**-row as the components of a vector, which we shall designate as $b_i^{(t)}$. It follows that the contribution to D_i associated with the tth nucleus is simply the dot product $b_i^{(t)} . x^{(t)}$, and that we may accordingly write

$$D_i = \sum_{t=1}^{t=N} (b_i^{(t)} . x^{(t)}). \tag{6.5}$$

Now consider the matrix product $\mathbf{BM^{-1}B^{\dagger}}$ which defines the **G** matrix. Just as the ith row of **B** has a set of three elements which form the vector $b_i^{(t)}$, so also will the jth column of $\mathbf{B^{\dagger}}$ have three elements forming the vector $b_j^{(t)}$. In evaluating the product $\mathbf{BM^{-1}B^{\dagger}}$ it will be seen that what we may call the contribution to G_{ij} associated with the tth nucleus will be $(b_i^{(t)} . b_j^{(t)})/m_t$, and that we may accordingly write

$$G_{ij} = \sum_t (b_i^{(t)} . b_j^{(t)}) \frac{1}{m_t}. \tag{6.6}$$

The special feature of the Wilson vector method of setting up the **G** matrix is the way that the use of eqn (6.6) is facilitated by first expressing the b-vectors in terms of unit vectors directed conveniently in the molecule (for example along the bonds). Dot products of the kind occurring in eqn (6.6) can thus be converted into dot products of these unit vectors, and evidently such products can be written down very easily from a knowledge of the geometry of the molecule.

To illustrate the elegance and power of the method, we shall now consider in detail its application to two commonly occurring types of internal coordinates, namely, bond stretching and bond-angle increase. Consider first the internal coordinate Δr, which is the increase in the distance r between two nuclei which we shall number as 1 and 2. We introduce a unit vector e_{21} directed along the bond in the sense (indicated by the order of the subscripts) from nucleus 2 towards nucleus 1. Only these two nuclei can make contributions to the row of **B** associated with Δr, so what we must do is to express the vectors $b_{\Delta r}^{(1)}$ and $b_{\Delta r}^{(2)}$ in terms of the unit vector e_{21}. The contribution to Δr due to a general displacement $x^{(1)}$ of the first nucleus will clearly be $e_{21}.x^{(1)}$. But in accordance with eqn (6.5) this contribution must be $b_{\Delta r}^{(1)}.x^{(1)}$. Thus we may write

$$b_{\Delta r}^{(1)}.x^{(1)} = e_{21}.x^{(1)},$$

whence
$$b_{\Delta r}^{(1)} = e_{21}. \tag{6.7}$$

As to nucleus 2, a general displacement $x^{(2)}$ will make a contribution to Δr which will be of magnitude $e_{21}.x^{(2)}$, but will clearly be of opposite sign to that of the corresponding contribution from nucleus 1; for if a certain displacement of nucleus 1 results in an extension of the bond, then a similar displacement of nucleus 2 will necessarily cause a shortening. We therefore deduce that

$$b_{\Delta r}^{(2)} = -e_{21}. \tag{6.8}$$

We have thus found the appropriate values of the vectors $b_{\Delta r}^{(1)}$ and $b_{\Delta r}^{(2)}$ for use in calculating **G**-matrix elements by means of eqn (6.6).

We shall now consider the case of an internal coordinate which represents the increase of a bond angle. Consider for example the angle β between the two bonds in Fig. 6.1. The internal coordinate could be chosen simply as $\Delta\beta$, but in practice it is usual to include a factor having the dimensions of length. In this way the coordinate representing the change in bond angle is made dimensionally the same as other internal coordinates of bond stretching. The factor is arbitrary, but often some related bond length is used. As the internal coordinate to be considered in the present instance we shall choose $r\Delta\beta$, where r is the equilibrium length of the bond between nuclei 2 and 1. We first

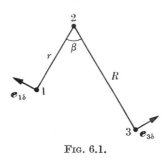

FIG. 6.1.

introduce a unit vector (see Fig. 6.1) which lies in the plane of the two bonds and is directed normally to the bond between nuclei 2 and 1 in the sense corresponding to an increase of the bond angle. We shall designate this unit vector by $e_{1\delta}$. Clearly the contribution to $r\Delta\beta$ from a displacement $x^{(1)}$ of nucleus 1 will be $e_{1\delta}.x^{(1)}$. As before, this must be $b_{r\Delta\beta}^{(1)}.x^{(1)}$. We have thus found that $b_{r\Delta\beta}^{(1)} = e_{1\delta}$; but for the purposes of the Wilson vector method it remains to express the unit vector $e_{1\delta}$ in terms of unit vectors along the directions of the bonds. These we now introduce as e_{21} (directed from nucleus 2 to nucleus 1) and e_{23} (directed from nucleus 2 to nucleus 3). To obtain $e_{1\delta}$ in terms of e_{21} and e_{23} we make use of the appropriate vector parallelogram, as shown in Fig. 6.2. We at once find that $e_{1\delta}$ is the vector sum of $e_{21}/\tan\beta$ and $-e_{23}/\sin\beta$. The desired expression for $b_{r\Delta\beta}^{(1)}$ is therefore

$$b_{r\Delta\beta}^{(1)} = \frac{1}{\sin\beta}(\cos\beta\, e_{21} - e_{23}). \tag{6.9}$$

We consider next nucleus 3 (in Fig. 6.1), for which the state of affairs is similar to that of nucleus 1, except that nucleus 3 is situated at the end of a bond whose length is R instead of r. We first introduce a unit vector $e_{3\delta}$ (analogous to $e_{1\delta}$) directed so as to correspond to increase in the angle β. Because of the different bond length, the contribution of a displacement $x^{(3)}$ of nucleus 3 to the coordinate $r\Delta\beta$ is $(r/R)(e_{3\delta}.x^{(3)})$, rather than simply $e_{3\delta}.x^{(3)}$. Since this contribution must be $b_{r\Delta\beta}^{(3)}.x^{(3)}$, we find that $b_{r\Delta\beta}^{(3)} = (r/R)e_{3\delta}$. We now have to express $e_{3\delta}$ in terms of the unit vectors e_{21} and e_{23} along the bond directions. From the vector parallelogram (see Fig. 6.3) we at once see that $e_{3\delta}$ is the vector sum of $e_{23}/\tan\beta$ and $-e_{21}/\sin\beta$. The desired expression for $b^{(3)}_{r\Delta\beta}$ is therefore

$$b^{(3)}_{r\Delta\beta} = \frac{r}{R}\cdot\frac{1}{\sin\beta}\,(\cos\beta\,e_{23}-e_{21}).\qquad(6.10)$$

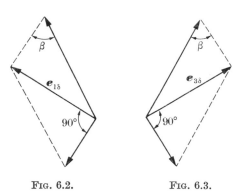

FIG. 6.2.   FIG. 6.3.

To complete our consideration of the coordinate of angle increase, $r\Delta\beta$, we still have to deal with the effect of displacements of nucleus 2, which (see Fig. 6.1) is common to the two bonds involved. In other words, we must express $b^{(2)}_{r\Delta\beta}$ in terms of the same unit vectors, e_{21} and e_{23}, as before. This is most neatly done by considering the situation in which all three nuclei have the same displacement, such that $x^{(1)} = x^{(2)} = x^{(3)} = x$, say. The resulting value of the internal coordinate will be

$$(b^{(1)}_{r\Delta\beta}.x)+(b^{(2)}_{r\Delta\beta}.x)+(b^{(3)}_{r\Delta\beta}.x).$$

But since the simultaneous displacements amount simply to a mere translation of the system as a whole, the value of the internal coordinate must be zero. We have thus proved that

$$b^{(1)}_{r\Delta\beta}+b^{(2)}_{r\Delta\beta}+b^{(3)}_{r\Delta\beta} = 0.\qquad(6.11)$$

Since we have already evaluated $b^{(1)}_{r\Delta\beta}$ and $b^{(3)}_{r\Delta\beta}$ in terms of e_{21} and e_{23}, we can use this result to obtain $b^{(2)}_{r\Delta\beta}$ in similar terms. Thus suppose that $b^{(2)}_{r\Delta\beta} = ae_{21}+be_{23}$. Substituting this value into eqn (6.11), as well as the values of the other two b vectors from eqns (6.9) and (6.10), and equating to zero the sum of the coefficients of e_{21} and the sum of the coefficients of e_{23}, we obtain

$$a = \frac{1}{\sin \beta} \left(\frac{r}{R} - \cos \beta \right),$$

$$b = \frac{1}{\sin \beta} \left(1 - \frac{r}{R} \cos \beta \right).$$

The desired result is thus

$$b_{r\Delta\beta}^{(2)} = \frac{1}{\sin \beta} \left\{ \left(\frac{r}{R} - \cos \beta \right) e_{21} + \left(1 - \frac{r}{R} \cos \beta \right) e_{23} \right\}.$$

$$(6.12)$$

Once the **b**-vectors for a molecule have been thus expressed in terms of unit vectors along the bond directions, it becomes a simple matter to write down the G-matrix elements by the use of eqn (6.6). This can best be appreciated by working out an actual case in detail. As an illustrative example we will take the phosphorus trichloride molecule PCl_3.

6.3. The PCl_3 model and the choice of internal coordinates

The phosphorus trichloride molecule is known to have a regular triangular pyramidal structure, with a bond between the

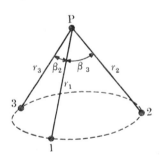

FIG. 6.4. Model of the PCl_3 molecule.

apical P atom and each of the three Cl atoms. All three bonds are, of course, identical. We shall assume that all three Cl nuclei are of the same isotopic species, i.e. have exactly the same mass. Fig. 6.4 shows the molecular model. The three point masses representing the Cl nuclei are numbered 1, 2, and 3, and the lengths of the bonds to them are numbered correspondingly. In fact, of course, $r_1 = r_2 = r_3 = r$ in the equilibrium configuration. The angles beween the pairs of bonds carry the same numbers, but it should be noted that β_1 is the angle between the bonds to nuclei 2 and 3, i.e. it is the bond angle *opposite* the bond labelled r_1. The other two bond angles β_2 and β_3 are numbered in a similar fashion.

Since the molecule contains 4 nuclei and is non-linear, the number of normal vibrational modes it possesses is $3N-6 = 6$. We therefore have to choose six mutually independent internal coordinates for the treatment of the vibrational problem. In considering the likely nature of the force field, it is reasonable to choose intuitively as internal coordinates the extensions of the three P–Cl bonds and the increases in the three Cl–P–Cl bond angles. In order to make the latter dimensionally equivalent to the former, we include for each a factor with the dimension of length. A natural choice is the common equilibrium bond length r. Our choice of internal coordinates is thus $D_1 = \Delta r_1$, $D_2 = \Delta r_2$, $D_3 = \Delta r_3$, $D_4 = r\Delta\beta_1$, $D_5 = r\Delta\beta_2$, and $D_6 = r\Delta\beta_3$.

The considerations of the preceding section have brought us to a position in which we can now proceed to set up the **G** matrix for the PCl_3 molecule. It will be a square matrix with 6 rows and 6 columns, giving a total of 36 elements. However, because it is necessarily a symmetric matrix, we need only evaluate the diagonal elements and half of the off-diagonal ones. This makes a total of $6+15 = 21$ elements. In fact we shall find that, as a result of the symmetry of the molecule, the number of distinct elements is considerably smaller than this.

6.4. Setting up G for PCl_3 by the vector method

We start by introducing three unit vectors directed respectively along the three bonds from the apical point mass P towards the point masses 1, 2, and 3. We shall call these unit vectors e_1, e_2, and e_3. We can now use the results of the previous section, namely, eqns (6.7), (6.8), (6.9), (6.10), and (6.12), to construct a table of the b vectors of the problem. The result is Table 6.1, which has 6 rows (one for each internal coordinate) and 4 columns (one for each nucleus). To save space, $\sin\beta$ and $\cos\beta$ are abbreviated to s and c respectively.

It is usually wise also to construct a table of the dot products of the e vectors. In the present example this is a trivial undertaking; in general, however, reference to a table of this kind, in the course of working out the G-matrix elements, may help to

*Setting up the **G** matrix*

TABLE 6.1

***b** vectors for* PCl_3 *in terms of unit vectors along bonds*

	1	2	3	P
$D_1 = \Delta r_1$	e_1	0	0	$-e_1$
$D_2 = \Delta r_2$	0	e_2	0	$-e_2$
$D_3 = \Delta r_3$	0	0	e_3	$-e_3$
$D_4 = r\Delta\beta_1$	0	$\frac{1}{s}(ce_2-e_3)$	$\frac{1}{s}(ce_3-e_2)$	$\frac{1-c}{s}(e_3+e_2)$
$D_5 = r\Delta\beta_2$	$\frac{1}{s}(ce_1-e_3)$	0	$\frac{1}{s}(ce_3-e_1)$	$\frac{1-c}{s}(e_1+e_3)$
$D_5 = r\Delta\beta_3$	$\frac{1}{s}(ce_1-e_2)$	$\frac{1}{s}(ce_2-e_1)$	0	$\frac{1-c}{s}(e_1+e_2)$

avoid errors. In the form given as Table 6.2, each entry represents the value of the dot product of the two vectors which stand respectively at the left-hand side of the row and the top of the column in question. Again $c = \cos\beta$. The array is, of course, symmetric.

TABLE 6.2

Dot products of unit vectors for PCl_3

	e_1	e_2	e_3
e_1	1	c	c
e_2	c	1	c
e_3	c	c	1

The next step requires rather more algebraic work, for we must construct tables of the dot products of the ***b*** vectors contained in Table 6.1. As reference back to eqn (6.6) will remind us, the **G**-matrix elements involve only dot products in which both ***b*** factors are associated with the same nucleus. The PCl_3 molecule contains 4 nuclei, for each of which we have six ***b*** vectors. It is convenient to construct a separate table for each nucleus, and to give it a square-array form similar to that of Table 6.2. For the first Cl the result is shown in Table 6.3. The entries are the dot products of ***b*** vectors, and the labels of the row and column at whose intersection a particular entry is situated denote the internal coordinates involved. Thus the entry

$c(1-c)/s^2$ is the value of the dot product $b_{r\Delta\beta_2}^{(1)} \cdot b_{r\Delta\beta_3}^{(1)}$. Clearly the array must be symmetric; therefore only the diagonal entries and the off-diagonal entries on one side of it need be specified. Because Table 6.3 is concerned solely with the first Cl, it follows that the only internal coordinates involved will be Δr_1, $r\Delta\beta_2$, and $r\Delta\beta_3$, and for this reason there are many zero entries in the table. Using the appropriate entries in the column of Table 6.1 headed 1, the reader should have no difficulty in verifying the entries in Table 6.3. A couple of illustrative examples may, however, be helpful. Thus consider the entry in the row labelled $r\Delta\beta_2$ and

TABLE 6.3

Dot products of **b** *vectors for* PCl_3

Cl of bond 1	Δr_1	Δr_2	Δr_3	$r\Delta\beta_1$	$r\Delta\beta_2$	$r\Delta\beta_3$
Δr_1	1	0	0	0	0	0
Δr_2		0	0	0	0	0
Δr_3			0	0	0	0
$r\Delta\beta_1$				0	0	0
$r\Delta\beta_2$					1	$\dfrac{c(1-c)}{s^2}$
$r\Delta\beta_3$						1

the column also labelled $r\Delta\beta_2$. Its value is obtained as follows.

$$b_{r\Delta\beta_2}^{(1)} \cdot b_{r\Delta\beta_2}^{(1)} = \frac{1}{s}(ce_1-e_3) \cdot \frac{1}{s}(ce_1-e_3)$$

$$= \frac{1}{s^2}\{c^2(e_1 \cdot e_1)-c(e_1 \cdot e_3)-c(e_1 \cdot e_3)+(e_3 \cdot e_3)\}$$

$$= \frac{1}{s^2}(c^2-c^2-c^2+1) = \frac{1-c^2}{s^2} = 1.$$

Secondly, consider the entry in the row labelled $r\Delta\beta_2$ and the column labelled $r\Delta\beta_3$. In this case

$$b_{r\Delta\beta_2}^{(1)} \cdot b_{r\Delta\beta_3}^{(1)} = \frac{1}{s}(ce_1-e_3) \cdot \frac{1}{s}(ce_1-e_2)$$

$$= \frac{1}{s^2}\{c^2(e_1 \cdot e_1)-c(e_1 \cdot e_2)-c(e_1 \cdot e_3)+(e_2 \cdot e_3)\}$$

$$= \frac{1}{s^2}(c^2-c^2-c^2+c) = \frac{c(1-c)}{s^2}.$$

The remaining entries may be verified in a similar manner.

The corresponding arrays for masses 2 and 3 are given respectively in Tables 6.4 and 6.5. The close relationship with Table 6.3 is, of course, a consequence of the symmetrical equivalence of

TABLE 6.4

Dot products of **b** *vectors for* PCl_3

Cl of bond 2	Δr_1	Δr_2	Δr_3	$r\Delta\beta_1$	$r\Delta\beta_2$	$r\Delta\beta_3$
Δr_1	0	0	0	0	0	0
Δr_2		1	0	0	0	0
Δr_3			0	0	0	0
$r\Delta\beta_1$				1	0	$\dfrac{c(1-c)}{s^2}$
$r\Delta\beta_2$					0	0
$r\Delta\beta_3$						1

TABLE 6.5

Dot products of **b** *vectors for* PCl_3

Cl of bond 3	Δr_1	Δr_2	Δr_3	$r\Delta\beta_1$	$r\Delta\beta_2$	$r\Delta\beta_3$
Δr_1	0	0	0	0	0	0
Δr_2		0	0	0	0	0
Δr_3			1	0	0	0
$r\Delta\beta_1$				1	$\dfrac{c(1-c)}{s^2}$	0
$r\Delta\beta_2$					1	0
$r\Delta\beta_3$						0

the three Cl-masses. The array for the point mass P (see Table 6.6) is more complicated. Because of the unique position of the P nucleus at the end of all three bonds and at the apex of all three bond angles, there are no zero entries. Verification of Table 6.6 is left as a simple exercise for the reader.

At last we can write down the elements of the **G** matrix, using eqn (6.6), which for convenience we repeat here:

$$G_{ij} = \sum_t (b_i^{(t)} \cdot b_j^{(t)}) \frac{1}{m_t}.$$

All the required information is contained in the tables of dot products of **b** vectors which have been given above. For example, consider the simple case of the element G_{11}. The

TABLE 6.6

*Dot products of **b** vectors for* PCl_3

P nucleus	Δr_1	Δr_2	Δr_3	$r\Delta\beta_1$	$r\Delta\beta_2$	$r\Delta\beta_3$
Δr_1	1	c	c	$\dfrac{-2c(1-c)}{s}$	$-s$	$-s$
Δr_2		1	c	$-s$	$\dfrac{-2c(1-c)}{s}$	$-s$
Δr_3			1	$-s$	$-s$	$\dfrac{-2c(1-c)}{s}$
$r\Delta\beta_1$				$2(1-c)$	$\dfrac{(1-c)^2(1+3c)}{s^2}$	$\dfrac{(1-c)^2(1+3c)}{s^2}$
$r\Delta\beta_2$					$2(1-c)$	$\dfrac{(1-c)^2(1+3c)}{s^2}$
$r\Delta\beta_3$						$2(1-c)$

internal coordinate D_1 is Δr_1. We therefore take the entries in Tables 6.3, 6.4, 6.5, and 6.6 which are situated in the first row and the first column, multiply in each case by the reciprocal of the mass of the nucleus in question, and then sum the products. The entries from the four tables are found to be 1, 0, 0, and 1. Writing m_{Cl} and m_P for the masses of the Cl and P respectively, we find at once that

$$G_{11} = \frac{1}{m_{Cl}} + \frac{1}{m_P}.$$

Some of the other elements of the **G** matrix are rather less simple in form, but the reader should have no difficulty in working them out. When he has done this he will find that, even after allowing for the symmetric nature of **G**, the elements are not all distinct. This, of course, is a consequence of the symmetry of the molecule. In fact the array which constitutes the matrix **G** may be presented in the convenient form of Table 6.7, in which we have labelled the rows and columns with the internal coordinates to which they relate, and have used the top left-hand corner to accommodate the symbol for the matrix.

TABLE 6.7

The form of the matrix **G** *for* PCl_3

G	Δr_1	Δr_2	Δr_3	$r\Delta\beta_1$	$r\Delta\beta_2$	$r\Delta\beta_3$
$D_1 = \Delta r_1$	A	B	B	E	F	F
$D_2 = \Delta r_2$		A	B	F	E	F
$D_3 = \Delta r_3$			A	F	F	E
$D_4 = r\Delta\beta_1$				C	D	D
$D_5 = r\Delta\beta_2$					C	D
$D_6 = r\Delta\beta_3$						C

To complete the specification of the **G** matrix we finally give a list of the six distinct element-values, these are

$$A = \frac{1}{m_{Cl}} + \frac{1}{m_P},$$

$$B = \cos\beta \frac{1}{m_P},$$

$$C = \frac{2}{m_{Cl}} + \frac{2(1-\cos\beta)}{m_P},$$

$$D = \left\{\frac{\cos\beta(1-\cos\beta)}{\sin^2\beta}\right\}\frac{1}{m_{Cl}} + \left\{\frac{(1-\cos\beta)^2(1+3\cos\beta)}{\sin^2\beta}\right\}\frac{1}{m_P}$$

$$= \left\{\frac{\cos\beta}{1+\cos\beta}\right\}\frac{1}{m_{Cl}} + \left\{\frac{(1-\cos\beta)(1+3\cos\beta)}{1+\cos\beta}\right\}\frac{1}{m_P},$$

$$E = -\left\{\frac{2\cos\beta(1-\cos\beta)}{\sin\beta}\right\}\frac{1}{m_P},$$

$$F = -\sin\beta\frac{1}{m_P}.$$

6.5. The F matrix for the PCl_3 molecule, using a general valency force field

A commonly used type of quadratic force field, and the one directly related to our choice of bond-length extensions and bond-angle increases as internal coordinates, is the *valency force field*.

On account of the symmetrical equivalence of the three bonds
on the one hand, and of the three bond angles on the other, the
general form of this field has only 6 distinct force constants, i.e.
6 distinct **F**-matrix elements to occupy the 21 available places.
(The occupancy of the remaining 15 places in the 6×6 array
follows from its symmetric nature.) The first of the distinct force
constants is concerned simply with the stretching of one bond.
It will necessarily be the same for each of the three bonds, and so
will occupy the first three places on the diagonal of **F**. Denoting
it by f_r, we have $F_{11} = F_{22} = F_{33} = f_r$. In a similar way the force
constant f_β for the increase of one bond angle will occupy the
remaining three diagonal places: $F_{44} = F_{55} = F_{66} = f_\beta$. The
non-diagonal elements are known as interaction force constants,
since they are concerned with contributions to the potential
energy which arise from the products of two different internal
coordinates. The stretching interaction constant $f_{rr'}$ will clearly
occupy 3 out of the 15 off-diagonal places for which we have to
account; in fact we may write $F_{12} = F_{23} = F_{13} = f_{rr'}$. The
analogous constant $f_{\beta\beta'}$, concerned with interaction between
changes of two different bond angles, will also occupy 3 places:
$F_{45} = F_{56} = F_{46} = f_{\beta\beta'}$. Finally we come to the interaction
constants, of which there are two, concerned with contributions
to the potential energy from simultaneous stretching of a bond
and change of a bond angle. The first, which we may call $f_{r\beta}$,
relates to the interaction between the stretching of a certain bond
and the change of one of the two bond angles of which the bond
in question forms a limb. This will clearly occupy 6 places in the
F array: $F_{15} = F_{16} = F_{24} = F_{26} = F_{34} = F_{35} = f_{r\beta}$. The last
remaining force constant, $f_{r\beta'}$, relates to the interaction between
the stretching of a certain bond and the change of the bond angle
opposite to it, i.e. $F_{14} = F_{25} = F_{36} = f_{r\beta'}$.

We have now accounted for all the elements of the **F** matrix.
It is shown in Table 6.8 in the same convenient form as we
adopted for the **G** matrix in Table 6.7.

It is interesting to note that the six distinct force constants are
arranged in the **F** matrix in a pattern identical with that of the
arrangement of the six distinct elements in the **G** matrix.

TABLE 6.8

The matrix **F** *for* PCl_3: *general valency force field*

F	Δr_1	Δr_2	Δr_3	$r\Delta\beta_1$	$r\Delta\beta_2$	$r\Delta\beta_3$
$D_1 = \Delta r_1$	f_r	$f_{rr'}$	$f_{rr'}$	$f_{r\beta'}$	$f_{r\beta}$	$f_{r\beta}$
$D_2 = \Delta r_2$		f_r	$f_{rr'}$	$f_{r\beta}$	$f_{r\beta}$	$f_{r\beta}$
$D_3 = \Delta r_3$			f_r	$f_{r\beta}$	$f_{r\beta}$	$f_{r\beta'}$
$D_4 = r\Delta\beta_1$				f_β	$f_{\beta\beta'}$	$f_{\beta\beta'}$
$D_5 = r\Delta\beta_2$					f_β	$f_{\beta\beta'}$
$D_6 = r\Delta\beta_3$						f_β

6.6. The secular equation for the vibrations of the PCl_3 molecule

Having set up the **G** matrix and the **F** matrix, we could go on to calculate the vibrational secular equation, which (as we have shown in Section 5.6) may be written in the form $|\mathbf{FG}-\lambda\mathbf{E}| = 0$. The process would involve multiplying the two matrices together; but even for such a small molecule as PCl_3, where each of them has only 6 rows and 6 columns, this would be rather a laborious undertaking. With increase in the number of nuclei the labour involved increases rapidly.

The secular determinant, like **G** and **F**, must have 6 rows and 6 columns, and when expanded must take the form of a polynomial of the sixth degree in λ. The solution of the secular equation in this form would again be a laborious matter, and we shall not pursue it here. Indeed, it would be pointless to do so; for, as we shall presently show, it is a consequence of the symmetry of the molecule that only four of the six roots can be distinct, and that the values of these four can be obtained, two at a time, by the trivial process of solving two quadratic equations. The two equations in question can be derived by taking symmetry into account in the mechanical treatment. The very great advantage of doing this in the case of the PCl_3 molecule (and others like it) arises from the circumstance that the number of nuclei is quite small, while at the same time the degree of symmetry is quite high; for other species of lower symmetry the advantage will not be so striking. However, for any molecule possessing any degree of symmetry, a valuable simplification of the vibrational problem can be achieved by taking this symmetry fully into account. This is true, not only of the classical, but also of the quantum-mechanical treatment. At this point, therefore, we shall pause to consider the theory of molecular symmetry and how we may use it to advantage in dealing with molecular vibrations.

7 Symmetry point groups and matrix representations of them

7.1. Symmetry elements and symmetry operations

PCl_3 will continue to serve us well as an example of a molecule possessing symmetry, it being assumed that all three Cl nuclei are of the same isotopic species. The regular triangular pyramidal model has been shown in Fig. 6.4, in which the point masses representing the three Cl nuclei were numbered 1, 2, and 3. In this connection it should be noted that, in the molecule itself, the three actual Cl nuclei are in principle indistinguishable, so that any such individual labelling would be improper. No such limitation arises, however, in the discussion of symmetry properties in terms of a molecular model, in which the Cl nuclei are represented by distinguishable point masses; nor shall we encounter any difficulty in applying the results of such discussion in the subsequent treatment of the vibrations of the molecule.

The symmetry properties of a molecule in its equilibrium configuration are described by saying that it possesses certain *symmetry elements*, with which are associated certain *symmetry operations*. The symmetry elements include reflection planes and rotational axes etc., with which are associated the operations of reflection in the planes, rotations through certain angles about the axes, etc. A symmetry operation is defined as one which, when performed upon the molecular model (i.e. upon the model of the molecule in its equilibrium configuration) produces a result which (except for the labelling of certain equivalent point masses) is indistinguishable from the original model.

Let us now consider the symmetry properties of the PCl_3 model. The first symmetry element is the three-fold axis of rotation, designated as C_3. This is shown in Fig. 7.1 as the vertical dotted line passing through the apical point mass P.

Associated with this element are two symmetry operations. The first, which is also (perhaps rather confusingly) called C_3, is a rotation of the model around the C_3 axis through the angle $\frac{2}{3}\pi$ (i.e. 120°) in an anticlockwise direction as viewed along the axis, looking from the point P towards the centre of mass of the trio of point masses representing the Cl nuclei. Clearly this operation leaves point mass P unmoved, but causes point mass 1 to move to the position originally occupied by point mass 2, point mass 2 to move to the position originally occupied by point mass 3, and point mass 3 to move to the position originally occupied by point mass 1. Thus each of the point masses 1, 2, and 3 becomes replaced by another which is indistinguishable from it, except for the number-label. The operation C_3 therefore satisfies the definition of a symmetry operation given above. It may be conveniently specified in terms of the replacements it effects:

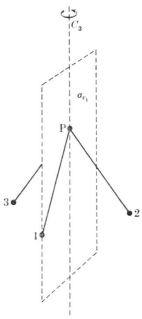

Fɪɢ. 7.1. Model of the PCl_3 molecule, showing the rotational axis and one of the three reflection planes.

$$C_3 \quad \begin{Bmatrix} 1 \text{ replaced by } 3 \\ 2 \text{ replaced by } 1 \\ 3 \text{ replaced by } 2 \end{Bmatrix}. \qquad (7.1)$$

The second symmetry operation associated with the three-fold axis of rotation, is a similar rotation but through the angle $\frac{4}{3}\pi$ (i.e. 240°). It is entirely equivalent to two successive C_3 operations and is designated as C_3^2. It leaves point mass P unmoved, and effects the following replacements amongst the other three:

$$C_3^2 \quad \begin{Bmatrix} 1 \text{ replaced by } 2 \\ 2 \text{ replaced by } 3 \\ 3 \text{ replaced by } 1 \end{Bmatrix}. \qquad (7.2)$$

101

We may here note an alternative nomenclature which is sometimes encountered, in which C_3 and C_3^2 are called C_3^+ and C_3^-
respectively, the superscripts $+$ and $-$ denoting anticlockwise
and clockwise respectively. Clearly an anticlockwise rotation
through $\frac{4}{3}\pi$ is entirely equivalent to a clockwise one through $\frac{2}{3}\pi$.

It might be thought reasonable to associate a third symmetry
operation with the axis C_3, namely, a rotation through 2π (i.e.
360°). This would, of course, be equivalent to doing nothing at
all; for it would not only leave the point mass P unmoved, but
would also bring each of the other masses back to its original
position. Indeed, this trivial operation, called the identity
operation, is included in the list of symmetry operations which
characterize the molecule; but it is regarded as distinct, and is
not associated with any particular symmetry element. We shall
denote the identity operation by the symbol I. This seems
preferable to the symbol E, which is probably more widely
current, but may tend to cause confusion, as it is also conventionally used in symmetry theory with another quite different
significance.

In addition to the C_3 axis, the PCl_3 model possesses three
others, namely, the reflection planes or planes of symmetry
(called σ_{v_1}, σ_{v_2}, and σ_{v_3}) which contain the C_3 axis and the point
masses labelled 1, 2, and 3 respectively. The σ_{v_1} plane is indicated
in Fig. 7.1. Each such plane has one symmetry operation
associated with it, namely, the reflection of the model in the
plane. These operations are denoted by the same symbols as
the symmetry elements themselves. The reflection σ_{v_1} leaves
point mass 1 (as well as point mass P) unmoved, but interchanges
point masses 2 and 3. We may specify it by the replacements it
effects:

$$\sigma_{v_1} \quad \begin{Bmatrix} 2 \text{ replaced by } 3 \\ 3 \text{ replaced by } 2 \end{Bmatrix}. \tag{7.3}$$

The remaining two reflections are correspondingly specified thus:

$$\sigma_{v_2} \quad \begin{Bmatrix} 1 \text{ replaced by } 3 \\ 3 \text{ replaced by } 1 \end{Bmatrix}, \tag{7.4}$$

and $$\sigma_{v_3} \quad \begin{Bmatrix} 1 \text{ replaced by } 2 \\ 2 \text{ replaced by } 1 \end{Bmatrix}. \tag{7.5}$$

This completes the list of the six symmetry operations for the PCl$_3$ model, namely I, C_3, C_3^2, σ_{v_1}, σ_{v_2}, and σ_{v_3}.

Other molecular models may possess other types of symmetry elements and associated symmetry operations. Besides further reflection planes denoted by slightly different symbols, we encounter twofold, fourfold, fivefold, and sixfold rotational axes (C_2, C_4, C_5, C_6 respectively) with associated operations of rotation through appropriate angles; a centre of symmetry i, whose associated operation is reflection of all point masses through it; and rotation–reflection axes S_n (also called improper axes) of which the associated operation consists of appropriate rotations followed by reflections in a plane at right angles to the axis.

7.2. Products of symmetry operations

The overall operation consisting of one symmetry operation R followed by another symmetry operation S is itself a symmetry operation, conventionally referred to as the *product SR*, which will of course have been included in the complete list applicable to the molecule under consideration. We may exemplify this by considering PCl$_3$ and the effect of C_3 followed by σ_{v_1}. Point mass P is left unmoved by both operations. The effects upon point masses 1, 2, and 3 is shown in Fig. 7.2, which is a plan seen from the side on which point mass P is situated. It should be noted in particular that the reflection plane σ_{v_1} is regarded as fixed in space in the position determined by the *initial* position of point mass 1, with no possibility of alteration as symmetry operations are performed upon the model. Thus after the performance of C_3 the plane σ_{v_1} no longer passes through point mass 1, but through point mass 3. The overall effect of the two consecutive operations is seen to be to replace 1 by 3 and 3 by 1, while bringing 2 back to its initial position. The product operation is thus identified as σ_{v_2} (see (7.4)), and we may write

$$\sigma_{v_1} C_3 = \sigma_{v_2}. \tag{7.6}$$

Here we note that, in writing the product, the symbol of the first operation is written on the right of that of the subsequent operation. Thus the order of writing (from left to right) is the

reverse of the order of performance. One value of this convention is that it distinguishes between a product SR and the
reverse product RS; for, as we shall now demonstrate, they are

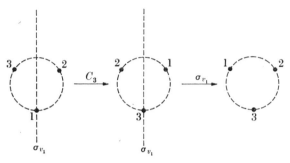

FIG. 7.2.   Effects of successive symmetry operations upon
the three equal point masses representing the three Cl
nuclei in the model of the PCl_3 molecule.

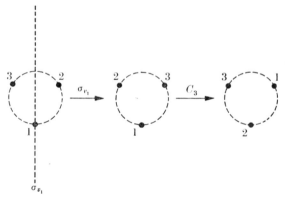

FIG. 7.3.   Effects of reversing the order of the operations,
as compared with Fig. 7.2.

not in general the same. Thus consider the effect of reversing
the order of performance of the operations C_3 and σ_{v_1} upon the
PCl_3 model. This is shown diagrammatically in Fig. 7.3, from
which we see that the product operation now effects the replacement of 1 by 2 and of 2 by 1. It is thus identified as σ_{v_3} (see
(7.5)). We thus have

$$C_3\sigma_{v_1} = \sigma_{v_3} \qquad (7.7)$$

which is to be contrasted with eqn (7.6).

104

Similar considerations apply to the successive performances of more than two symmetry operations, and the same convention is followed in writing the symbol for the product. Thus if the operations R, S, and T are carried out in that order, the product is written as TSR.

Two successively performed symmetry operations need not necessarily be different. Thus $C_3 C_3 = C_3^2$, a result which incidentally indicates the reasonable basis for the use of the index 2 in the symbol of the operation C_3^2. Two successive performances of the reflection σ_{v_1} obviously bring us back identically to the original situation. Accordingly we write $\sigma_{v_1} \sigma_{v_1} = I$.

Given the complete list of symmetry operations in any particular case, it is possible to construct a table of all the products of them two at a time. A convenient form in which this may be done is shown in Table 7.1 which applies to our example, the PCl_3 molecule. (For the meaning of 'point group C_{3v}' in the title of Table 7.1, see the next section.) It is necessary to explain that the entry at the intersection of a certain column and a certain

TABLE 7.1

Multiplication table for the point group C_{3v}

C_{3v}	I	C_3	C_3^2	σ_{v_1}	σ_{v_2}	σ_{v_3}
I	I	C_3	C_3^2	σ_{v_1}	σ_{v_2}	σ_{v_3}
C_3	C_3	C_3^2	I	σ_{v_3}	σ_{v_1}	σ_{v_2}
C_3^2	C_3^2	I	C_3	σ_{v_2}	σ_{v_3}	σ_{v_1}
σ_{v_1}	σ_{v_1}	σ_{v_2}	σ_{v_3}	I	C_3	C_3^2
σ_{v_2}	σ_{v_2}	σ_{v_3}	σ_{v_1}	C_3^2	I	C_3
σ_{v_3}	σ_{v_3}	σ_{v_1}	σ_{v_2}	C_3	C_3^2	I

row is the product obtained by performing first the operation shown at the head of the column, and then the operation shown at the left-hand end of the row. Since in general a product SR is not the same as the reverse product RS, the table is not symmetrical about the diagonal from the top left-hand corner to the bottom right-hand corner.

Inspection of the table shows that the entry I occurs just once in each column and once in each row. Thus for any symmetry operation R there exists one other symmetry operation S such that $SR = I$. This means that S entirely undoes the effect of R, and accordingly we refer to S as the inverse of R and write $S = R^{-1}$. It follows also that $R = S^{-1}$.

7.3. The symmetry operations of a molecule as a point group

The full set of symmetry operations appropriate to the model of any molecule in its equilibrium configuration is said to constitute a *mathematical point group*. Each such group is designated by a conventional symbol. In the so-called Schoenflies notation, which we shall use in this book, the symbol for our example (the PCl_3 molecule) is C_{3v}. In speaking of the mathematical point group whose elements are the symmetry operations of a molecular model, the word 'point' is used because there is always at least one point (not necessarily occupied by a mass) which is left unaffected by every one of the operations. Clearly the centre of mass of a system is always a point of this kind. In our example of PCl_3 all points lying on the C_3 axis, including in particular the point representing the P nucleus, are unmoved by all the six symmetry operations.

A set of elements (such as the symmetry operations for a molecular model) must, in order that they shall constitute a group, satisfy the following four requirements:

(1) the set must contain the identity operation I with the property that, if R be any element of the group, then $RI = IR = R$;

(2) the product SR of any two elements of the group must itself be an element of the group;

(3) each element R must have an inverse R^{-1} which is also an element of the group, satisfying the relation
$$RR^{-1} = I = R^{-1}R;$$

(4) products of elements must satisfy the associative law, as in $T(SR) = (TS)R$.

We have already seen that requirements 1, 2, and 3 are satisfied by the complete set of symmetry operations for PCl_3.

That requirement 4 is also satisfied may be verified by the use of the binary products in Table 7.1. For example, we find

$$C_3^2(\sigma_{v_1} C_3) = C_3^2 \sigma_{v_2} = \sigma_{v_3},$$

and
$$(C_3^2 \sigma_{v_1})C_3 = \sigma_{v_2} C_3 = \sigma_{v_3},$$

thus verifying that

$$C_3^2(\sigma_{v_1} C_3) = (C_3^2 \sigma_{v_1})C_3.$$

In common parlance, the situation is summed up by saying that the PCl_3 molecule belongs to the point group C_{3v} or that it possesses C_{3v} symmetry.

Molecules of other shapes belong to other point groups, of which there are only a limited number, however. We shall have occasion to refer to some of them later.

7.4. Representation of symmetry operations by matrices

The symmetry operations we have been discussing are those appropriate to the model of a molecule *in its equilibrium configuration*, and we specified their effects by listing the replacements they produced amongst the point masses representing symmetrically equivalent, identical nuclei of the molecule. We may now go on to consider the effects of these same symmetry operations upon the *displacements* of the nuclei when the molecule is in a distorted (non-equilibrium) configuration, such as occurs during molecular vibrations.

We have already seen in Chapter 4 that nuclear displacements may be regarded as vectors, and (in Section 4.7) that the performance of a symmetry operation upon a vector may be represented by the pre-multiplication of a single-column matrix by a square matrix. We shall now consider this more closely, using the PCl_3 molecule as an example. In the interests of simplicity a general displacement of the apical phosphorus nucleus will first be considered. As a set of three basis vectors in terms of which the nuclear displacement under consideration may be expressed we choose the mutually orthogonal set of unit vectors e_x, e_y, e_z. Let e_z be directed along the C_3 axis so as to point away from the trio of Cl nuclei and towards the P nucleus. Our special choices of e_x and e_y are indicated in Fig. 7.4. This shows the xy plane, the e_z vector being directed towards the reader. The three P–Cl bonds are situated on the far side of the

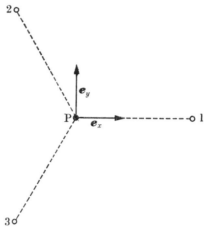

Fig. 7.4.   Particular choice of unit basis vectors e_x and e_y in considering a general displacement of the P nucleus in PCl_3.

plane, their projections upon it being indicated by the dotted lines. The arbitrary elements of the choice may be summed up by saying that e_x is chosen so as to lie in the plane containing the C_3 axis and a particular site occupied by a Cl nucleus in the equilibrium configuration of the molecule. While continuing to accept (as indeed we must) that all three identical Cl nuclei are in principle indistinguishable, so that it would be improper to label them with individual numbers, we may nevertheless distinguish in this way between the different *sites* occupied by these nuclei; for to distinguish between the sites in no way involves any claim to know which of the three Cl nuclei is, at any particular moment, occupying any particular one of them. Accordingly we shall henceforth use the numbers 1, 2, and 3 to label the sites of the Cl nuclei in the equilibrium configuration of the PCl_3 molecule. The reflection planes σ_{v_1}, σ_{v_2}, and σ_{v_3} may now be defined as containing the C_3 axis and the sites 1, 2, and 3 respectively. Incidentally, the reason for using the subscripts

v_1, v_2, and v_3 to designate the reflection planes is that, by convention, the rotation axis of highest order of a molecule (in PCl$_3$
there is only one rotation axis, C_3) is regarded as being vertical.
Planes containing it are therefore referred to as vertical planes
and the symbol for them is given the subscript v. Having chosen
the basis vector $\boldsymbol{e}_z$ in the direction of the C_3 axis, our choice of $\boldsymbol{e}_x$
may thus be specified by saying that it lies in the σ_{v_1} plane.
Thereafter the choice of $\boldsymbol{e}_y$ is uniquely determined (see Fig.
7.4).

Having used the numbers 1, 2, and 3 to label the sites of the Cl
nuclei, we shall use the number 4 to label the site of the P nucleus.
A general displacement of this nucleus will thus be described by
the coordinates x_4, y_4, z_4 in our chosen basis. Each of the six
symmetry operations of PCl$_3$ will now be representable by a 3×3
matrix, the elements of which will be determined by the particular basis choice we have made. Since a symmetry operation
leaves the length of the displacement vector unchanged, it follows
from Section 3.8 that all the matrices in question will be orthogonal ones. We have already deduced in Section 4.7 the matrix
representing the operation of rotation through the general
angle θ—see eqn (4.34). Using $\theta = 120°$, so that $\sin\theta = \frac{1}{2}\sqrt{3}$
and $\cos\theta = -\frac{1}{2}$, we find that the transformation in question
may be expressed as

$$\begin{bmatrix} x_4' \\ y_4' \\ z_4' \end{bmatrix} = \begin{bmatrix} -\frac{1}{2} & -\frac{1}{2}\sqrt{3} & 0 \\ \frac{1}{2}\sqrt{3} & -\frac{1}{2} & 0 \\ 0 & 0 & 1 \end{bmatrix} \begin{bmatrix} x_4 \\ y_4 \\ z_4 \end{bmatrix}. \tag{7.8}$$

Here x_4', y_4', and z_4' are the coordinates after the operation C_3.
Similarly using $\theta = 240°$, so that $\sin\theta = -\frac{1}{2}\sqrt{3}$ and $\cos\theta = -\frac{1}{2}$,
we obtain

$$\begin{bmatrix} x_4' \\ y_4' \\ z_4' \end{bmatrix} = \begin{bmatrix} -\frac{1}{2} & \frac{1}{2}\sqrt{3} & 0 \\ -\frac{1}{2}\sqrt{3} & -\frac{1}{2} & 0 \\ 0 & 0 & 1 \end{bmatrix} \begin{bmatrix} x_4 \\ y_4 \\ z_4 \end{bmatrix}. \tag{7.9}$$

The square matrix represents C_3^2.

We shall now deduce the corresponding matrices for the
remaining four symmetry operations of PCl$_3$. The treatment

of the identity operation I is trivial, and the matrix representing it is obviously the unit matrix,

$$\begin{bmatrix} 1 & 0 & 0 \\ 0 & 1 & 0 \\ 0 & 0 & 1 \end{bmatrix}.$$

The reflection σ_{v_1} leaves x_4 and z_4 unchanged, and simply reverses the sign of y_4. The matrix representing it is thus

$$\begin{bmatrix} 1 & 0 & 0 \\ 0 & -1 & 0 \\ 0 & 0 & 1 \end{bmatrix}.$$

The operations σ_{v_2} and σ_{v_3} are most easily dealt with by using the binary-product relations in Table 7.1. Thus $\sigma_{v_2} = C_3^2 \sigma_{v_1}$. The matrix representing σ_{v_2} is therefore obtained as follows:

$$\begin{bmatrix} -\frac{1}{2} & \frac{1}{2}\sqrt{3} & 0 \\ -\frac{1}{2}\sqrt{3} & -\frac{1}{2} & 0 \\ 0 & 0 & 1 \end{bmatrix} \begin{bmatrix} 1 & 0 & 0 \\ 0 & -1 & 0 \\ 0 & 0 & 1 \end{bmatrix} = \begin{bmatrix} -\frac{1}{2} & -\frac{1}{2}\sqrt{3} & 0 \\ -\frac{1}{2}\sqrt{3} & \frac{1}{2} & 0 \\ 0 & 0 & 1 \end{bmatrix}.$$

Likewise from $\sigma_{v_3} = C_3 \sigma_{v_1}$ we find that the matrix representing σ_{v_3} is

$$\begin{bmatrix} -\frac{1}{2} & \frac{1}{2}\sqrt{3} & 0 \\ \frac{1}{2}\sqrt{3} & \frac{1}{2} & 0 \\ 0 & 0 & 1 \end{bmatrix}.$$

We have thus arrived at a complete set of orthogonal matrices, one for each of the symmetry operations appropriate to the PCl_3 molecule. As we have already pointed out, the actual elements of them are determined by the particular basis to which they are related. An unlimited number of alternative choices of basis is possible, and for each the matrix elements would be different. Nevertheless, all the 3×3 matrices representing a particular symmetry operation, are essentially *equivalent*, and as we saw in Section 4.8 they are related to one another by so-called similarity transformations.

Furthermore, the representation of a symmetry operation by a 3×3 matrix is limited to showing the effect of the operation upon a vector in 3-dimensional space. In order to represent such an effect upon a vector in n-dimensional space, an $n \times n$ matrix is required. Thus the number of matrices that can represent a particular symmetry operation is unlimited for two reasons: on the one hand an unlimited number of choices of basis is possible for a given value of n, and on the other the number of n-values is unlimited. We shall presently see how the theory of groups has brought order to this complicated situation.

7.5. The effect of a symmetry operation upon a vector in configuration space

For the purposes of illustration we shall continue to use the PCl_3 molecule, considering now a general distortion in which all four nuclei are displaced from their equilibrium sites. For each site we choose three basis unit vectors, their directions being the same as we chose in the preceding section for the site of the P nucleus. We now combine all the four 3-dimensional bases into a single orthonormal basis for 12-dimensional configuration space. The vector representing the molecular distortion in this space will have 12 coordinates: x_1, y_1, z_1 for the displacement from site 1; x_2, y_2, z_2 for the displacement from site 2; and so on.

Consider, for example, the effect upon these coordinates (and so upon the displacement vector in configuration space) of performing the symmetry operation C_3 upon the distorted molecule in ordinary physical space. Note that, although it is performed upon the molecule in a non-equilibrium configuration, the symmetry operation is one which is characteristic of the symmetry of the molecule in its equilibrium configuration. As far as site 4 is concerned, we have already dealt with the effect of operation C_3 upon the general displacement described by $x_4, y_4,$ and z_4. In fact we found that, if the coordinate-values after the operation are $x'_4, y'_4,$ and z'_4, then

$$\begin{bmatrix} x'_4 \\ y'_4 \\ z'_4 \end{bmatrix} = \begin{bmatrix} -\frac{1}{2} & -\frac{1}{2}\sqrt{3} & 0 \\ \frac{1}{2}\sqrt{3} & -\frac{1}{2} & 0 \\ 0 & 0 & 1 \end{bmatrix} \begin{bmatrix} x_4 \\ y_4 \\ z_4 \end{bmatrix}.$$

The displacement vector in question remains associated with site 4 under operation C_3, and is merely rotated about the axis of symmetry.

The effects on the displacements originally associated with the

111

three Cl-sites are not so simple. In Section 7.1 we saw that, when applied to the molecular model in its equilibrium configuration, operation C_3 carried the point masses originally at the sites 1, 2, and 3 into the new sites 2, 3, and 1 respectively. We note that the sites, unlike the associated point masses, are fixed in space and so are unaffected by any symmetry operation. It now follows that when the operation C_3 is performed upon the distorted molecule, the displacement vectors originally associated with the sites 1, 2, and 3 must be carried round so that they become associated with the new sites 2, 3, and 1 respectively. Simultaneously each will suffer a rotation through 120° about the axial direction.

The reader should here note in particular that the statements in the last two sentences, although based upon consideration of a model in which the atomic nuclei are represented by distinguishable point masses, remain meaningful and true for the actual molecule, where the three identical Cl nuclei are in principle absolutely indistinguishable. In fact the statements, being about nuclear displacements, do not in any way imply any knowledge of the identities of the individual nuclei at any particular moment. They consequently remain unimpaired by any conceivable (though in principle unobservable) exchanges amongst these nuclei. We do not commit the impropriety of labelling the nuclei; we merely label the nuclear displacements. Considerations of this kind apply quite generally to the applications of symmetry theory in the treatment of molecular vibrations.

Returning now to the details of our example, we see that the components z_1, z_2, and z_3 are carried by operation C_3 to the new sites 2, 3, and 1 respectively. We can therefore write $z_1' = z_3$, $z_2' = z_1$, and $z_3' = z_2$. The effect of C_3 in transferring the components x_3 and y_3, for example, from site 3 to site 1 can be seen in Fig. 7.5. This represents the plane containing the three Cl-sites, the rotational axis being directed towards the reader. For clarity, only the components originally associated with the site 3 are shown. From the figure we easily see that $x_1' = -\frac{1}{2}x_3 - \frac{1}{2}\sqrt{3}\,y_3$, and $y_1' = \frac{1}{2}\sqrt{3}\,x_3 - \frac{1}{2}y_3$. The transformations for the x and y

components associated with the other two Cl-sites can be similarly deduced.

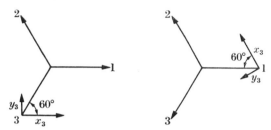

FIG. 7.5. Effect of the symmetry operation C_3 in moving the displacements x_3 and y_3 from the Cl site 3 of PCl_3 to the Cl site 1.

It follows that the transformation for the whole configuration vector has the form:

$$
\begin{bmatrix} x_1' \\ y_1' \\ z_1' \\ x_2' \\ y_2' \\ z_2' \\ x_3' \\ y_3' \\ z_3' \\ x_4' \\ y_4' \\ z_4' \end{bmatrix}
=
\begin{bmatrix}
& & & & & & -\tfrac{1}{2} & -\tfrac{1}{2}\sqrt{3} & 0 & & & \\
& & & & & & \tfrac{1}{2}\sqrt{3} & -\tfrac{1}{2} & 0 & & & \\
& & & & & & 0 & 0 & 1 & & & \\
-\tfrac{1}{2} & -\tfrac{1}{2}\sqrt{3} & 0 & & & & & & & & & \\
\tfrac{1}{2}\sqrt{3} & -\tfrac{1}{2} & 0 & & & & & & & & & \\
0 & 0 & 1 & & & & & & & & & \\
& & & -\tfrac{1}{2} & -\tfrac{1}{2}\sqrt{3} & 0 & & & & & & \\
& & & \tfrac{1}{2}\sqrt{3} & -\tfrac{1}{2} & 0 & & & & & & \\
& & & 0 & 0 & 1 & & & & & & \\
& & & & & & & & & -\tfrac{1}{2} & -\tfrac{1}{2}\sqrt{3} & 0 \\
& & & & & & & & & \tfrac{1}{2}\sqrt{3} & -\tfrac{1}{2} & 0 \\
& & & & & & & & & 0 & 0 & 1
\end{bmatrix}
\begin{bmatrix} x_1 \\ y_1 \\ z_1 \\ x_2 \\ y_2 \\ z_2 \\ x_3 \\ y_3 \\ z_3 \\ x_4 \\ y_4 \\ z_4 \end{bmatrix}
$$

$$(7.10)$$

All elements which are not shown are zero in value.

The square matrix is a 12×12 representation of the operation C_3. A corresponding 12×12 matrix could be derived for each of the other symmetry operations of the molecular point group. Other choices of basis would have given rise to other equivalent matrices. We could equally well have considered only the displacements of the three Cl nuclei of PCl_3, just as we previously considered only that of the P nucleus in obtaining 3×3 matrices representing the group elements. This would have resulted in a set of 9×9 matrices for each choice of basis. Moreover, PCl_3

is by no means the only type of molecule belonging to the point group C_{3v}. Examples of other types are $POCl_3$, $P(CN)_3$, and $P(NCO)_3$, containing respectively 5, 7, and 10 nuclei, and having the structures shown in Fig. 7.6. Considerations similar to those above concerning PCl_3 would give respectively 15×15, 21×21, and 30×30 matrices representing the symmetry operations of the group C_{3v}.

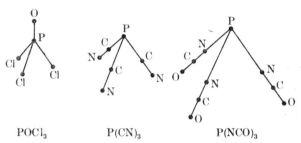

POCl₃   P(CN)₃   P(NCO)₃

FIG. 7.6.   Models of molecules belonging to the point group C_{3v}.

7.6. Matrix representations of a group

Let R and S be two symmetry operations of a group, and let their product SR (operation R followed by operation S) be the operation T, i.e.

$$SR = T. \tag{7.11}$$

With a certain choice of basis, let the matrices representing these operations be **R**, **S**, and **T** respectively. Suppose that **R** transforms the single-column matrix **X** of a general displacement vector into the single-column matrix **X'**, and that **S** transforms **X'** into **X''**. Then we may write

$$\mathbf{X'} = \mathbf{RX},$$

and $$\mathbf{X''} = \mathbf{SX'} = \mathbf{SRX}.$$

But we know that $\mathbf{SRX} = \mathbf{TX}$, so it follows that

$$\mathbf{SR} = \mathbf{T}. \tag{7.12}$$

Comparison of eqns (7.12) and (7.11) shows that the product relation for the matrices representing the symmetry operations exactly parallels that for the operations themselves.

Since the complete set of symmetry operations constitutes a mathematical group, it follows that a complete set of matrices representing these operations also forms such a group, for it likewise satisfies the four requirements listed in Section 7.3. Just as the individual matrices represent the individual symmetry operations, so also the complete set of matrices is said to constitute a *representation of the group*.

7.7. Reducible and irreducible group representations

We have already seen that it is possible to obtain an unlimited number of matrix representations of a group, according not only to the dimensions of the vector space but also to the choice of basis vectors. It is an important achievement of general group theory that it enables us to escape from this *embarras de choix* by showing how all the representations of a particular group may be 'decomposed' into a quite small number of underlying types. In the case of the group C_{3v} of the PCl_3 molecule, for example, this number is three.

For purposes of exposition, it will be advantageous to proceed in what is admittedly a rather illogical fashion. Instead of presenting an argued development of the theory from first principles, we shall begin by giving a descriptive sketch of the results of simplification of group representations, using the group C_{3v} for illustrative purposes. In doing this we shall have to invoke certain findings of group theory, and for the time being take their validity for granted without proof. Having thus reviewed the general situation, and in doing so become acquainted with the sort of consequences that follow from the application of the theory, we shall afterwards complete the picture by discussing how the lacunae in the logical framework can be filled in.

Consider a set of $n \times n$ matrices, one for each symmetry operation, which constitute a representation of a point group with a particular choice of basis vectors. In general, the effect of a symmetry operation upon a configuration vector is to transform each of its coordinates into a linear combination of a number of those coordinates. We may conveniently express this by saying that the operation causes *mixing* of the coordinates. Consider now the change to a new basis. As follows from Section 4.8, the

115

new group representation is obtained by subjecting each of the matrices of the original representation to the same appropriate similarity transformation. It may be that, using these new matrices, the extent of mixing of the new coordinates is less than before. If this is so, we say that the original representation was *reducible*, and that the change of basis has effected a certain measure of *reduction*.

The possible extent of reduction is, however, limited. When the result of a change to a new basis is such as to achieve the maximum degree of reduction, the representation so obtained (still of course consisting of $n \times n$ matrices) is said to be *completely reduced*. In general, the choice of a basis which will achieve complete reduction is not unique. Every such choice is characterized, however, by the fact that the associated displacement coordinates form a number of distinct sets, in the sense that the members of any one set may mix with one another under a symmetry operation, but that in no case can a member of one set mix with a member of another set. An alternative way of expressing this is to say that complete reduction brings about the resolution of the n-dimensional configuration space into the largest possible number of *invariant sub-spaces*, each with the property that any symmetry operation transforms any vector of a particular sub-space into another vector which also belongs to that same sub-space. It also transpires that, no matter what the point group, the number of coordinates in a distinct set (or, otherwise expressed, the dimension of the corresponding invariant subspace) is either 1, 2, or 3. We shall see later that the maximum of 3 is connected with the 3-dimensional nature of the physical space in which the symmetry operation is actually carried out.

Let us now suppose that, in writing all these displacement coordinates as the elements of the single-column matrix **X**, we have arranged them in successive distinct sets. Because the members of a particular set may mix together as a result of the symmetry operations, in each of the square matrices representing these operations there will be a non-zero square block of the same dimensions, and this block will be situated on the diagonal.

For example, a distinct set containing two coordinates will give rise to a 2×2 block, while a set containing only one co-ordinate will give rise to a 1×1 block, i.e. to a single non-zero diagonal element. Moreover the non-mixing of coordinates belonging to different sets will mean that all matrix elements not included in the above-mentioned diagonal blocks will be zero. We say that the effect of a complete reduction is to bring each matrix of a group representation into a *diagonal-block form*. It is important to note that the pattern of diagonal blocks will be the same for all the matrices of the completely reduced group representation, although, of course, corresponding blocks in matrices representing different symmetry operations will in general have different entries, some of which may happen to be zero.

An illustrative example is discussed in the first section of the next chapter, with a diagram showing the diagonal-block form of the matrices of a 12×12 representation of the point group C_{3v}.

8 Irreducible representations and the character table of a group

8.1. The completely reduced form of a group representation

B Y way of illustration, consider any reducible 12×12 representation of the point group C_{3v}, such as could be derived by choosing an arbitrary basis and then writing down (as in Section 7.5) the matrices representing the effects of the different symmetry operations upon the 12 coordinates describing a general non-equilibrium configuration of the PCl_3 molecule. Complete reduction is possible by a suitable change of basis, which means subjecting each of the matrices of the reducible representation to the same suitably chosen similarity transformation. At a later stage we shall discuss in detail how the choice of a suitable new basis may be determined. For the present we merely state the fact (which will be substantiated later) that the effect of a complete reduction will be to cause each of the matrices to assume the diagonal-block form shown in Fig. 8.1, consisting of a diagonal array of four 1×1 blocks (i.e. single diagonal elements) and four 2×2 blocks. All other elements will be zero. Moreover, in the matrix representing any chosen symmetry operation, three of the 1×1 blocks (labelled A_1 in the figure) will necessarily be identical, whereas the fourth (labelled A_2) may be different. Also all four of the 2×2 blocks (labelled E) will necessarily be identical. The order in which the blocks appear on the diagonal merely depends upon the arbitrary order in which the distinct coordinate sets are written in the column matrix representing the displacement vector. In the case of the point group C_{3v}, with which we are at present concerned, it is conventional to regard them as having the order shown in the figure, i.e. A_1 blocks first, followed by the A_2 block and then by the E blocks.

As we have previously noted, the choice of basis effecting a complete reduction is not unique. Nevertheless, we shall see that the value of an element in an A_1 block is independent of the choice made and must be unity for all the symmetry operations.

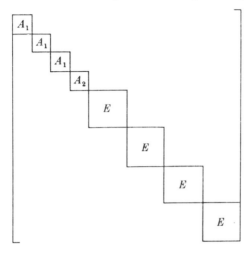

FIG. 8.1. The diagonal-block form of the matrices of a completely reduced 12×12 representation of the point group C_{3v}.

The value of the element in each A_2 block is likewise independent of the choice of basis, and is $+1$ or -1, according to the operation concerned. On the other hand, the four elements in each E block are not uniquely fixed, but vary with the choice of the basis which has effected the complete reduction.

Of great importance is the fact that, for any matrix representation of the point group C_{3v} (no matter what the dimension n of the relevant configuration space), the resulting diagonal-block form contains only blocks of the types A_1, A_2, and E. Of course, the numbers of blocks of each type will vary from case to case, according to the value of n.

8.2 Irreducible representations of a group

Let us now refer back to Section 7.4, in which we deduced a set of 3×3 matrices, one for each symmetry operation of the group C_{3v}, by considering a general displacement of the P nucleus of the PCl_3 molecule and using a special choice of basis for the 3-dimensional vector space. For convenience we reproduce the matrices in Table 8.1, which shows for each the symmetry operation which it represents.

We notice that all six matrices have the same diagonal-block pattern (indicated by the dashed lines), with a 2×2 block in the top left-hand position and a 1×1 block in the bottom right-hand position. Thus the particular basis which we chose was such as to effect a reduction, as compared with other choices which would have given rise to 3×3 matrices not possessing the diagonal-block form. In fact, as will become clear later in this section, the reduction is complete.

In passing, we may remark that the relative positions of the 2×2 and 1×1 blocks is a trivial matter, depending simply upon the order in which the displacement coordinates are written in the single-column matrix representing the displacement vector. In fact, in deriving the matrix representation shown in Table 8.1. we adopted the order x, y, z. By adopting instead the order z, x, y, we could obviously cause the blocks to exchange positions.

In Section 3.4 we saw that the product of two matrices, both of the same diagonal-block form, is a matrix of the same form, in which each block is simply the product of the corresponding blocks of the original matrices. It follows that, like the set of 3×3 matrices, the set of 1×1 blocks must itself form a representation of the group C_{3v}, for its members satisfy the four requirements set out in Section 7.3. Likewise the set of 2×2 blocks must form another representation of the group. Our special choice of basis has thus caused the 'decomposition' of the 3×3 group representation into a 1×1 representation and a 2×2 representation.

Obviously the representation formed by the 1×1 blocks (i.e. by simple numbers) cannot be further reduced. It is said to be an *irreducible representation* of the group, and is designated by the symbol A_1. The corresponding 1-dimensional invariant subspace has the unit basis vector e_z, to which corresponds the displacement coordinate z. This coordinate is therefore said to belong to the *symmetry species* A_1. The 1×1 matrices of the

120

TABLE 8.1

A matrix representation of the point group C_{3v}

$$
\begin{array}{cccccc}
I & C_3 & C_3^2 & \sigma_{v_1} & \sigma_{v_2} & \sigma_{v_3} \\[4pt]
\begin{bmatrix} 1 & 0 & 0 \\ 0 & 1 & 0 \\ 0 & 0 & 1 \end{bmatrix} &
\begin{bmatrix} -\frac{1}{2} & -\frac{\sqrt{3}}{2} & 0 \\ \frac{\sqrt{3}}{2} & -\frac{1}{2} & 0 \\ 0 & 0 & 1 \end{bmatrix} &
\begin{bmatrix} -\frac{1}{2} & \frac{\sqrt{3}}{2} & 0 \\ -\frac{\sqrt{3}}{2} & -\frac{1}{2} & 0 \\ 0 & 0 & 1 \end{bmatrix} &
\begin{bmatrix} 1 & 0 & 0 \\ 0 & -1 & 0 \\ 0 & 0 & 1 \end{bmatrix} &
\begin{bmatrix} -\frac{1}{2} & -\frac{\sqrt{3}}{2} & 0 \\ -\frac{\sqrt{3}}{2} & \frac{1}{2} & 0 \\ 0 & 0 & 1 \end{bmatrix} &
\begin{bmatrix} -\frac{1}{2} & \frac{\sqrt{3}}{2} & 0 \\ \frac{\sqrt{3}}{2} & \frac{1}{2} & 0 \\ 0 & 0 & 1 \end{bmatrix}
\end{array}
$$

TABLE 8.2

Matrices of an irreducible representation (species E) of C_{3v}

C_{3v}	I	C_3	C_3^2	σ_{v_1}	σ_{v_2}	σ_{v_3}
E	$\begin{bmatrix} 1 & 0 \\ 0 & 1 \end{bmatrix}$	$\begin{bmatrix} -\frac{1}{2} & -\frac{\sqrt{3}}{2} \\ \frac{\sqrt{3}}{2} & -\frac{1}{2} \end{bmatrix}$	$\begin{bmatrix} -\frac{1}{2} & \frac{\sqrt{3}}{2} \\ -\frac{\sqrt{3}}{2} & -\frac{1}{2} \end{bmatrix}$	$\begin{bmatrix} 1 & 0 \\ 0 & -1 \end{bmatrix}$	$\begin{bmatrix} -\frac{1}{2} & -\frac{\sqrt{3}}{2} \\ -\frac{\sqrt{3}}{2} & \frac{1}{2} \end{bmatrix}$	$\begin{bmatrix} -\frac{1}{2} & \frac{\sqrt{3}}{2} \\ \frac{\sqrt{3}}{2} & \frac{1}{2} \end{bmatrix}$

associated irreducible representation represent the symmetry operations in so far as their effects upon the vector z are concerned. In fact, all of them leave this vector unchanged. A vector which is transformed into itself (i.e. left unchanged) by an operation, is said to be *symmetric* with respect to that operation. The vector z (and the corresponding coordinate z) are therefore described as *totally symmetric*, in that no operation of the group changes them. The species A_1 to which they belong is likewise called the totally symmetric species. The fact that all the elements of the irreducible representation of species A_1 are unity is of course in accord with the general requirement that all matrices representing symmetry operations must be orthogonal matrices. For a 1×1 matrix (i.e. a simple number) this means that the square of the number must equal unity. The only possible values of the number are therefore $+1$ and -1. In the A_1 irreducible representation all the elements have the value $+1$.

We come now to the group representation formed by the 2×2 blocks. Here the invariant sub-space is 2-dimensional and has the basis vectors e_x and e_y, which we chose in a particular way. An unlimited number of alternative choices was, however, open to us, for the x and y directions (unlike the z direction along the C_3 axis) are not uniquely determined by symmetry. Any pair of mutually orthogonal unit vectors lying in the plane normal to e_z would have served equally well to span the sub-space. Whatever the choice made, there would always have been an inevitable mixing of the corresponding coordinates x and y under the symmetry operations of the group. In other words, it is a consequence of the C_{3v} symmetry that the 2×2 representation in question is an *irreducible* one. By convention it is designated by the symbol E. The two coordinates x and y are said to belong to the symmetry species E, and this species is said to be *doubly degenerate*. We note again in passing that the actual elements of the 2×2 matrices forming the irreducible representation belonging to this species will depend upon the particular choice of the x and y directions.

It is important to realize that the existence of A_1 and E symmetry species for the point group C_{3v} is a direct consequence of the fact that one direction in space (namely, that of the C_3 axis) is differentiated from the other two by symmetry, and that the other two are symmetrically equivalent. Symmetry species which are 3-fold degenerate can only occur where all three directions in physical space are symmetrically equivalent. This is the case for the so-called cubic point groups, to which (for example) belong the regular tetrahedral CCl_4 molecule (point group T_d) and the regular octahedral SF_6 molecule (point group O_h). Obviously, since physical space is only 3-dimensional, symmetry species with degeneracies greater than 3-fold are impossible.

We return now to the non-degenerate species A_1 for the point group C_{3v}. The irreducible representation belonging to this species is a rather trivial one. It may be presented in convenient tabular form as follows:

C_{3v}	I	C_3	C_3^2	σ_{v_1}	σ_{v_2}	σ_{v_3}
A_1	[1]	[1]	[1]	[1]	[1]	[1]

The base vector e_z and coordinate z of species A_1 refer to a simple translation of the point considered in the z-direction. Now the simple translation of any system as a whole in this direction is described in exactly the same way by a z-coordinate giving the displacement of a special point of the system, namely, its centre of mass. We have denoted this overall translation by T_z, and we can now say that T_z belongs to the totally symmetric species A_1 of the group C_{3v}. For similar reasons it will be clear that we can say that the pair of translations T_x and T_y belong to the doubly degenerate species E. The irreducible representation belonging to this species and based upon our special choice of x- and y-directions, is shown in Table 8.2 (p. 121).

Now, as stated earlier, and as we shall presently prove, the number of irreducible representations of the group C_{3v} is three. We have so far dealt with two of them. The remaining one is non-degenerate and is designated by the symbol A_2. As a simple way of deriving it, consider the rotation R_z of the PCl_3 molecule about the C_3 axis. The P nucleus is unmoved by it, and stays so when the rotationally displaced molecule is subjected to any of the symmetry operations of the group. As to the three Cl nuclei, inspection of Fig. 8.2 shows that the displacement vectors remain unaffected by the operations I, C_3, and C_3^2 but are reversed

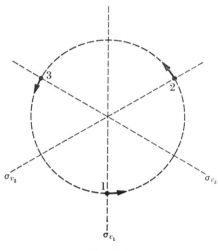

FIG. 8.2.

by the operations σ_{v_1}, σ_{v_2}, and σ_{v_3}. The corresponding irreducible representation is therefore as follows:

C_{3v}	I	C_3	C_3^2	σ_{v_1}	σ_{v_2}	σ_{v_3}
A_2	[1]	[1]	[1]	[−1]	[−1]	[−1]

In connection with the rotations, we note that R_x and R_y form a pair belonging to the doubly degenerate species E. This is again a consequence of the complete symmetrical equivalence of the x- and y-directions in space.

8.3. The characters of the matrices of irreducible representations

In Section 4.9 we became acquainted with the character χ of a square matrix as the sum of its diagonal elements, and proved that χ is invariant under similarity transformations. It follows that all matrices that are equivalent, in the sense that they all represent the same symmetry operation but with different

124

choices of basis, must have the same character. Now it will transpire that, for the purposes of applying symmetry theory to the problem of molecular vibrations, it suffices to know the characters of the matrices forming the irreversible representations, without knowing the individual matrix elements for any particular basis choice. The enormously valuable simplification that group theory introduces will now be appreciated. It shows, first, that all the matrix representations of a particular group (no matter what the dimensions of the matrices or what the choice of basis) can be decomposed into irreducible representations belonging to a small number of symmetry species; and secondly, that the non-uniqueness of matrices in these irreducible representations is immaterial, since only their unique characters need be known.

For reference purposes in applying group theory to molecular vibrational problems, it is of advantage to tabulate, for each point group, the values of the characters of the matrices in the irreducible representations belonging to the different symmetry species. A preliminary form of such a group character table is shown in Table 8.3. The columns are headed by the symbols for

TABLE 8.3

A form of character table for the point group C_{3v}

C_{3v}	I	C_3	C_3^2	σ_{v_1}	σ_{v_2}	σ_{v_3}	
A_1	1	1	1	1	1	1	T_z
A_2	1	1	1	-1	-1	-1	R_z
E	2	-1	-1	0	0	0	$(T_x, T_y);\quad (R_x, R_y)$

the symmetry operations and the rows are preceded by the symbols of the symmetry species. The top left-hand corner is used to accommodate the symbol of the point group. An extra column to the right of the main table serves to indicate the symmetry species to which the three translations and the three rotations belong.

In connection with Table 8.3 the following rather trivial points may be noted. For a 1×1 matrix the character is obviously identical with the single element. For any irreducible representation the identity operation has a character which is equal to the dimension of the associated sub-space (this being the same as the degeneracy of the symmetry species). Thus for A_1 and A_2 the value is unity, while for E it is 2. For the three-fold degenerate species, which occur among other point groups, the corresponding value is 3.

8.4. Classes of symmetry operations and the group character table

Inspection of Table 8.3 shows that the two rotational operations C_3 and C_3^2 have the same characters as one another in all the irreducible representations. A similar state of affairs is apparent for the three reflection operations. We say that C_3 and C_3^2 form a distinct *class* of operations, and that σ_{v_1}, σ_{v_2}, and σ_{v_3} form another such class. In any point group the identity operation forms a class of its own, of which it is the sole member. The mathematical definition of a class and the derivation of its membership will be discussed towards the end of this section.

In consequence of the existence of these mutually exclusive classes of symmetry operations, it is possible to effect an economy of space in presenting the group character table. Instead of setting out all the characters of all the operations, it will suffice to give only the characters for the different classes and to indicate at the head of each column of the table the number and nature

TABLE 8.4

Character table for the point group C_{3v}

C_{3v}	I	$2C_3$	$3\sigma_v$	
A_1	1	1	1	T_z
A_2	1	1	-1	R_z
E	2	-1	0	(T_x, T_y) ; (R_x, R_y)

of the operations in each class. In this way we arrive at the form of group character table shown in Table 8.4, which is more compact than that of Table 8.3. This is indeed the form usually encountered in reference works which give sets of character tables for the various point groups. An additional column on the right is usually found, containing information as to the species or other quantities of special spectroscopic interest, which we shall discuss later. Group theory shows that the number of symmetry species for any group is equal to the number of classes of symmetry operations. For our example, the group C_{3v}, this number is 3. It follows that, in the compact form shown in Table 8.4, the main part of the character table of a group must always be a square array.

The division of the symmetry species of C_{3v} into classes seems natural and plausible one, the two rotations forming one class and the three reflections another; for other groups also the divisions into classes generally have this plausible appearance. Nevertheless a more rigorous, general definition of a class is clearly necessary. This will now be discussed. The element B of a group of symmetry operations is said to be *conjugate* to another element A of the group if there is some other element R of the group such that

$$B = R^{-1}AR. \qquad (8.1)$$

We see at once that, if B is conjugate to A, then A is conjugate to B, for it follows from eqn (8.1) that

$$A = RBR^{-1}. \qquad (8.2)$$

Now since R is an element of the group, R^{-1} must also be such an element. Therefore eqn (8.2) is of the same form as eqn (8.1). We note that the matrices representing conjugate symmetry operations must have the same character, because they are related to one another by a similarity transformation. Now suppose that another element C of the group is generated from A by a transformation similar to that in eqn (8.1), but with some other group element S in place of R, so that we may write

$$C = S^{-1}AS. \qquad (8.3)$$

The situation may be expressed by saying that both B and C are conjugate to the same operation A. We can now show that B and C must therefore be conjugate to one another. The proof is quite simple. By substituting the value of A from eqn (8.2) into eqn (8.3), we obtain

$$C = S^{-1}RBR^{-1}S. \qquad (8.4)$$

The product $R^{-1}S$ must be an element of the group, say the element T, so that
$$R^{-1}S = T, \qquad (8.5)$$

whence
$$T^{-1} = S^{-1}R. \qquad (8.6)$$

Substitution of these relations into eqn (8.4) at once gives

$$C = T^{-1}BT. \qquad (8.7)$$

This proves that C and B are conjugate to one another. Starting with some selected element A of a group we can therefore deduce all the other elements which are conjugate to A. All that is necessary is to form all the products of the type $R^{-1}AR$, letting R be each element of the group in turn.

We can now define a class as consisting of all those elements which are conjugate to the same element, and therefore conjugate to one another. Clearly by the procedure just outlined we can derive all the classes by letting all the group elements in succession play the role of A. The classes so obtained are mutually exclusive, i.e. no element can belong to more than one class. The proof of this is quite simple, and is left as an exercise for the reader.

By way of illustration, the above definition of a class may be applied to the point group C_{3v} in order to find the number of symmetry classes and their membership. In working out the expressions of the form $R^{-1}AR$, use may be made of the binary products given in Table 7.1. Let us start with the identity operation I. We see at once that, for every operation R of the group, the product $R^{-1}IR$ is simply equal to I itself. It follows therefore that (as stated earlier) the identity operation constitutes a distinct class on its own. This will clearly be so in any point group. Consider next the operation C_3 of the group C_{3v}, and from it form the products of the type $R^{-1}C_3R$, in which R is successively I, C_3, C_3^2, σ_{v_1}, σ_{v_2}, and σ_{v_3}. With the help of Table 7.1, the reader will easily verify that the result is always either C_3 or C_3^2. This proves that the two rotations form a distinct class.

Of course, the same conclusion would have been reached by forming products of the type $R^{-1}C_3^2 R$. By the same method the reader may verify that the three reflection operations form a third distinct class.

9 Orthogonality properties of irreducible representations

9.1. Relations between matrix elements in irreducible representations

BEFORE we can proceed with the application of group theory to the molecular vibrational problem, we must become acquainted with a fundamental theorem concerning the relations between the elements of matrices which form the various irreducible representations of a group. We shall exemplify these relations for the group C_{3v}, using the irreducible representations of species A_1, A_2, and E which we obtained in Section 8.2. We shall also state the theorem in a general form applicable to all point groups.

For convenience, the three irreducible representations of Section 8.2 are brought together here in the form of Table 9.1. Although the elements of the matrices of the irreducible representation belonging to species E, as shown in Table 9.1, depend upon the special choice of basis adopted in their derivation, the general properties we shall now be considering will hold for any basis. Consider any particular group operation R. The total number of elements in all the three matrices representing R in the three irreducible representations is clearly equal to the sum of the squares of the dimensions of the matrices, i.e. $1^2 + 1^2 + 2^2 = 6$. We may distinguish between these elements by appending to the element-symbol the customary two subscripts, denoting respectively the row and the column in which the element occurs, and also one superscript (in parenthesis) denoting the species of the representation in which the matrix occurs. For example, suppose R is C_3^2. Then $R_{11}^{(A_1)} = 1$ and $R_{12}^{(E)} = \frac{1}{2}\sqrt{3}$. The general symbol is $R_{ij}^{(\mu)}$, where μ denotes a symmetry species. Clearly for non-degenerate species the values of i and j are limited to unity,

TABLE 9.1

Irreducible representations of the group C_{3v}

C_{3v}	I	C_3	C_3^2	σ_{v_1}	σ_{v_2}	σ_{v_3}
A_1	$[1]$	$[1]$	$[1]$	$[1]$	$[1]$	$[1]$
A_2	$[1]$	$[1]$	$[1]$	$[-1]$	$[-1]$	$[-1]$
E	$\begin{bmatrix} 1 & 0 \\ 0 & 1 \end{bmatrix}$	$\begin{bmatrix} -\frac{1}{2} & -\frac{\sqrt{3}}{2} \\ \frac{\sqrt{3}}{2} & -\frac{1}{2} \end{bmatrix}$	$\begin{bmatrix} -\frac{1}{2} & \frac{\sqrt{3}}{2} \\ -\frac{\sqrt{3}}{2} & -\frac{1}{2} \end{bmatrix}$	$\begin{bmatrix} 1 & 0 \\ 0 & -1 \end{bmatrix}$	$\begin{bmatrix} -\frac{1}{2} & -\frac{\sqrt{3}}{2} \\ -\frac{\sqrt{3}}{2} & \frac{1}{2} \end{bmatrix}$	$\begin{bmatrix} -\frac{1}{2} & \frac{\sqrt{3}}{2} \\ \frac{\sqrt{3}}{2} & \frac{1}{2} \end{bmatrix}$

while for a doubly degenerate species either i or j may have the values 1 and 2.

Now for any particular operation R choose two matrix elements from *different* sites. The sites may be both in one matrix, or they may be in different matrices. Suppose the chosen elements are $R_{ij}^{(\mu)}$ and $R_{kl}^{(\nu)}$, where ν denotes a symmetry species. Then because their sites are different, at least one of the conditions $i \neq k$, $j \neq l$, or $\mu \neq \nu$ must be satisfied. Now multiply the two elements together and form the sum of all the corresponding products for all the operations of the group. For instance, referring to Table 9.1, we find

$$\sum_R R_{11}^{(A_1)} R_{12}^{(E)}$$

$$= (1 \times 0) - (1 \times \tfrac{1}{2}\sqrt{3}) + (1 \times \tfrac{1}{2}\sqrt{3}) - (1 \times 0) + (1 \times \tfrac{1}{2}\sqrt{3}) - (1 \times \tfrac{1}{2}\sqrt{3})$$

$$= 0.$$

Similarly,

$$\sum_R R_{11}^{(E)} R_{12}^{(E)}$$

$$= (1 \times 0) - (\tfrac{1}{2} \times \tfrac{1}{2}\sqrt{3}) + (\tfrac{1}{2} \times \tfrac{1}{2}\sqrt{3}) + (1 \times 0) + (\tfrac{1}{2} \times \tfrac{1}{2}\sqrt{3}) - (\tfrac{1}{2} \times \tfrac{1}{2}\sqrt{3})$$

$$= 0.$$

It is in fact a fundamental and general property of the matrix elements in a complete set of irreducible representations of any point group that all sums of the type in question must always vanish. The reader may verify the truth of this for all the possible pairs of sites in Table 9.1.

Next, instead of choosing two different matrix sites, choose just one, form the square of the matrix element, and then sum all such squares over all the operations of the group. For example, using Table 9.1 we obtain

$$\sum_R R_{11}^{(A_2)} R_{11}^{(A_2)} = 1^2 + 1^2 + 1^2 + (-1)^2 + (-1)^2 + (-1)^2$$

$$= 6.$$

Similarly

$$\sum_R R_{12}^{(E)} R_{12}^{(E)} = 0^2 + (-\tfrac{1}{2}\sqrt{3})^2 + (\tfrac{1}{2}\sqrt{3})^2 + 0^2 + (-\tfrac{1}{2}\sqrt{3})^2 + (\tfrac{1}{2}\sqrt{3})^2$$

$$= 3.$$

It is in fact a fundamental and general property of all groups that all sums of the type in question must always be equal to g/n_μ, where g is the *order of the group* (i.e. the number of symmetry operations it contains) and n_μ is the dimension (or the degeneracy) of the irreducible representation to which the matrix elements in question belong. In our example of the group C_{3v}, the order g is 6, $n_{A_1} = n_{A_2} = 1$, and $n_E = 2$. The values for the choices $R_{11}^{(A_1)}$ and $R_{12}^{(E)}$, as worked out above, are seen to be in accordance with the general statement. The reader may check the truth of this statement for all other possible choices of $R_{ij}^{(\mu)}$.

The basic general theorem, which covers all the cases we have considered, may now be stated in the following conveniently compact form:

$$\sum_R R_{ij}^{(\mu)} R_{kl}^{(\nu)} = \delta_{ik}\delta_{jl}\delta_{\mu\nu}\frac{g}{n_\mu}, \tag{9.1}$$

where each of the three Kronecker deltas on the right-hand side has the value unity when its two subscripts are equal, but the value zero when they are not. The proof of this very important and fundamental theorem is rather lengthy and involved, and will not be given in this book. It may be found in standard works on group theory, such as that by E. Wigner.†

We may remark in passing that the theorem is consistent with the fact (which we have previously noted) that all the matrices of irreducible representations must be orthogonal matrices. However it goes much further than this. For example, consider the matrix representing an operation R in an irreducible representation belonging to the species E. Because the matrix is orthogonal, we know that, for every R,

$$R_{11}^{(E)} R_{12}^{(E)} + R_{21}^{(E)} R_{22}^{(E)} = 0$$

and hence, summing over all the group operations, that

$$\sum_R R_{11}^{(E)} R_{21}^{(E)} + \sum_R R_{12}^{(E)} R_{22}^{(E)} = 0. \tag{9.2}$$

The theorem of eqn (9.1) goes further, in that it states that each of the two sums on the left-hand side of eqn (9.2) is *separately*

† **Wigner, E.** (1959). *Group theory.* Academic Press, New York.

zero. Likewise, from the orthogonal nature of the matrices, we know that for every R

$$R_{11}^{(E)} R_{11}^{(E)} + R_{12}^{(E)} R_{12}^{(E)} = 1,$$

and hence, summing as before, that

$$\sum_R R_{11}^{(E)} R_{11}^{(E)} + \sum_R R_{12}^{(E)} R_{12}^{(E)} = g. \qquad (9.3)$$

The theorem of eqn (9.1) goes further, in that it states that each of the two sums on the left-hand side of eqn (9.3) is *separately* equal to $g/n_E = \tfrac{1}{2}g$. The scope of the theorem also extends to sums of matrix-element products where $\mu \neq \nu$, i.e. to cases where its requirements cannot possibly be foreseen from the mere knowledge that each of the matrices concerned is orthogonal.

An interesting and instructive way of looking at the general theorem is as follows. In the irreducible representation belonging to any chosen species μ there are g matrix elements occupying corresponding sites (i.e. sites characterized by the same pair of values of i and j). We now regard these g elements (one for each symmetry operation) as constituting the components (or coordinates) of a vector in a purely conceptual g-dimensional space. For each choice of symmetry species and of site type (i.e. for each choice of μ, i, and j) there will be one such vector. The content of the general theorem of eqn (9.1) is then expressible by saying that all these vectors are mutually orthogonal, while the sum of the squares of the components of each of them is equal to g/n_μ. We shall have occasion later to make use of this point of view in applying group theory to molecular vibrations.

9.2. Orthogonality properties of matrix characters in irreducible representations

From the basic theorem of eqn (9.1) we can deduce another very useful theorem which concerns the characters of the matrices in a complete set of irreducible representations of a group. By definition, the character χ_R of a matrix representing the symmetry operation R is the sum of its diagonal elements.

Thus for the character of a matrix of an irreducible representation belonging to the species μ, we write

$$\chi_R^{(\mu)} = \sum_i R_{ii}^{(\mu)}.$$

Because of the 3-dimensional nature of physical space, the largest dimension for a matrix in an irreducible representation is 3. The species is then triply degenerate, and we may write

$$\chi_R^{(\mu)} = R_{11}^{(\mu)} + R_{22}^{(\mu)} + R_{33}^{(\mu)}. \tag{9.4}$$

When dealing with a doubly degenerate species, we simply delete R_{33} in eqn (9.4); for a non-degenerate species we delete both R_{22} and R_{33}. The square of the character is given by

$$(\chi_R^{(\mu)})^2$$
$$= (R_{11}^{(\mu)})^2 + (R_{22}^{(\mu)})^2 + (R_{33}^{(\mu)})^2 + 2(R_{11}^{(\mu)} R_{22}^{(\mu)} + R_{22}^{(\mu)} R_{33}^{(\mu)} + R_{33}^{(\mu)} R_{11}^{(\mu)}).$$

Now consider the sum of the squares of the characters over all the operations R of the group. We have

$$\sum_R (\chi_R^{(\mu)})^2 = \sum_R (R_{11}^{(\mu)})^2 + \sum_R (R_{22}^{(\mu)})^2 + \sum_R (R_{33}^{(\mu)})^2 +$$
$$+ 2\left(\sum_R R_{11}^{(\mu)} R_{22}^{(\mu)} + \sum_R R_{22}^{(\mu)} R_{33}^{(\mu)} + \sum_R R_{33}^{(\mu)} R_{11}^{(\mu)} \right). \tag{9.5}$$

This equation covers doubly degenerate and non-degenerate species as well as triply degenerate ones. When $n_\mu = 2$ or $n_\mu = 1$ all that is necessary is to delete those terms which do not apply.

The basic theorem of eqn (9.1) tells us that each of the first n_μ sums on the right-hand side of eqn (9.5) must be equal to g/n_μ, and that the remaining sums must all vanish. We have thus proved that

$$\sum_R (\chi_R^{(\mu)})^2 = g, \tag{9.6}$$

i.e. for any irreducible representation the sum of the squares of the characters of all the matrices is simply equal to the order of the group.

Next, instead of $\sum_R (\chi_R^{(\mu)})^2$, let us consider $\sum_R \chi_R^{(\mu)} \chi_R^{(\nu)}$ where μ and ν are different symmetry species. By proceeding as before, the reader will easily find that the value is again the sum of a

135

number of terms, each of which is itself a summation of matrix elements with different superscripts. Invoking the basic theorem of eqn (9.1), we find that all these terms must vanish. We therefore conclude that, when μ and ν are not identical,

$$\sum_R \chi_R^{(\mu)} \chi_R^{(\nu)} = 0. \tag{9.7}$$

This case and the case when $\mu = \nu$ may be covered simultaneously by writing

$$\sum_R \chi_R^{(\mu)} \chi_R^{(\nu)} = \delta_{\mu\nu} g. \tag{9.8}$$

The significance of this very important and useful conclusion may be expressed as follows. For the particular irreducible representation belonging to the species μ, the g values of χ_R, of which there is one for each symmetry operation of the group, may be regarded as constituting the components (or coordinates) of a vector in a conceptual g-dimensional space. Then all such vectors, of which there is one for each irreducible representation or symmetry species, are mutually orthogonal, and for each of them the sum of the squares of its components has the value g. The present theorem has obvious affinities with the general theorem of eqn (9.1) from which it was deduced; but it has the important practical advantage that the matrix characters (with which it is alone concerned) differ from the matrix elements in being independent of the particular choice of a basis.

By referring to the characters for the point group C_{3v}, as given in Table 8.3, we may easily verify the validity of eqn (9.8) for this particular group. For example we find

$$\sum_R (\chi_R^{(A_1)})^2 = 1^2 + 1^2 + 1^2 + 1^2 + 1^2 + 1^2 = 6;$$

$$\sum_R (\chi_R^{(E)})^2 = 2^2 + (-1)^2 + (-1)^2 + 0^2 + 0^2 + 0^2 = 6;$$

$$\sum_R \chi_R^{(A_1)} \chi_R^{(A_2)} = (1 \times 1) + (1 \times 1) + (1 \times 1) - (1 \times 1) - $$
$$- (1 \times 1) - (1 \times 1) = 0;$$

$$\sum_R \chi_R^{(A_1)} \chi_R^{(E)} = (1 \times 2) - (1 \times 1) - (1 \times 1) + 0 + 0 + 0 = 0.$$

The reader may easily check the remaining cases.

9.3. The number and dimensions of the symmetry species of a group

The conclusion of the foregoing section may be extended by taking into account the facts that (as we have seen in Section 8.4) the symmetry operations of a group belong to mutually exclusive classes and that in any irreducible representation all the matrices representing operations in the same class must have the same character. Let k be the number of classes and g_t the number of symmetry operations in the tth class; then

$$\sum_{t=1}^{t=k} g_t = g,$$

where g is the order of the group. Let us designate by $\chi_t^{(\mu)}$ the common character of all the matrices of the class t in an irreducible representation belonging to the species μ. Instead of summing over the individual symmetry operations (as we did in the foregoing section) we may reach the same result by summing over the classes. Thus

$$\sum_R \chi_R^{(\mu)}\chi_R^{(\nu)} = \sum_{t=1}^{t=k} g_t \chi_t^{(\mu)}\chi_t^{(\nu)} = \delta_{\mu\nu}g. \tag{9.9}$$

At this point we may note that this result may be used to determine the characters for certain irreducible representations without the necessity of first deriving actual matrices. For example, suppose that we already know, for each of the three classes of symmetry operation of the group C_{3v}, the characters in the irreducible representations belonging to the two non-degenerate species A_1 and A_2. Numbering the classes as 1, 2, and 3 in the conventional order in which they are arranged in Table 8.4, we write $g_1 = 1$, $g_2 = 2$, and $g_3 = 3$. In addition to knowing that $\chi_1^{(A_1)} = \chi_2^{(A_1)} = \chi_3^{(A_1)} = 1$ and that $\chi_1^{(A_2)} = \chi_2^{(A_2)} = 1$ while $\chi_3^{(A_2)} = -1$, we also know that (as always for a doubly degenerate species) $\chi_1^{(E)} = 2$. We can now make use of eqn (9.9) to obtain the unknown values of $\chi_2^{(E)}$ and $\chi_3^{(E)}$. We have

$$\sum_{t=1}^{t=3} g_t \chi_t^{(A_1)}\chi_t^{(E)} = \chi_1^{(E)}+2\chi_2^{(E)}+3\chi_3^{(E)} = 0 \tag{9.10}$$

and $$\sum_{t=1}^{t=3} g_t \chi_t^{(A_2)}\chi_t^{(E)} = \chi_1^{(E)}+2\chi_2^{(E)}-3\chi_3^{(E)} = 0. \tag{9.11}$$

137

From eqns (9.10) and (9.11) it follows at once that $\chi_3^{(E)} = 0$ and that $\chi_2^{(E)} = -\frac{1}{2}\chi_1^{(E)} = -1$. We have thus been able to calculate the values of $\chi_2^{(E)}$ and $\chi_3^{(E)}$ without ever having had to derive matrices of the kind given in Table 8.2.

We revert now to the discussion of eqn (9.9), which we may rewrite in the equivalent form

$$\sum_{t=1}^{t=k} (g_t^{\frac{1}{2}} \chi_t^{(\mu)})(g_t^{\frac{1}{2}} \chi_t^{(\nu)}) = \delta_{\mu\nu} g. \qquad (9.12)$$

Its significance may now be expressed in the following way. For each symmetry species regard the set of numbers $g_t^{\frac{1}{2}} \chi_t$, of which there will be one for each of the k classes of operation, as the components (or coordinates) of a vector in a purely conceptual k-dimensional space. Then eqn (9.12) states that all these vectors are mutually orthogonal and that for each of them the sum of the squares of its components has the value g. Now if γ is the number of irreducible representations (or symmetry species) possessed by the group, γ will also be the number of vectors of the kind in question. Since the number of mutually orthogonal vectors in a k-dimensional space cannot exceed k,

$$\gamma \leqslant k.$$

It may in fact be shown that it is the equality which holds, i.e. that

$$\gamma = k. \qquad (9.13)$$

Expressed in words, this means that the number of symmetry species possessed by a group is equal to the number of classes of symmetry operations. We had occasion to mention this property in Section 8.4. The proof is not simple and will not be given here. In this connection, however, see Section 9.4 for a further comment on eqn (9.13). As an example of the validity of this equation we note that for the group C_{3v} the number of classes of symmetry operation is 3, and that therefore the number of symmetry species must also be 3. These are, of course, the ones we have designated as A_1, A_2, and E.

The dimensions of the species of a group can also be deduced by a further application of the basic theorem of eqn (9.1). We adopt the point of view described near the end of Section 9.2, regarding the set of g matrix elements $R_{ij}^{(\mu)}$, all of which have the same i, j, and μ, as being the components of a vector in a conceptual g-dimensional space. The basic theorem states that all the vectors of this kind must be mutually orthogonal, from which it follows that the number of them cannot exceed g. Now for a species μ of dimension (or degeneracy) n_μ, the number of vectors is obviously n_μ^2, so that the total number must be $\sum n_\mu^2$, where the summation is made over all the symmetry species. If, as above, we let γ be the number of species, we conclude that

$$n_1^2 + n_2^2 + \ldots + n_\gamma^2 \leqslant g.$$

Once again it can be proved that it is the equality which holds, though the details of the proof will not be given here. Finally, therefore,

$$n_1^2 + n_2^2 + \ldots + n_\gamma^2 = g. \tag{9.14}$$

This equation, taken in conjunction with eqn (9.13), enables us to deduce, for any point group, the dimensions of all its symmetry species. For example, we have seen that the number of species for the point group C_{3v} is 3. Also we know that $g = 6$. Let n_1, n_2, and n_3 be the dimensions of the three symmetry species. Then

$$n_1^2 + n_2^2 + n_3^2 = 6. \tag{9.15}$$

Obviously all the three n-values must be integral, and we see at once that the only three integers which satisfy eqn (9.15) are 1, 1, and 2. We therefore must have two non-degenerate species (these we have designated as A_1 and A_2) and one doubly-degenerate species (designated as E).

As a further example we may take the point group T_d, to which belong regular tetrahedral molecules like CCl_4. This group has 24 operations, divided into 5 classes. For the dimensions (or degeneracies) of the species, of which eqn (9.13) shows that there must necessarily be 5, we seek 5 integers such that the sum of their squares is equal to 24. A little thought shows that the unique set is 1, 1, 2, 3, and 3. We have thus proved that the group T_d possesses two non-degenerate species, one doubly degenerate one, and two triply degenerate ones. The first three are in fact designated as A_1, A_2, and E respectively. The remaining two are usually called F_1 and F_2. Sometimes (though rarely in discussing molecular vibrations) the triply degenerate species are called T_1 and T_2, instead of F_1 and F_2.

9.4. The structures of completely reduced group representations

In what has gone before we have been concerned with the irreducible representations of a group into which any reducible representation can be decomposed; but we have not shown how such a decomposition (i.e. a complete reduction) can be effected. We shall do this in Chapter 11. First, however, we shall consider and exemplify the convenient fact that it is always possible to derive the *structure* of the completely reduced form (i.e. the pattern of the diagonal blocks) without the necessity of actually carrying out the complete reduction at all. As we shall see in Chapter 10, this is of great value in connection with the problem of molecular vibrations.

For a molecule containing N nuclei, the configuration space is $3N$-dimensional, and each choice of orthonormal basis vectors gives rise to a group representation whose matrices are $3N \times 3N$. In general this representation will be reducible. Complete reduction is effected by changing to an appropriate new basis, whereupon all the matrices of the original reducible representation assume the same diagonal-block pattern, in which each set of corresponding blocks forms an irreducible representation of the group. Whatever the transformation that produces this result, we know that all the matrices will have been subjected to a similarity transformation, and that therefore the character of each of them will have remained unchanged. For each matrix, therefore, the original character must be equal to the sum of the characters of all the blocks in its completely reduced, diagonal-block form. From this very simple fact it is possible to deduce the structure of this form, i.e. the number of times that each distinct irreducible representation occurs in it.

Let Γ denote the original reducible representation, so that $\chi_R^{(\Gamma)}$ will denote the character of the matrix representing the operation R in it. Let the γ symmetry species (or distinct irreducible representations) be numbered $1, 2, ..., \gamma$, so that the characters of the matrices representing R in them are respectively $\chi_R^{(1)}, \chi_R^{(2)}, ..., \chi_R^{(\gamma)}$. If now $a_1, a_2, ..., a_\gamma$ are respectively the numbers of times that these irreducible representations are contained in the completely reduced form of Γ, we can write

$$\chi_R^{(\Gamma)} = a_1 \chi_R^{(1)} + a_2 \chi_R^{(2)} + \ldots + a_\gamma \chi_R^{(\gamma)}. \tag{9.16}$$

It might at first sight be thought that the number of equations of this kind would be the number g of operations in the group; but we must remember that these operations are divided into classes, and that the character is the same for all the operations in a class. In fact, therefore, the number of independent equations of the type of eqn (9.16) is equal only to the number k of classes. Treating $a_1, a_2, \ldots, a_\gamma$ as unknowns, the number of them is γ, and this is just equal to the number of independent equations connecting them; for, as we have seen in the preceding section (eqn (9.13)), $\gamma = k$. Thus by solving the simultaneous equations we can obtain the values of all the unknowns, i.e. the structure of the completely reduced form of the representation Γ. The result is commonly expressed in the form

$$\Gamma = a_1 \Gamma_1 + a_2 \Gamma_2 + \ldots + a_\gamma \Gamma_\gamma, \tag{9.17}$$

where Γ_1, Γ_2, etc. stand for the distinct irreducible representations. The right-hand side of eqn (9.17) is often referred to as a *direct sum*.

Alternatively, we might start out by regarding the number γ of symmetry classes as an unknown quantity. Then the assumption that the completely reduced form of Γ contains certain definite numbers of each of the distinct irreducible representations, i.e. the assumption that the numbers $a_1, a_2, \ldots, a_\gamma$ are determinate, would lead to the conclusion that the number of equations must be equal to the number k of classes. But the number of equations is equal to the number γ of the symmetry species. We should thus reach the conclusion that γ must be equal to k. This is the result given (without proof) in eqn (9.13).

The solution of the γ simultaneous equations may be elegantly performed by invoking the orthogonality properties of the characters of the matrices in the irreducible representations, as expressed in eqn (9.9). Let us multiply the eqn (9.16) by $\chi_R^{(1)}$

and then sum over all the operations R of the group. We thus obtain

$$\sum_R \chi_R^{(1)} \chi_R^{(\Gamma)} = a_1 \sum_R \chi_R^{(1)} \chi_R^{(1)} + a_2 \sum_R \chi_R^{(1)} \chi_R^{(2)} + \dots + a_\gamma \sum_R \chi_R^{(1)} \chi_R^{(\gamma)}. \quad (9.18)$$

By the theorem of eqn (9.9), all the terms on the right-hand side of eqn (9.18) must vanish with the sole exception of the first, whose value must be $a_1 g$, where g is the order of the group. Thus we have shown that

$$a_1 = \frac{1}{g} \sum_R \chi_R^{(1)} \chi_R^{(\Gamma)}.$$

In exactly the same way, by multiplying eqn (9.16) by $\chi_R^{(2)}$ and summing over all operations R, we obtain

$$a_2 = \frac{1}{g} \sum_R \chi_R^{(2)} \chi_R^{(\Gamma)}.$$

In general, for the number of times a_i that the ith irreducible representation is contained in the completely reduced structure of Γ, we have

$$a_i = \frac{1}{g} \sum_R \chi_R^{(i)} \chi_R^{(\Gamma)}. \quad (9.19)$$

9.5. An example of a reducible representation and the determination of its structure

To illustrate the application of eqn (9.19), consider again the nuclear displacements of the PCl_3 molecule, the point group of which is C_{3v}. We shall adopt the same choice of basis for the 12-dimensional configuration space as in Section 7.5, where the matrix representing the operation C_3 was deduced and shown in eqn (7.10). In order to apply eqn (9.19) we need to know the characters $\chi_R^{(\Gamma)}$ of the matrices representing all the symmetry operations. In fact, because of the division of the operations into classes, it suffices to know the characters for each of the classes, for which purpose it is not necessary to determine the whole 12×12 matrix for any operation. As inspection of the C_3 matrix in eqn (7.10) at once shows, only the displacements associated with nuclear site 4 make any contribution to the character. This is because this is the only site for which the displacements are not shifted by the operation to some other site. The same sort of consideration holds for all the symmetry operations. For the reflection σ_{v_1}, for example, only the displacements associated with sites 1 and 4 remain unshifted, and therefore they are the only sites to contribute to the character of the 12×12 matrix representing this symmetry operation. The contribution from each of them is simply the character of the 3×3 matrix representing σ_{v_1} in Table 8.1, namely unity. Since the contribution from any site to the character of any matrix in the 12×12 representation Γ is independent of the choice of basis in determining it, we are at liberty to choose whatever basis is most convenient for the purpose. This, together with the fact that characters are the same for all operations belonging to the same class, considerably lightens the task of evaluating the values of $\chi_R^{(\Gamma)}$.

In dealing with calculations of this kind, it is helpful to set out the quantities involved in the form of a simple table like Table 9.2, which applies to the example in hand.

TABLE 9.2

Data for the calculation of the structure of a 12×12 representation of the group C_{3v}

PCl$_3$ molecule (C_{3v})	I	$2C_3$	$3\sigma_v$
Contribution to $\chi_R^{(\Gamma)}$ per site with unshifted displacements	3	0	1
Number of such sites	4	1	2
Value of $\chi_R^{(\Gamma)}$	12	0	2
Value of $\chi_R^{(A_1)}$	1	1	1
Value of $\chi_R^{(A_2)}$	1	1	-1
Value of $\chi_R^{(E)}$	2	-1	0

The lowest three rows of the table, which are included for convenience, are merely the character table of the group (taken from Table 8.4).

If a_{A_1}, a_{A_2}, and a_E are the numbers of times that the respective irreducible representations are contained in the completely

143

reduced form of Γ, we may use Table 9.2 to set up three independent equations of the type of eqn (9.16), one for each class of symmetry operations:

$$\left.\begin{array}{l} 12 = a_{A_1}+a_{A_2}+2a_E \\ 0 = a_{A_1}+a_{A_2}-a_E \\ 2 = a_{A_1}-a_{A_2} \end{array}\right\}. \tag{9.20}$$

The solution of these simultaneous equations is easily performed, and gives the following result:

$$a_{A_1} = 3, \quad a_{A_2} = 1, \quad a_E = 4. \tag{9.21}$$

The same result may be obtained by applying eqn (9.19), though in such a simple case the advantage is minimal. In this way we find that

$$a_{A_1} = \frac{1}{g} \sum_R \chi_R^{(A_1)} \chi_R^{(\Gamma)}$$

$$= \tfrac{1}{6}\{(1\times 12)+2(1\times 0)+3(1\times 2)\} = 3,$$

$$a_{A_2} = \tfrac{1}{6}\{(1\times 12)+2(1\times 0)-3(1\times 2)\} = 1,$$

$$a_E = \tfrac{1}{6}\{(2\times 12)-2(1\times 0)+3(0\times 2)\} = 4.$$

In more complicated cases the use of eqn (9.19) is less troublesome than the procedure which involves solving a number of simultaneous linear equations.

The result we have obtained may be expressed by writing the structure of Γ as a direct sum,

$$\Gamma = 3A_1+A_2+4E. \tag{9.22}$$

We have thus provided the theoretical justification for the structure previously shown (without proof) in Fig. 8.1. All that remains now, in order to fulfil earlier promises, is to develop a procedure whereby a basis can be chosen such that a change to it will bring about the complete reduction of the reducible 12×12 matrix representation Γ. We shall take this up in Chapter 11.

9.6. Sets of symmetrically equivalent nuclear sites

While we have the example of the PCl_3 molecule before us, attention may be directed to one consideration which will be of interest at a later stage in connection with vibrations (particularly in Section 10.3). In a molecule possessing symmetry elements, it is possible to divide up the equilibrium nuclear sites into sets which are mutually exclusive, in the sense that, under all the symmetry operations of the molecular point group, the displacement coordinates associated with the members of any one set are exclusively shifted amongst the members of that set, and never 'mix' with the coordinates associated with any member of any other set. For the PCl_3 molecule we clearly have two such sets, set (a) consisting solely of the site of the P nucleus (site 4), and set (b) consisting of the three sites (1, 2, and 3) of the symmetrically equivalent Cl nuclei. The matrices of the reducible representation Γ must therefore consist of two diagonal blocks (one 3×3 and the other 9×9) corresponding to the non-mixing sets.

The considerations of the preceding section are evidently applicable, not only to the molecule as a whole, but also just as well to each of the individual sets of which it is composed. This is expressed in Table 9.3, the entries in which should be compared with the corresponding ones in Table 9.2.

TABLE 9.3

Data for the calculation of the structures of representations arising from sets of symmetrically equivalent sites

PCl_3 molecule (C_{3v})		I	$2C_3$	$3\sigma_v$
Number of sites with	Set (a)	1	1	1
unshifted coordinates	Set (b)	3	0	1
Value of χ_R	Set (a)	3	0	1
	Set (b)	9	0	1

Let $\Gamma_{(a)}$ and $\Gamma_{(b)}$ be the representations (respectively 3×3 and 9×9) arising from the two sets. Using the data in Table 9.3 and applying exactly the same method as we used for the whole PCl_3 molecule in the preceding section, we find

$$\left. \begin{aligned} \Gamma_{(a)} &= A_1 + E \\ \Gamma_{(b)} &= 2A_1 + A_2 + 3E \end{aligned} \right\}. \tag{9.23}$$

Of course, by adding together the two direct sums we necessarily get the structure of Γ for the whole molecule, i.e.

$$\Gamma = \Gamma_{(a)} + \Gamma_{(b)} = 3A_1 + A_2 + 4E.$$

This agrees with eqn (9.22).

9.7. Note on the uniqueness of the structure of a representation

This note merely reiterates the more important points in remarks which have already been made in the course of our discussion of irreducible representations. Its object is to emphasize that the structures given in eqns (9.22) and (9.23) are independent of the particular choice of basis for the original reducible representation Γ. In fact the only quantities relating to Γ that were used in calculating the structures were the characters $\chi_R^{(\Gamma)}$, and these (unlike the matrices themselves) are the same for all bases. Thus all equivalent 12×12 representations of the group C_{3v} must have the unique structure given in eqn (9.22). Moreover, the particular new basis which effects the complete reduction is also immaterial. No matter how the complete reduction is achieved, the resulting *structure* will always be the same.

We have used only PCl_3 and the point group C_{3v} as examples throughout, but it will, of course, be realized that these considerations apply quite generally to the structures of reducible representations of all point groups.

10 The number of distinct normal vibrational frequencies of a molecule and their symmetry species

10.1. Symmetry species of normal coordinates

THE considerations on the basis of group theory in the preceding chapter are of great importance in the treatment of the problem of molecular vibrations. The underlying reason for this is that, when based upon the complete set of normal coordinates (including translations and rotations as well as genuine vibrations), the $3N$-dimensional representation of the point group of any N-atomic molecule is completely reduced. In other words, we can say that the normal coordinates belong to the various symmetry species of the molecular point group. We shall now see why this is so.

We recall that the normal coordinates Q_k (where k may take the values 1 to $3N$) are defined (in Section 1.5) in such a way that the expressions for the kinetic energy T and the potential energy V in terms of such coordinates contain no cross-terms, i.e.

$$2T = \sum_k \dot{Q}_k^2, \tag{10.1}$$

$$2V = \sum_k \lambda_k Q_k^2. \tag{10.2}$$

Here λ_k is the kth root of the secular equation, with the value $4\pi^2 \nu_k^2$ where ν_k is the associated normal frequency.

We now consider the mixing of the normal coordinates which occurs as the result of the performance of a symmetry operation. It is not difficult to see that if a translational normal coordinate mixes with another normal coordinate, then the latter must necessarily also be a translational coordinate. Similarly a rotational coordinate, if it mixes at all, can only do so with another rotational coordinate. In the particular case of the group C_{3v}, we have seen how the translational and rotational

147

coordinates (either singly or in pairs) belong to the symmetry species of the group; this information is included in the character table which was presented as Table 8.4. We thus come to the genuine vibrational normal coordinates with $\lambda_k \neq 0$ which, if they mix at all, must do so with one another and not with any translational or rotational coordinates.

The important thing to realize in this connection is that both the kinetic energy and the potential energy are invariant under any symmetry operation of the molecular point group. It is true that, in general, such an operation will affect the nuclear displacements by altering their directions and possibly by shifting them between equivalent sites; but regarding the molecule as a whole, the displacements present after the operation will be identical with those present before it, i.e. there will be identical internuclear distances, bond angles, etc. Since the potential energy depends upon these displacement coordinates, its value will therefore remain unchanged. Similarly, if we imagine the operation carried out instantaneously at a certain moment in the motion of the molecule, it is clear that, in a similar fashion, the same time derivatives of these displacements will be present after the operation as were present before. Since the kinetic energy is dependent upon these time derivatives, its value will also remain unchanged.

Consider now two vibrational normal coordinates Q_1 and Q_2 with the roots λ_1 and λ_2 respectively. To express their mixing (if any) under the symmetry operation R we may write

$$\left. \begin{aligned} Q_1 &\xrightarrow{R} Q_1' = aQ_1 + bQ_2 \\ \text{and} \quad Q_2 &\xrightarrow{R} Q_2' = cQ_1 + dQ_2 \end{aligned} \right\} . \tag{10.3}$$

Analogous relations will also hold for the time derivatives

$$\left. \begin{aligned} \dot{Q}_1 &\xrightarrow{R} \dot{Q}_1' = a\dot{Q}_1 + b\dot{Q}_2 \\ \text{and} \quad \dot{Q}_2 &\xrightarrow{R} \dot{Q}_2' = c\dot{Q}_1 + d\dot{Q}_2 \end{aligned} \right\} . \tag{10.4}$$

Now take a particular state of affairs in which only the mode with the normal coordinate Q_1 is present. The absence of the other mode is expressed by writing $Q_2 = 0$ and $\dot{Q}_2 = 0$. The kinetic energy before the operation R is given by $2T = \dot{Q}_1^2$. After the operation its value must be $2T = (\dot{Q}_1')^2 + (\dot{Q}_2')^2$ whence, substituting the values of $\dot{Q}_1'$ and $\dot{Q}_2'$ from eqns (10.4) and remembering that, in the special case we are considering, $\dot{Q}_2 = 0$, we obtain $2T = (a^2 + c^2)\dot{Q}_1^2$. In view of the invariance of the kinetic energy, it follows that

$$\dot{Q}_1^2 = (a^2 + c^2)\dot{Q}_1^2,$$

so that

$$a^2 + c^2 = 1. \tag{10.5}$$

Now consider the potential energy in the same special case. Its value before the operation is $\lambda_1 Q_1^2$, and its value after is $\lambda_1(Q_1')^2 + \lambda_2(Q_2')^2 = (\lambda_1 a^2 + \lambda_2 c^2)Q_1^2$. In view of the invariance of the potential energy, it follows that

$$\lambda_1 Q_1^2 = (\lambda_1 a^2 + \lambda_2 c^2)Q_1^2,$$

so that

$$a^2 + \frac{\lambda_2}{\lambda_1} c^2 = 1. \tag{10.6}$$

Elimination of a^2 by combining eqn (10.5) with eqn (10.6) gives

$$\left(\frac{\lambda_2}{\lambda_1} - 1\right)c^2 = 0. \tag{10.7}$$

Thus we can have one or other of two cases:

(a) $\lambda_1 = \lambda_2$ with $c^2 \neq 0$,

or   (b) $\lambda_1 \neq \lambda_2$ with $c^2 = 0$.

In case (b) it follows from eqn (10.6) that $a^2 = 1$. By exactly similar reasoning, if we take a particular state of affairs in which only the mode of normal coordinate Q_2 is present, so that both Q_1 and $\dot{Q}_1$ are zero, we conclude that one or other of the following two cases may arise:

(a) $\lambda_1 = \lambda_2$ with $b^2 \neq 0$,

or   (b) $\lambda_1 \neq \lambda_2$ with $b^2 = 0$ and $d^2 = 1$.

We have thus proved that if $\lambda_1 \neq \lambda_2$ then $a = d = \pm 1$ and $b = c = 0$. Referring to eqn (10.3) we see that this means that, if the two normal vibrational modes have different frequencies, there can be no mixing under any symmetry operation. Therefore a vibrational mode that has a frequency different from that of any other, must be such that its normal coordinate belongs to a non-degenerate symmetry species. Under any operation R of the molecular group, such a non-degenerate vibrational normal coordinate must either remain unaffected or must change in sign. In the first of these cases the coordinate is said to be symmetric with respect to the symmetry operation; in the second case it is said to be antisymmetric. A coordinate that is symmetric with respect to one symmetry operation may be anti-symmetric with respect to another. Only when it is symmetric with respect to all the operations of the group is it (and its normal coordinate) said to belong to the totally symmetric symmetry species (e.g. A_1 for C_{3v}); otherwise both it and its normal co-ordinate are said to belong to an antisymmetric species (e.g. A_2 for C_{3v}), even though the mode may be antisymmetric with respect to only some of the group operations and symmetric with respect to the others.

If, on the other hand, the roots λ_1 and λ_2 are equal, i.e. if the two modes in question have the same frequency, then mixing is possible, and therefore the two normal coordinates Q_1 and Q_2 belong to the same invariant subspace or, in other words, to the same degenerate symmetry species. If there is no third root which is identical with λ_1 and λ_2, then the pair of coordinates (Q_1, Q_2) belongs to a doubly degenerate species E. Where three roots are identical, the three associated normal coordinates belong to a triply degenerate species F.

10.2. The number of distinct normal vibrational frequencies of a molecule possessing symmetry

From what has been said in the preceding section, it follows that the representation of a molecular point group based upon the complete set of normal coordinates must be completely reduced. Certain of the diagonal blocks will be concerned with the translational and rotational coordinates. Knowing the species to which these belong (and this information is customarily obtainable from the character table of the group), we may remove the appropriate numbers and types of blocks from the structure, thus leaving only those concerned with the genuine vibrations. Each of these remaining blocks will yield one distinct root, i.e. one distinct frequency-value. The point of great interest and practical value is that, in order to know the structure of the completely reduced representation in question, we do not have to introduce the normal coordinates at all, for this structure is independent of the choice of basis (as explained in Section 9.7). We may therefore choose any convenient set of coordinates (e.g. Cartesian displacement coordinates) upon which to base an original reducible representation and then, by the method of Section 9.4, calculate its structure. This must be identical with the structure of the representation based upon the normal coordinates, no knowledge of which is necessary. Thus we are able to find, not only the number of distinct normal vibrational frequencies, but also the symmetry species to which each of them belongs.

An example will make the method clear. We may with advantage choose the PCl_3 molecule again, as we have already deduced for it the structure of the relevant representation of dimension $3N = 12$ (in Section 9.5). The result, previously given in eqn (9.22), is

$$\Gamma = 3A_1 + A_2 + 4E. \tag{10.8}$$

We have also found that the three translations account for $A_1 + E$, and that the three rotations account for $A_2 + E$ (see Section 8.2 and the character table of the point group C_{3v}, Table 8.4). Subtracting off $A_1 + A_2 + 2E$ from the structure given in eqn (10.8), we are left with the symmetry species of the genuine vibrations. We may call the residual representation $\Gamma_{(\text{vib})}$ and write

$$\Gamma_{(\text{vib})} = 2A_1 + 2E. \tag{10.9}$$

We have thus shown (without requiring any knowledge of the normal coordinates) that the PCl_3 molecule possesses just *four* distinct normal vibrational frequencies. We have also shown that two of the associated normal coordinates belong to the totally symmetric species A_1 and that the other four form two doubly degenerate pairs belonging to the species E.

151

The method, which is seen to be of great power and elegance, is applicable to any molecule of known point group. Later in this chapter we shall apply it to the BF_3 molecule, which belongs to a point group different from that of PCl_3.

At this point it is convenient to describe the convention which is now widely followed in numbering the distinct frequencies of a molecule. To begin with, they are arranged by species, usually starting with the totally symmetric and continuing with the remaining non-degenerate species, the doubly degenerate, and finally the triply degenerate. Then within a particular species the frequencies are numbered in order of descending magnitude. Thus for PCl_3 we have $\nu_1(A_1)$, $\nu_2(A_1)$, $\nu_3(E)$, and $\nu_4(E)$ with $\nu_1 > \nu_2$ and $\nu_3 > \nu_4$. The corresponding normal coordinates are numbered analogously, the two members of a degenerate pair being distinguished by the subscripts a and b. Thus for PCl_3 we have: $Q_1(A_1)$; $Q_2(A_1)$; Q_{3a}, $Q_{3b}(E)$; and Q_{4a}, $Q_{4b}(E)$. Four distinct fundamental frequencies are indeed observed experimentally. We shall later see how they can be assigned as ν_1, ν_2, ν_3, and ν_4.

10.3. Use of internal coordinates in determining the number of distinct normal frequencies

In the preceding section we started with a full set of $3N$ displacement coordinates, and later subtracted off the translations and rotations so as to leave only the genuine vibrations. However, we may with advantage exclude the translations and rotations at the start, by using a set of $3N-6$ (or in the case of linear molecules $3N-5$) purely internal coordinates. Indeed, this procedure can provide extra information, for we can without difficulty choose our internal coordinates in such a way that they fall into sets that are mutually exclusive, in the sense that the members of any one set may mix amongst themselves under the symmetry operations of the molecular group, but that no member of one set can mix with a member of any other set. The internal coordinates belonging to a set of this kind will be referred to as being symmetrically equivalent.

The nature of such sets may be illustrated by reference again to the PCl$_3$ molecule, the model of which has been shown in Fig. 6.4. The three equal bond lengths are called r_1, r_2, and r_3 and the three equal bond angles are called β_1, β_2, and β_3, these angles being respectively the ones opposite the bonds of lengths r_1, r_2, and r_3. The number of independent internal coordinates required is $3N-6 = 6$. We shall choose them in the same way as we did in Section 6.3 when deriving the G matrix, namely as Δr_1, Δr_2, Δr_3, $r\Delta\beta_1$, $r\Delta\beta_2$, and $r\Delta\beta_3$. Here r is the equilibrium bond length and is introduced into the three coordinates of angle change in order to make them dimensionally the same as the three coordinates of bond stretch. We see that the first three of our internal coordinates form one distinct set, and that the last three form another; for whereas coordinates of bond stretch may mix with one another, and similarly coordinates of angle change may also mix with one another, it is obvious that a coordinate of bond stretch can never mix with a coordinate of angle change.

The mutually exclusive nature of the two sets means that the 6×6 group representation based upon them is already partially reduced, in that each matrix must consist of two 3×3 diagonal blocks, one for each of the non-mixing sets of symmetrically equivalent coordinates. Complete reduction can be achieved by completely reducing each of these blocks separately. (The reader will recognize that the situation is just like that previously discussed in Section 9.6.) The total structure of $\Gamma_{(\text{vib})}$ will be the direct sum of the structures of $\Gamma_{(\Delta r)}$ and $\Gamma_{(r\Delta\beta)}$.

Let us first apply the general equation (9.19) to determine the structure of $\Gamma_{(\Delta r)}$. The relevant data are shown in Table 10.1, which is of the same form as the upper part of Table 9.2. The character of a matrix of the representation based on the set of Δr-coordinates is denoted by $\chi_R^{(\Delta r)}$. Using eqn (9.19), we obtain $a_{A_1} = 1$, $a_{A_2} = 0$, and $a_E = 1$. The desired structure is thus

$$\Gamma_{(\Delta r)} = A_1 + E. \qquad (10.10)$$

Consideration of the set of symmetrically equivalent coordinates of angle change shows that $r\Delta\beta_1$ transforms in exactly the same way as does the change Δr_1 of the length of the bond directly opposite to β_1. Similarly $r\Delta\beta_2$ and $r\Delta\beta_3$ transform in exactly the same way as do Δr_2 and Δr_3 respectively. It therefore follows at once that

$$\Gamma_{(r\Delta\beta)} = \Gamma_{(\Delta r)} = A_1 + E. \tag{10.11}$$

Adding eqns (10.10) and (10.11) we obtain

$$\Gamma_{(vib)} = 2A_1 + 2E,$$

which, of course, is in agreement with the previous finding expressed in eqn (10.9).

TABLE 10.1

Data for the calculation of the structure of the representation based on the set of coordinates of bond stretch

PCl$_3$ molecule (C_{3v})	I	$2C_3$	$3\sigma_v$
Contribution to $\chi_R^{(\Delta r)}$ per unshifted internal coordinate	1	1	1
Number of such coordinates	3	0	1
Value of $\chi_R^{(\Delta r)}$	3	0	1

Now, however, we have obtained a somewhat deeper insight into the vibrational problem, in that we have found that one of the two A_1 representations may be based upon a coordinate of bond stretch and the other upon a coordinate of angle change. Similarly we have found that one of the two E representations may be based on a pair of coordinates of bond stretch and the other upon a pair of coordinates of angle change. In fact, as we shall see later, each of the actual normal modes of species A_1 (and likewise each of those of species E) will consist of a mixture of bond-stretching and angle-change motions. The extent of the mixing will depend upon the masses and the force constants involved. In general we can say that the extent of mixing will tend to be large when the frequency-values to be expected for hypothetical pure bond-stretching and hypothetical pure angle-change motions are nearly equal; whereas when these expected frequencies are very different in magnitude, the extent of mixing will be correspondingly small. Now it is a matter of experience that bond-stretch force constants for single bonds are considerably larger (roughly by factors of 5 to 10) than those for angle change. Thus if the masses involved are not very different, pure bond-stretching frequencies will be expected to be considerably higher than pure angle-change frequencies. Consequently in the actual normal modes the extent of mixing of the two types of motion will be expected to be rather small. In a rough qualitative manner of speaking, therefore, we may describe one of the two A_1 modes of PCl_3 (the one with the higher frequency) as a bond-stretching mode, meaning thereby only that (although we recognize that we are dealing with a mixture) it is the bond-stretch contribution that predominates. In the same approximate way we may describe the other A_1 mode as an angle-change mode, though in doing so we must not lose sight of the fact that it contains a contribution (believed to be a minor one) from bond stretching. Bearing in mind the rough order of the ratio of the two kinds of force constant, we might estimate that the so-called bond-stretch frequency of the species A_1 is likely to be 2 or 3 times greater than the so-called angle-change frequency of the same species. Exactly similar considerations apply to the two frequencies of species E. On the whole, therefore, we are led to anticipate that two of the four distinct normal frequencies of PCl_3 will be relatively high, as compared with the other two.

In fact this expectation is in fair qualitative agreement with the results of experimental observations, which show that the four frequencies (in order of descending magnitude) are 511, 484, 258, and 190 cm^{-1}. The first two relatively high frequencies, involving predominantly bond stretching, are thus $\nu_1(A_1)$ and $\nu_3(E)$; but at present we cannot say which is ν_1 and which is ν_3. The other two relatively lower frequencies, involving predominantly angle changes, are $\nu_2(A_1)$ and $\nu_4(E)$; but again we are not at present able to say which is which. When we come to discuss vibrational spectra, we shall learn how the proper assignments can be made by taking account of experimental observations of other properties besides the mere frequency-values.

10.4. Concerning redundancies

The number and species of the distinct frequencies of an N-atomic molecule can always be determined without difficulty by making use of a full set of $3N$ Cartesian displacement coordinates to derive the structure of the group representation of dimension $3N$, and finally subtracting off the translational and rotational contributions. The alternative method, however, in which the translations and rotations are excluded at the start by using only internal coordinates, may in certain cases involve new features which are not always easy to deal with. We shall exemplify this by the case of the phosphorus oxychloride molecule $POCl_3$, the model of which was shown in Fig. 7.6. Like PCl_3, it belongs to the point group C_{3v}.

If we adopt the perfectly straightforward method and make use of a full set of $3N = 15$ Cartesian displacement coordinates, we find (by considerations similar to those discussed for PCl_3 in Section 10.2) that

$$\Gamma = 4A_1 + A_2 + 5E.$$

We know that for the group C_{3v} the translations and rotations account for $A_1 + A_2 + 2E$. Subtracting off these contributions to Γ, we obtain

$$\Gamma_{(vib)} = 3A_1 + 3E. \tag{10.12}$$

The $POCl_3$ molecule therefore has 6 distinct frequencies. This conclusion is in agreement with experimental findings. But now let us try to use internal coordinates only, as we did for PCl_3 in Section 10.3. The number of these we require is $3N - 6 = 9$. From the model we see that there are two sets of symmetrically equivalent bond stretches and two sets of symmetrically equivalent angle changes, these being

(a) 3 P–Cl bond stretches, Δr_1, Δr_2, and Δr_3;

(b) 1 P–O bond stretch, Δr_4,

(c) 3 ClPCl angle changes, $r\Delta\beta_1$, $r\Delta\beta_2$, and $r\Delta\beta_3$;

(d) 3 ClPO angle changes, $r\Delta\alpha_1$, $r\Delta\alpha_2$, and $r\Delta\alpha_3$.

The notation for sets (a) and (c) is the same as for PCl_3; for sets (b) and (d) it is self-explanatory. The total number of internal coordinates contained in all four sets is 10, which is one in excess of the number required for the vibrational problem. We say that our choice must involve one redundancy. Nevertheless we carry on and use the excessively large number of coordinates of bond stretch and angle change, as indeed we must in order to find the characters of the matrices in the associated group representation. Proceeding exactly as for PCl_3 in Section 10.3, we reach the result

$$\Gamma_{(vib)} = 4A_1 + 3E. \tag{10.13}$$

This differs from the correct result, as given in eqn (10.12), which is $\Gamma_{(vib)} = 3A_1 + 3E$. Thus by using a set of internal coordinates containing one redundancy we have arrived (as was indeed to be expected) at a result giving one too many vibrational normal coordinates.

In the case of $POCl_3$ it is not difficult to see the nature of the redundancy. There is one combination of the coordinates $r\Delta\alpha_1$, $r\Delta\alpha_2$, and $r\Delta\alpha_3$, namely $r\Delta\alpha_1 + r\Delta\alpha_2 + r\Delta\alpha_3$, which obviously belongs to the totally symmetric species A_1, for, under any symmetry operation of the group, the coordinates either remain unchanged or have their order altered in the combination. In exactly the same way, we see that the second combination $r\Delta\beta_1 + r\Delta\beta_2 + r\Delta\beta_3$ also belongs to the species A_1. These combinations account for two of the four totally symmetric normal coordinates required by eqn (10.13), but they are not independent. From the geometry of the molecule it follows, in the approximation being used (i.e. that the displacements are small), that

$$\Delta\alpha_1 + \Delta\alpha_2 + \Delta\alpha_3 + \Delta\beta_1 + \Delta\beta_2 + \Delta\beta_3 = 0.$$

Thus one of the two combinations in question is merely the other with its sign reversed. We must therefore retain only one or the other of them, but not both. The redundancy is thus eliminated, and we reach agreement with the correct result (as expressed in eqn (10.12)) previously found by the trouble-free method which used a full set of $3N$ Cartesian displacement coordinates.

In this simple example, the one redundancy was quite easily formulated. In general, however, redundancies may not be so simple to detect, so that the method which uses only internal coordinates is not always so straightforward.

157

10.5. The point group D_{3h} to which the BF₃ molecule belongs

As a further example of the use of the methods described in Sections 10.2 and 10.3, we may deduce the number and species of the distinct normal frequencies of the boron trifluoride molecule, BF₃. This has a different structure from that of PCl₃, and consequently belongs to a different point group. Before

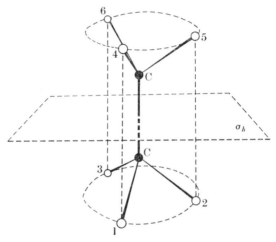

FIG. 10.1. Model of the C₂Cl₆ molecule in the eclipsed configuration.

going on to consider BF₃, however, we shall discuss this group, which is called D_{3h}, in terms of a more general model. For this purpose we may choose a conceivable model of the hexachloro-ethane molecule C₂Cl₆, in which it is assumed to have the so-called 'eclipsed' configuration. (The fact that this is not the structure which the molecule actually possesses in no way affects what we shall have to say.) The model in question is represented diagrammatically in Fig. 10.1. The molecule consists of two identical CCl₃ groups (each with the same regular pyramidal structure as the PCl₃ molecule) joined back to back by a bond between the two carbon atoms. The line of this bond obviously constitutes a three-fold rotation axis C_3, and following the usual convention we take it as the vertical z-direction. The so-called eclipsed configuration is characterized by the fact that

at equilibrium each of the upper set of three Cl-masses is vertically over one in the lower set. In Fig. 10.1 the vertical dotted
lines are intended to indicate that the Cl-masses numbered 4,
5, and 6 are vertically over those numbered 1, 2, and 3 respectively. Like the PCl_3 model, with which we are familiar, the
present one is seen to possess three vertical planes of symmetry
(or vertical reflection planes) σ_v. Because of the eclipsed configuration, it has in addition a reflection plane passing through
the centre of mass of the system and normal to the vertical C_3
axis. This is called a horizontal reflection plane, and is denoted
by the symbol σ_h. Associated with it is the single reflection
operation, for which the same symbol is used. Furthermore, it
is not difficult to see that the model also necessarily possesses
three two-fold rotation axes, C_2. They coincide respectively
with the intersections of the three σ_v planes with the σ_h plane,
and are thus all at right angles to the C_3 axis. For instance, the
first of these C_2 axes coincides with the intersection of σ_{v_1} with
σ_h. The single symmetry operation associated with it is rotation
through $\pi = 180°$. From Fig. 10.1 (or more conveniently by
handling an actual model) it can be seen that the effects of this
operation are to interchange the two point masses representing
the C nuclei, at the same time interchanging mass 1 and mass 4,
mass 2 and mass 6, and mass 3 and mass 5. The rotational
operations associated with the other two C_2 axes bring about
corresponding interchanges.

At first sight it might be thought that all the symmetry elements of the model have now been accounted for, and that the
list of associated symmetry operations, i.e. I, $2C_3$, $3C_2$, σ_h, and
$3\sigma_v$, is the complete one for the molecular point group. However,
this is not in fact the case. The inadequacy of the list can easily
be demonstrated by attempting to compile a binary-product
table, like that of Table 7.1 for C_{3v}. By doing this we find that
the products $\sigma_h C_3$ and $\sigma_h C_3^2$ (and the respectively identical ones
$C_3 \sigma_h$ and $C_3^2 \sigma_h$) are not equal to any single elements in the list we
have so far given. This means that the symmetry operations in
this list fail to satisfy the second requirement of Section 7.3, i.e.
they do not constitute a group. What we require in order to

159

complete the group (or, as is said, to close it) is some new symmetry element of the model which will generate the two single operations that are so far missing. We accordingly attribute to the model what is called a three-fold rotation–reflection axis S_3 (sometimes called an improper rotation axis). This new symmetry element is in fact coincident with the C_3 axis, but the basic symmetry operation S_3 associated with it is equivalent to a rotation C_3 followed by a reflection σ_h, or the corresponding, reverse procedure, σ_h followed by C_3.

Of course S_3^2, S_3^3, S_3^4, etc. must also be elements of the group. The reader may easily verify that, since

$$S_3 = \sigma_h C_3 = C_3 \sigma_h,$$

it follows that

$$S_3^2 = \sigma_h^2 C_3^2 = C_3^2,$$

$$S_3^3 = \sigma_h^3 C_3^3 = \sigma_h,$$

$$S_3^4 = \sigma_h^4 C_3^4 = C_3,$$

$$S_3^5 = \sigma_h^5 C_3^5 = \sigma_h C_3^2 = C_3^2 \sigma_h,$$

and

$$S_3^6 = \sigma_h^6 C_3^6 = I.$$

Thereafter we have $S_3^7 = S_3 I = S_3$; $S_3^8 = S_3^2 I = S_3^2$; etc. The cycle is thus repeated. We see that, except for S_3 and S_3^5, all the operations S_3^n are identical with operations already associated with other symmetry elements. Thus the rotation–reflection axis S_3 contributes just two new operations, and in fact just the two which were needed to close the group. The reader may easily verify that the order g of the group (i.e. the total number of symmetry operations in it) is 12, and that these fall into 6 classes (i.e. $k = 6$). It follows from eqn (9.13) that the number γ of symmetry species or irreducible representations must also be 6. Using eqn (9.14), we find that the only six integers whose squares add up to 12 are 1, 1, 1, 1, 2, and 2. There must therefore be 4 non-degenerate species and 2 doubly degenerate ones—in each case just twice as many as for the point group C_{3v}. Now the essential difference between C_{3v} and D_{3h} is the possession by D_{3h} of the horizontal reflection plane. In fact the presence of this plane means that for D_{3h} each of the symmetry species of C_{3v} has given rise to two, one symmetric and the other antisymmetric with respect to σ_h. For this reason the same symbols are used as for C_{3v}, but the behaviour with respect to σ_h is denoted by the addition of one prime (symmetric) or two primes (anti-symmetric).

Table 10.2 gives the character table for D_{3h} in the same conventional form as that of Table 8.4 for C_{3v}.

TABLE 10.2

Character table for the point group D_{3h}

D_{3h}	I	$2C_3$	$3C_2$	σ_h	$2S_3$	$3\sigma_v$	
A_1'	1	1	1	1	1	1	
A_2'	1	1	−1	1	1	−1	R_z
E'	2	−1	0	2	−1	0	(T_x, T_y)
A_1''	1	1	1	−1	−1	−1	
A_2''	1	1	−1	−1	−1	1	T_z
E''	2	−1	0	−2	1	0	(R_x, R_y)

10.6. The model of the BF_3 molecule

The BF_3 molecule has a planar structure, with a central B atom joined by an identical bond to each of the three F atoms, which therefore occupy sites at the corners of an equilateral triangle. Fig. 10.2 shows the molecular model. The plane of the figure is the xy plane, the C_3 axis (z-direction) being normal to this plane and directed towards the reader.

The molecule belongs to the point group D_{3h}, which was discussed in the preceding section. However, it represents a special case in that, in the equilibrium configuration which the model represents, all the four nuclei lie in the σ_h plane. In conse-quence, when applied to the model, the operation σ_h is identical with the identity operation I, and the operations S_3 and S_3^5 are respectively identical with C_3 and C_3^2. This, of course, is because the terms symmetric and antisymmetric with respect to σ_h lose their meaning when all the point masses remain unaffected, i.e. when, by virtue of their position in the reflection plane, they are simply 'reflected into themselves'. However, in dealing with displacements of the point masses (as we have to do in the treat-ment of the normal modes of the molecule) it is essential to retain all the elements of the point group D_{3h}, including of course σ_h and $2S_3$. The translation T_z, for example, involves displacements of the nuclei out of the σ_h plane of the equilibrium configuration, and belongs to the species A_2'', i.e. to one of the species characterized by being antisymmetric with respect to the horizontal

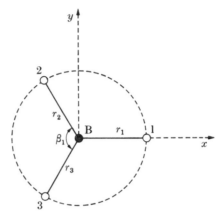

FIG. 10.2. Model of the planar BF_3 molecule.

reflection plane. In the rotation R_z, on the other hand, all the nuclei remain in this plane, and this external mode belongs to the species A_2', i.e. to one of the species characterized by being symmetric with respect to σ_h. The reader will be able without difficulty to confirm these species assignments, and also those of the degenerate pairs (T_x, T_y) and (R_x, R_y) as given in the character table, Table 10.2.

10.7. Calculation of the number and species of the distinct frequencies of BF_3, using Cartesian displacement coordinates

Choosing the x- and y-directions in physical space in the manner indicated in Fig. 10.2, and remembering that the z-direction is that of the C_3 axis, we follow the same course as we did for PCl_3 in Sections 9.5 and 10.2. Thus we must derive, for one operation R from each of the six classes, the number of sites with unshifted displacements, and the contribution from each such site to the character $\chi_R^{(\Gamma)}$ of the matrix representing R in the 12-dimensional group representation. In the case of the identity operation I, the number of sites with unshifted displacements is obviously 4, and the character contribution from each is obviously 3. For the operation C_3, only the central B site comes into the question, and we find that its contribution is zero, as in Table 9.2 for PCl_3. Amongst the C_2 operations, consider rotation through 180° about the bond from B to the site marked 1 in Fig. 10.2. The number of sites with unshifted displacements is found to be 2 (those of the B nucleus and the F nucleus on the two-fold axis under consideration). The transformation for either of them is of the form

$$\begin{bmatrix} x' \\ y' \\ z' \end{bmatrix} = \begin{bmatrix} 1 & 0 & 0 \\ 0 & -1 & 0 \\ 0 & 0 & -1 \end{bmatrix} \begin{bmatrix} x \\ y \\ z \end{bmatrix}.$$

Thus the contribution to $\chi_R^{(\Gamma)}$ from each of the two sites is -1. The next class of symmetry operations consists solely of σ_h. Here all four sites make character contributions. For each the transformation is of the form

$$\begin{bmatrix} x' \\ y' \\ z' \end{bmatrix} = \begin{bmatrix} 1 & 0 & 0 \\ 0 & 1 & 0 \\ 0 & 0 & -1 \end{bmatrix} \begin{bmatrix} x \\ y \\ z \end{bmatrix},$$

so that the contribution per site is 1. Coming now to the rotation–reflection operation S_3, only the central site of the B nucleus will contribute. The transformation in question is of the form

$$\begin{bmatrix} x' \\ y' \\ z' \end{bmatrix} = \begin{bmatrix} -\frac{1}{2} & -\frac{1}{2}\sqrt{3} & 0 \\ \frac{1}{2}\sqrt{3} & -\frac{1}{2} & 0 \\ 0 & 0 & -1 \end{bmatrix} \begin{bmatrix} x \\ y \\ z \end{bmatrix},$$

and the contribution to $\chi_R^{(\Gamma)}$ is therefore -2. Lastly we must consider the operation σ_{v_1}. The number of sites with unshifted displacements is 2 (the central site and the F site on the reflection plane being considered). The contribution from each is 1, as in Table 9.2 for PCl_3. The above information may be conveniently presented in a table like the upper part of Table 9.2 for PCl_3. This is done in Table 10.3.

We can now find the structure of the reducible group representation Γ based upon the full set of 12 displacement coordinates for the BF_3 molecule. This we do by making use of equation

TABLE 10.3

Data for the calculation of the structure of a 12×12 representation of the group D_{3h}

BF_3 molecule (D_{3h})	I	$2C_3$	$3C_2$	σ_h	$2S_3$	$3\sigma_v$
Contribution to $\chi_R^{(\Gamma)}$ per site with unshifted displacements	3	0	-1	1	-2	1
Number of such sites	4	1	2	4	1	2
Value of $\chi_R^{(\Gamma)}$	12	0	-2	4	-2	2

(9.19). In the case in hand $g = 12$, and the remaining necessary information is contained in the character table (Table 10.2) and in Table 10.3. We obtain the result

$$\Gamma = A_1' + A_2' + 3E' + 2A_2'' + E''. \tag{10.14}$$

We next subtract off the species for the translations and vibrations, which (see the character table and the discussion in Section 10.6) account for $A_2' + E' + A_2'' + E''$. This gives

$$\Gamma_{(vib)} = A_1' + 2E' + A_2''. \tag{10.15}$$

The BF_3 molecule thus has four distinct normal frequencies. Three of them belong to normal modes which are symmetric with respect to reflection in the σ_h plane, i.e. the plane of the molecule in its equilibrium configuration. These are the modes whose species are denoted by symbols with a single prime. In all these three modes all the nuclear motions must occur in the plane of the molecule. One of the three is totally symmetric (species A_1'); the other two are doubly degenerate (E'). The fourth frequency belongs to a normal mode which is anti-symmetric with respect to σ_v, and in which all the nuclear motions must therefore be normal to the plane of the molecule in its equilibrium configuration. Further insight into the nature of the normal vibrational modes is obtained when, as in the next section, they are considered in relation to internal coordinates.

10.8. Use of internal coordinates in deriving the number and species of the distinct frequencies of BF_3

As with PCl_3, which was treated in Section 10.3, we must begin by choosing suitable internal coordinates and arranging them into sets that are mutually exclusive in the sense that, under the symmetry operations of the group, the members of any one set may mix with one another, but not with any member of any other set. Following the course successfully taken for PCl_3, we choose as the first of such sets the three bond stretches Δr_1, Δr_2, and Δr_3. (The molecular model will be found in Fig. 10.2.) For this set, considered on its own, we can very easily derive the information contained in Table 10.4. Along with the group

TABLE 10.4

Data for the calculation of the structure of the representation based on the set of coordinates of bond stretch

BF_3 molecule (D_{3h})	I	$2C_3$	$3C_2$	σ_h	$2S_3$	$3\sigma_v$
Contribution to $\chi_E^{(\Delta r)}$ per unshifted internal coordinate	1	1	1	1	1	1
Number of such coordinates	3	0	1	3	0	1
Value of $\chi_E^{(\Delta r)}$	3	0	1	3	0	1

character table (Table 10.2), this table enables us to work out the structure of the group representation based upon the three bond stretches. The result is

$$\Gamma_{(\Delta r)} = A_1' + E'. \tag{10.16}$$

From eqn (10.15) we see that the A_1' frequency is the only one belonging to this species, and it therefore follows that the associated normal mode must be a purely bond-stretching one. Being totally symmetric, it must obviously be a mode in which the central boron atom remains stationary while the three fluorine atoms move in and out radially and synchronously in what has been aptly called a 'breathing' motion.

Again as for PCl_3, we choose as the second set of internal coordinates the three inter-bond angle changes $r\Delta\beta_1$, $r\Delta\beta_2$, and $r\Delta\beta_3$. (The numbering of the angles β is analogous to that for

165

PCl_3; see also Fig. 10.2.) Exactly as for the bond-stretch set, we obtain
$$\Gamma_{(r\Delta\beta)} = A_1' + E'.$$

Here, however, we note a redundancy of a kind that did not occur in the case of PCl_3. Obviously the sum of the three angles $\beta_1 + \beta_2 + \beta_3$ must be constant and equal to 2π, so that the combination $r(\Delta\beta_1 + \Delta\beta_2 + \Delta\beta_3)$ must vanish. But this combination is obviously totally symmetric, i.e. of species A_1'. Removing it from the structure $\Gamma_{(r\Delta\beta)}$ derived above, we obtain

$$\Gamma_{(r\Delta\beta)} = E'. \tag{10.17}$$

We have now accounted for three of the distinct frequencies of the BF_3 molecule. They are all frequencies of motion within the plane, as is indicated by the single primes on the symbols of their symmetry species. Referring back to the total structure $\Gamma_{(vib)}$ as expressed in eqn (10.15), we see that there remains just one distinct frequency to be accounted for, and that it belongs to the species A_2''. We have already exhausted all the species to which bond stretches and changes of the β-angles can give rise. The remaining frequency must therefore be associated with a normal mode which does not involve either of these types of internal coordinate. Being antisymmetric with respect to σ_h, it is identified uniquely as an out-of-plane deformation frequency. Accordingly it must be derivable from a set of symmetrically equivalent coordinates such as $r\Delta\alpha_1$, $r\Delta\alpha_2$, and $r\Delta\alpha_3$, where the angles α are those made by the respective bonds with the plane of the molecule in its equilibrium configuration. Of course, coordinates of this kind did not arise in the non-planar PCl_3 molecule. We shall have more to say about the set which they form when (in Chapter 13) we take up again the problem of the vibrations of the BF_3 molecule.

The expectation of four distinct frequencies is in agreement with experimental observation which, for the molecule $^{11}BF_3$ containing the heavier of the two boron isotopes, gives the values 1454, 888, 691, and 480 cm^{-1}. Our theoretical treatment has shown that two of the frequencies must belong to in-plane modes of bond stretching (species A_1' and E'), one to an in-plane mode of angle change (E') and one to the out-of-plane deformation mode (A_2''). For the modes of species A_1' and A_2'' the descriptions just given are exact, because each of these modes is the sole representative of its species; but for the two modes of species E' the descriptions are only approximate, it being admitted that each mode will have contributions from both types of motion, bond-stretch and angle-change. We shall certainly expect that the highest of the four observed frequencies will belong to one of the stretching modes, though we have no basis as yet for deciding which. Also we may think it likely that the lowest will belong to the mode of angle change within the plane. However, the assignments of all four frequencies must remain a matter of speculation until, from a study of vibrational spectra, we learn how they may be determined from further spectroscopic evidence.

11 Projection operators and their use for the complete reduction of group representations

11.1. Definition and properties of projection operators

WHEN a configuration vector p of a molecule is subjected to a symmetry operation R of the molecular point group, the transformed vector may be written (as in Section 4.6) in the form $p' = \mathscr{R}p$, where $\mathscr{R}$ is the operator of the operation R. Let us multiply $\mathscr{R}p$ by $\chi_R^{(\mu)}$, the character of the matrix representing R in the irreducible representation belonging to the particular symmetry species μ, and then sum over all the operations of the group. The result is expressible as $\mathscr{P}^{(\mu)}p$, where $\mathscr{P}^{(\mu)}$ is a new operator called the *projection operator* for the species μ. The equation defining it is thus

$$\mathscr{P}^{(\mu)} = \sum_R \chi_R^{(\mu)} \mathscr{R}. \tag{11.1}$$

There is evidently one such operator for each symmetry species of the group. Any such species must be either non-degenerate, doubly degenerate, or triply degenerate, i.e. the matrices of the associated irreducible representation must have either 1, 2, or 3 diagonal elements. When the number is 3, eqn (11.1) becomes

$$\mathscr{P}^{(\mu)} = \sum_R (R_{11}^{(\mu)} + R_{22}^{(\mu)} + R_{33}^{(\mu)})\mathscr{R}, \tag{11.2}$$

where $R_{11}^{(\mu)}$, $R_{22}^{(\mu)}$, and $R_{33}^{(\mu)}$ are the diagonal elements in question. The case in which μ is doubly degenerate is covered by eqn (11.2), $R_{33}^{(\mu)}$ being then simply set equal to zero. Similarly the non-degenerate case is also covered, if both $R_{22}^{(\mu)}$ and $R_{33}^{(\mu)}$ are struck out.

Our first objective is to examine the effect of applying the projection operator $\mathscr{P}^{(\mu)}$ to a basis vector of one of the irreducible representations, for instance the one belonging to the species ν.

168

At most this species will have the degeneracy 3, i.e. the basis of the corresponding invariant sub-space of configuration space will have 3 basis vectors, $e_1^{(\nu)}$, $e_2^{(\nu)}$, and $e_3^{(\nu)}$. The effect of the symmetry operation R upon them can be expressed in the following matrix equation, which is exactly analogous to eqn (4.19):

$$[\mathscr{R}e_1^{(\nu)} \quad \mathscr{R}e_2^{(\nu)} \quad \mathscr{R}e_3^{(\nu)}] = [e_1^{(\nu)} \quad e_2^{(\nu)} \quad e_3^{(\nu)}] \begin{bmatrix} R_{11}^{(\nu)} & R_{12}^{(\nu)} & R_{13}^{(\nu)} \\ R_{21}^{(\nu)} & R_{22}^{(\nu)} & R_{23}^{(\nu)} \\ R_{31}^{(\nu)} & R_{32}^{(\nu)} & R_{33}^{(\nu)} \end{bmatrix}.$$

Fixing our attention on $e_1^{(\nu)}$, for example, we can write

$$\mathscr{R}e_1^{(\nu)} = R_{11}^{(\nu)} e_1^{(\nu)} + R_{21}^{(\nu)} e_2^{(\nu)} + R_{31}^{(\nu)} e_3^{(\nu)}. \tag{11.3}$$

Here, as before, the case in which the species ν has the degeneracy 2 is covered, if we set equal to zero all quantities in eqn (11.3) which bear a subscript 3. Likewise the case of non-degeneracy is also covered, if we strike out all quantities bearing a subscript 3 or a subscript 2.

From eqns (11.2) and (11.3) we have

$$\mathscr{P}^{(\mu)}e_1^{(\nu)} = \sum_R \{(R_{11}^{(\mu)} + R_{22}^{(\mu)} + R_{33}^{(\mu)})(R_{11}^{(\nu)} e_1^{(\nu)} + R_{21}^{(\nu)} e_2^{(\nu)} + R_{31}^{(\nu)} e_3^{(\nu)})\}. \tag{11.4}$$

We see that, when multiplied out, the right-hand side of eqn (11.4) consists of the sum of nine terms, each of which consists of one of the basis vectors multiplied by a coefficient of the type $\sum_R R_{ii}^{(\mu)} R_{kl}^{(\nu)}$. We now invoke the basic theorem, expressed in eqn (9.1), concerning the orthogonality properties of the elements of the matrices in irreducible representations. It tells us that, if $\mu \neq \nu$, every one of the terms making up the right-hand side of eqn (11.4) in its multiplied-out form is necessarily zero. The same result would clearly have been obtained if the projection operator $\mathscr{P}^{(\mu)}$ had been applied to $e_2^{(\nu)}$ or to $e_3^{(\nu)}$. On the other hand, the basic theorem tells us that, if $\mu = \nu$, only one of the terms survives, and that its value is $(g/n_\mu)e_1^{(\mu)}$. Thus we have

$$\mathscr{P}^{(\mu)}e_1^{(\mu)} = \frac{g}{n_\mu} e_1^{(\mu)}. \tag{11.5}$$

In exactly the same way, we see that

$$\mathcal{P}^{(\mu)}e_2^{(\mu)} = \frac{g}{n_\mu}e_2^{(\mu)}, \tag{11.6}$$

and

$$\mathcal{P}^{(\mu)}e_3^{(\mu)} = \frac{g}{n_\mu}e_3^{(\mu)}. \tag{11.7}$$

It is well to repeat that the above arguments are by no means restricted to the case where μ and ν are three-fold degenerate. As has already been pointed out, they hold equally well when n_μ (or n_ν) is 1 or 2, when it is simply necessary to delete the parts which do not apply.

We may therefore sum up our conclusions so far by saying that in the projection operator $\mathcal{P}^{(\mu)}$ we have an operator whose application will annihilate any of the basis vectors of species other than μ, but will simply multiply by the constant factor g/n_μ any of the basis vectors which do belong to this species.

We now go on to consider any vector $p^{(\nu)}$ in the particular sub-space corresponding to the symmetry species ν. This general vector must, of course, be expressible as a linear combination (i.e. vector sum) of the basis vectors of this sub-space, i.e.

$$p^{(\nu)} = a_1^{(\nu)}e_1^{(\nu)} + a_2^{(\nu)}e_2^{(\nu)} + a_3^{(\nu)}e_3^{(\nu)},$$

in which the $a^{(\nu)}$s are numerical coefficients. If $n_\nu = 2$, the last term is to be deleted; if $n_\nu = 1$, only the first term is to be retained. The effect of applying the projection operator $\mathcal{P}^{(\mu)}$ may be written as

$$\mathcal{P}^{(\mu)}p^{(\nu)} = a_1^{(\nu)}\mathcal{P}^{(\mu)}e_1^{(\nu)} + a_2^{(\nu)}\mathcal{P}^{(\mu)}e_2^{(\nu)} + a_3^{(\nu)}\mathcal{P}^{(\mu)}e_3^{(\nu)}.$$

From the conclusions already reached above it follows that

$$(a) \text{ if } \mu \neq \nu, \quad \mathcal{P}^{(\mu)}p^{(\nu)} = 0, \tag{11.8}$$

and (b) if $\mu = \nu$, $$\mathcal{P}^{(\mu)}p^{(\nu)} = \frac{g}{n_\mu}p^{(\mu)}. \tag{11.9}$$

Thus the effect of $\mathcal{P}^{(\mu)}$ upon a general vector $p^{(\nu)}$ belonging to the species ν is the same as its effect upon a basis vector $e^{(\nu)}$ of this species.

11.2. The use of projection operators for the complete reduction of a group representation

Consider a reducible representation Γ with some conveniently chosen orthonormal basis vectors $e_i^{(\Gamma)}$, where i runs from unity up to a number equal to the dimension of the configuration space in question. For our purposes it will suffice to restrict our treatment to the case where no irreducible representation occurs more than once in the completely reduced structure of Γ. Each of the basis vectors of Γ must of course be expressible as a linear combination of all the basis vectors of all the irreducible representations. Let us arbitrarily select $e_1^{(\Gamma)}$ and accordingly write

$$e_1^{(\Gamma)} = \sum_\nu (a_1^{(\nu)}e_1^{(\nu)}+a_2^{(\nu)}e_2^{(\nu)}+a_3^{(\nu)}e_3^{(\nu)}), \qquad (11.10)$$

where the summation extends over all the irreducible representations (i.e. all the symmetry species) of the group. (Here again the cases where $n_\nu = 2$ or 1 are covered, if appropriate deletions are made.) We see that eqn (11.10) is of the form

$$e_1^{(\Gamma)} = \sum_\nu p_1^{(\nu)}, \qquad (11.11)$$

where $p_1^{(\nu)}$ is a particular vector belonging to the species ν.

Application of the projection operator $\mathscr{P}^{(\mu)}$ to the selected basis vector $e_1^{(\Gamma)}$ of our reducible representation therefore gives

$$\mathscr{P}^{(\mu)}e_1^{(\Gamma)} = \sum (\mathscr{P}^{(\mu)}p_1^{(\nu)}). \qquad (11.12)$$

From the conclusions reached in the preceding section—summarized in eqns (11.8) and (11.9)—it now follows that the only term that can survive in the summation on the right-hand side of eqn (11.12) is the one for which $\nu = \mu$, and that its value is $(g/n_\mu)p_1^{(\mu)}$. Application of $\mathscr{P}^{(\mu)}$ to $e_1^{(\Gamma)}$ has thus produced a vector belonging to the symmetry species μ. Moreover, since $\mathscr{P}^{(\mu)}$ is by definition a known linear combination of the known operators $\mathscr{R}$ (eqn (11.2)) the vector produced is a known linear combination of the known basis vectors $e_i^{(\Gamma)}$ of our reducible representation Γ.

By applying to $e_1^{(\Gamma)}$ the projection operator for each of the symmetry species in turn, we can thus produce one vector belonging to each. We can do this for each of the n unit vectors

171

$e_i^{(\Gamma)}$, and so generate n vectors belonging to each species. In general, of course, not all the n vectors for any particular species will be independent. If the species is a non-degenerate one, the invariant sub-space associated with it will be only 1-dimensional. In consequence, the n vectors in this space, will differ at most by mere numerical factors. In the case of a doubly or triply degenerate species, the number of independent linear combinations will be only 2 or 3 respectively.

We see that, by using the projection operators $\mathscr{P}^{(\mu)}$, we are able to obtain a set of vectors which will furnish a new basis of such a kind that in it the original representation will be completely reduced. Of course, in general, none of the vectors we obtain will be unit vectors, but if, as is usual, we wish the new basis to be an orthonormal one, we can very easily normalize each of the basis vectors to unity. Thus suppose that one of them is expressed as the linear combination $\sum_i c_i\, e_i^{(\Gamma)}$, where the c_is are numerical coefficients. To normalize (i.e. to obtain the corresponding unit vector) all that is necessary is to divide by the square root of $\sum_i c_i^2$. The ambiguity introduced by the square root is trivial, for a reversal of sign merely reverses the sense of the unit vector, without in any way impairing its suitability as a basis vector.

11.3. An example of the complete reduction of a representation of the point group C_{3v}

The method explained in the two preceding sections may be illustrated by considering the 3-dimensional representation of the point group C_{3v} which is generated by the complete set of symmetrically equivalent internal coordinates Δr_1, Δr_2, and Δr_3 of bond stretches in the PCl_3 molecule (fully described in Section 10.3). We have previously shown that this representation, which we called $\Gamma_{(\Delta r)}$, has the structure $A_1 + E$. This result is expressed in eqn (10.10). We shall now proceed to find a basis in which the representation is completely reduced.

We first introduce the unit vectors of the original orthonormal basis associated with the coordinates Δr_1, Δr_2, and Δr_3. We may call these basis vectors d_1, d_2, and d_3. Under the symmetry operations of the group, their behaviour is very simple; each of them merely remains unchanged or becomes one of the others. (This is, of course, the essential characteristic of the members of a complete set of symmetrically equivalent internal co-ordinates.) Under the operation C_3, for instance, $d_1 \to d_2$, $d_2 \to d_3$, and $d_3 \to d_1$, while under the operation σ_{v_1} we find $d_1 \to d_1$, $d_2 \to d_3$, and $d_3 \to d_2$. The full 3×3 reducible representation $\Gamma_{(\Delta r)}$ is easily seen to be that given in Table 11.1, which also includes the matrix characters for each symmetry operation.

TABLE 11.1

A reducible representation of the group C_{3v}

	I	C_3	C_3^2	σ_{v_1}	σ_{v_2}	σ_{v_3}
$\Gamma_{(\Delta r)}$	$\begin{bmatrix}1&0&0\\0&1&0\\0&0&1\end{bmatrix}$	$\begin{bmatrix}0&0&1\\1&0&0\\0&1&0\end{bmatrix}$	$\begin{bmatrix}0&1&0\\0&0&1\\1&0&0\end{bmatrix}$	$\begin{bmatrix}1&0&0\\0&0&1\\0&1&0\end{bmatrix}$	$\begin{bmatrix}0&0&1\\0&1&0\\1&0&0\end{bmatrix}$	$\begin{bmatrix}0&1&0\\1&0&0\\0&0&1\end{bmatrix}$
$\chi_R^{(\Gamma)}$	3	0	0	1	1	1

We now arbitrarily choose the particular basis vector d_1 as the one from which, by the application in turn of each of the projection vectors, we shall generate one vector belonging to each of the symmetry species μ of the group. We thus have to evaluate, for each species, the vector

$$\mathscr{P}^{(\mu)} d_1 = \sum_R \chi_R^{(\mu)} \mathscr{R} d_1.$$

The relevant values of $\mathscr{R} d_1$ for each symmetry operation R are given in the first row of Table 11.2. The next three rows give the values of $\chi_R^{(\mu)}$ for each of the three species of the group; they thus merely constitute the group character table, which is reproduced here for convenience. The last three rows of Table 11.2 give the values of the corresponding products $\chi_R^{(\mu)} \mathscr{R} d_1$.

Performing the summations, we obtain the following values for the vectors belonging to the species A_1, A_2, and E respectively:

$$\sum_R \chi_R^{(A_1)} \mathscr{R} d_1 = 2d_1 + 2d_2 + 2d_3; \qquad (11.13)$$

$$\sum_R \chi_R^{(A_2)} \mathscr{R} d_1 = 0; \tag{11.14}$$

$$\sum_R \chi_R^{(E)} \mathscr{R} d_1 = 2d_1 - d_2 - d_3. \tag{11.15}$$

The zero result for the species A_2 is, of course, in agreement with our earlier finding that this species does not appear in the structure $\Gamma_{\text{vib}} = A_1 + E$. For the species A_1 it is easy to see that exactly the same result would have been obtained if we had

TABLE 11.2

Details in the use of projection operators

C_{3v}	I	C_3	C_3^2	σ_{v_1}	σ_{v_2}	σ_{v_3}
$\mathscr{R} d_1$	d_1	d_2	d_3	d_1	d_3	d_2
$\chi_R^{(A_1)}$	1	1	1	1	1	1
$\chi_R^{(A_2)}$	1	1	1	-1	-1	-1
$\chi_R^{(E)}$	2	-1	-1	0	0	0
$\chi_R^{(A_1)} \mathscr{R} d_1$	d_1	d_2	d_3	d_1	d_3	d_2
$\chi_R^{(A_2)} \mathscr{R} d_1$	d_1	d_2	d_3	$-d_1$	$-d_3$	$-d_2$
$\chi_R^{(E)} \mathscr{R} d_1$	$2d_1$	$-d_2$	$-d_3$	0	0	0

applied $\mathscr{P}^{(A_1)}$ to d_2 or to d_3, instead of to d_1. The invariant sub-space of this non-degenerate species is only one-dimensional, so that two or more linearly independent vectors belonging to it would be an impossibility. As to the species E we have so far generated only one vector, whereas the 2-dimensional invariant sub-space requires two for its basis.

Before considering an additional vector for the species E, we shall normalize the ones which we already have. From eqn (11.13) we see that the normalizing factor for the vector belonging to the species A_1 is $1/(2^2 + 2^2 + 2^2)^{\frac{1}{2}} = 1/(2\sqrt{3})$. The normalized vector, which we will call $s^{(A_1)}$, is therefore

$$s^{(A_1)} = \frac{1}{\sqrt{3}} (d_1 + d_2 + d_3). \tag{11.16}$$

This is the unique unit vector belonging to the totally symmetric species, and therefore must be the basis vector of this species in any orthonormal basis in which $\Gamma_{(\Delta r)}$ is completely reduced. The normalizing factor in the case of the vector belonging to the species E (eqn (11.15)) is $1/(2^2+1^2+1^2)^{\frac{1}{2}} = 1/\sqrt{6}$. The corresponding unit vector is thus

$$s_a^{(E)} = \frac{1}{\sqrt{6}}(2\boldsymbol{d}_1 - \boldsymbol{d}_2 - \boldsymbol{d}_3). \tag{11.17}$$

We shall choose this as one of the two vectors of an orthonormal basis spanning the 2-dimensional sub-space, using the subscript a to distinguish it from the other, which we shall denote by $s_b^{(E)}$. This choice is not a uniquely determined one, for an unlimited number of pairs of mutually orthogonal unit vectors would serve equally well as a basis. Once we have made our arbitrary choice of this one basis vector, however, the second one becomes uniquely determined (except as to the trivial matter of sign).

As we saw in Section 4.5, a change of basis may be expressed in matrix form. Thus the contents of eqns (11.16) and (11.17) may be embodied in the following equation, which is analogous to eqn (4.19):

$$\begin{bmatrix} s^{(A_1)} & s_a^{(E)} & s_b^{(E)} \end{bmatrix} = \begin{bmatrix} \boldsymbol{d}_1 & \boldsymbol{d}_2 & \boldsymbol{d}_3 \end{bmatrix} \begin{bmatrix} \dfrac{1}{\sqrt{3}} & \dfrac{2}{\sqrt{6}} & b_1 \\[2mm] \dfrac{1}{\sqrt{3}} & -\dfrac{1}{\sqrt{6}} & b_2 \\[2mm] \dfrac{1}{\sqrt{3}} & -\dfrac{1}{\sqrt{6}} & b_3 \end{bmatrix}. \tag{11.18}$$

The second basis vector $s_b^{(E)}$ of the species E has the value $b_1\boldsymbol{d}_1 + b_2\boldsymbol{d}_2 + b_3\boldsymbol{d}_3$, where b_1, b_2, and b_3 are as yet unknown coefficients. Now since both the new basis and the original basis are orthonormal, the square matrix in eqn (11.18) must be an orthogonal one. We therefore deduce the three following relations connecting the coefficients in question:

$$b_1^2 + b_2^2 + b_3^2 = 1;$$

$$\frac{1}{\sqrt{3}}b_1 + \frac{1}{\sqrt{3}}b_2 + \frac{1}{\sqrt{3}}b_3 = 0;$$

$$\frac{2}{\sqrt{6}}b_1 - \frac{1}{\sqrt{6}}b_2 - \frac{1}{\sqrt{6}}b_3 = 0.$$

From the second and third of these equations we find at once that $b_1 = 0$ and $b_2 = -b_3$. From the first equation it then follows that $b_2 = 1/\sqrt{2}$ and $b_3 = -1/\sqrt{2}$. The required second unit basis vector belonging to the species E is thus found to be

$$s_b^{(E)} = \frac{1}{\sqrt{2}}(d_2 - d_3). \tag{11.19}$$

There is another way of arriving at this result, although it is essentially equivalent to the above. For the sake of completeness the details will be given. The basis vector $s_a^{(E)}$ was obtained by applying the projection operator $\mathscr{P}^{(E)}$ to the arbitrarily selected original basis vector d_1. If instead we had applied it to d_2 or d_3, we should have generated the corresponding vectors which, after normalization, would have been respectively $(1/\sqrt{6})(2d_2 - d_3 - d_1)$ and $(1/\sqrt{6})(2d_3 - d_1 - d_2)$. Let us call the first of these u, and the second w. Having decided upon $(1/\sqrt{6})(2d_1 - d_2 - d_3)$ as one of the basis vectors, $s_a^{(E)}$, neither u nor w is in itself acceptable as the second, $s_b^{(E)}$. Instead we must choose a linear combination of the two, say $\alpha u + \beta w$, which shall be orthogonal to $s_a^{(E)}$. In order to determine $s_b^{(E)}$, we must find the values of the coefficients α and β. We have

$$s_b^{(E)} = \frac{\alpha}{\sqrt{6}}(2d_2 - d_3 - d_1) + \frac{\beta}{\sqrt{6}}(2d_3 - d_1 - d_2)$$

$$= -\frac{\alpha + \beta}{\sqrt{6}}d_1 + \frac{2\alpha - \beta}{\sqrt{6}}d_2 + \frac{2\beta - \alpha}{\sqrt{6}}d_3. \tag{11.20}$$

Because this must be orthogonal to the vector

$$s_a^{(E)} = \frac{2}{\sqrt{6}}d_1 - \frac{1}{\sqrt{6}}d_2 - \frac{1}{\sqrt{6}}d_3,$$

the following equation must hold:

$$\frac{-2(\alpha + \beta)}{6} - \frac{(2\alpha - \beta)}{6} - \frac{(2\beta - \alpha)}{6} = 0;$$

i.e. $$\alpha + \beta = 0. \tag{11.21}$$

By using this result to eliminate β from the expression for $s_b^{(E)}$ in eqn (11.20), we obtain

$$s_b^{(E)} = \frac{3\alpha}{\sqrt{6}}\, d_2 - \frac{3\alpha}{\sqrt{6}}\, d_3.$$

The requirement that $s_b^{(E)}$ shall be a unit vector determines the value of α as $1/\sqrt{3}$ and the value of $s_b^{(E)}$ as $(1/\sqrt{2})(d_2 - d_3)$. This, of course, is exactly the same result as previously obtained and expressed in eqn (11.19).

We have thus obtained all the elements of the square matrix which, in the manner shown in eqn (11.18), will transform the set of basis vectors of the original reducible representation $\Gamma_{(\text{vib})}$ into the new set, in relation to which the representation is completely reduced. Let us denote this matrix by $\mathbf{A}$. Then, as we proved in Section 4.8, the effect of the change of basis upon a matrix $\mathbf{R}$ of the original representation is obtained by subjecting it to the similarity transformation $\mathbf{A^{-1}RA}$. Because $\mathbf{A}$ is an orthogonal matrix, we can write this transformation as $\mathbf{A^{\dagger}RA}$. As an example, take the symmetry operation C_3. The matrix $\mathbf{R}$ representing it before reduction is given in Table 11.1. The matrix $\mathbf{A}$ is the square matrix in eqn (11.18), with $b_1 = 0$, $b_2 = 1/\sqrt{2}$, and $b_3 = -1/\sqrt{2}$. The matrix representing C_3 after changing to the new basis is therefore given by

$$\begin{bmatrix} \dfrac{1}{\sqrt{3}} & \dfrac{1}{\sqrt{3}} & \dfrac{1}{\sqrt{3}} \\ \dfrac{2}{\sqrt{6}} & -\dfrac{1}{\sqrt{6}} & -\dfrac{1}{\sqrt{6}} \\ 0 & \dfrac{1}{\sqrt{2}} & -\dfrac{1}{\sqrt{2}} \end{bmatrix} \begin{bmatrix} 0 & 0 & 1 \\ 1 & 0 & 0 \\ 0 & 1 & 0 \end{bmatrix} \begin{bmatrix} \dfrac{1}{\sqrt{3}} & \dfrac{2}{\sqrt{6}} & 0 \\ \dfrac{1}{\sqrt{3}} & -\dfrac{1}{\sqrt{6}} & \dfrac{1}{\sqrt{2}} \\ \dfrac{1}{\sqrt{3}} & -\dfrac{1}{\sqrt{6}} & -\dfrac{1}{\sqrt{2}} \end{bmatrix},$$

which, as the reader may easily verify, is

$$\begin{bmatrix} 1 & 0 & 0 \\ 0 & -\tfrac{1}{2} & -\tfrac{1}{2}\sqrt{3} \\ 0 & \tfrac{1}{2}\sqrt{3} & -\tfrac{1}{2} \end{bmatrix}.$$

As expected, it is seen to have the diagonal-block form

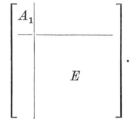

The same diagonal-block pattern is found for all the matrices, after changing to the new basis consisting of the vectors $s^{(A_1)}$, $s_a^{(E)}$, and $s_b^{(E)}$. The change has thus achieved the desired complete reduction of the original reducible 3×3 representation $\Gamma_{(\Delta r)}$.

11.4. Internal symmetry coordinates

In the example considered in the preceding section, the coordinates associated with the reducible representation $\Gamma_{(\Delta r)}$ were the internal coordinates Δr_1, Δr_2, and Δr_3 of the PCl_3 molecule. Let $S^{(A_1)}$, $S_a^{(E)}$, and $S_b^{(E)}$ be the coordinates associated with the new basis in which the representation is completely reduced. We now invoke the rule which we proved in Section 4.5, i.e. that when the matrix which effects a change of basis is an orthogonal one, the coordinates of a vector are transformed by the change in precisely the same way as are the basis vectors. Referring back to eqn (11.18), we see that this means that the S-coordinates are related to the Δr-coordinates by the matrix equation

$$\begin{bmatrix} S^{(A_1)} & S_a^{(E)} & S_b^{(E)} \end{bmatrix} = \begin{bmatrix} \Delta r_1 & \Delta r_2 & \Delta r_3 \end{bmatrix} \begin{bmatrix} \dfrac{1}{\sqrt{3}} & \dfrac{2}{\sqrt{6}} & 0 \\[2mm] \dfrac{1}{\sqrt{3}} & -\dfrac{1}{\sqrt{6}} & \dfrac{1}{\sqrt{2}} \\[2mm] \dfrac{1}{\sqrt{3}} & -\dfrac{1}{\sqrt{6}} & -\dfrac{1}{\sqrt{2}} \end{bmatrix}.$$

(11.22)

In accordance with our convention that sets of coordinates shall be written as single-column matrices, rather than as single-row matrices, we prefer to transpose both sides of eqn (11.22) in order to bring it into the conventionally acceptable (though entirely equivalent) form

$$
\begin{bmatrix}
S^{(A_1)} \\[2ex]
S_a^{(E)} \\[2ex]
S_b^{(E)}
\end{bmatrix}
=
\begin{bmatrix}
\dfrac{1}{\sqrt{3}} & \dfrac{1}{\sqrt{3}} & \dfrac{1}{\sqrt{3}} \\[2ex]
\dfrac{2}{\sqrt{6}} & -\dfrac{1}{\sqrt{6}} & -\dfrac{1}{\sqrt{6}} \\[2ex]
0 & \dfrac{1}{\sqrt{2}} & -\dfrac{1}{\sqrt{2}}
\end{bmatrix}
\begin{bmatrix}
\Delta r_1 \\[2ex]
\Delta r_2 \\[2ex]
\Delta r_3
\end{bmatrix}.
\qquad (11.23)
$$

It becomes clear that, in order to arrive at the square matrix which effects the change of basis required to reduce a representation completely, it is not necessary explicitly to introduce the basis vectors. Instead we can apply the projection vectors to the initial coordinates, and so produce linear combinations of them which belong to the respective symmetry species (or irreducible representations) of the group. In order that the square transformation matrix shall be an orthogonal one, it is necessary to normalize each of these linear combinations in just the same way as we normalized to unity each of the basis vectors. This, indeed, is the procedure that is usually followed in applications to the problem of molecular vibrations.

From a set of internal coordinates it is thus possible to generate new coordinates, each of which belongs to a particular symmetry species of the molecular point group. Such coordinates are usually designated by the symbol S and are called *internal symmetry coordinates*. Where no confusion can arise, we shall refer to them simply as symmetry coordinates. In one context however (in Section 14.2), we shall have occasion to distinguish them from the so-called external symmetry coordinates $\bar{S}$ which are similarly related to Cartesian nuclear displacement coordinates.

The matrix which transforms the internal coordinates D into the (internal) symmetry coordinates S is conventionally known as U. Thus writing the internal coordinates as elements of a single-column matrix **D**, and the symmetry coordinates as elements of a corresponding column **S**, we have the matrix relationship

$$
\mathbf{S} = \mathbf{UD}. \qquad (11.24)
$$

For a non-linear molecule containing N nuclei the number of D-coordinates (and hence of S-coordinates) will be $3N-6$. The matrix U will therefore be an orthogonal one with $3N-6$ rows and $3N-6$ columns.

Complete sets of symmetrically equivalent internal coordinates may be treated individually, as was done above for the set of coordinates Δr_1, Δr_2, and Δr_3 of PCl_3. Being mutually exclusive with respect to mixing under symmetry operations, the different sets will contribute distinct parts of the total U matrix. We may illustrate this by going on to consider the set of internal coordinates of angle change $r\Delta\beta_1$, $r\Delta\beta_2$, and $r\Delta\beta_3$ of the PCl_3 molecule. In this connection, we recall that the inter-bond angle β_1 is the one opposite bond 1. Similarly β_2 is opposite bond 2, and β_3 opposite bond 3 (Fig. 6.4). In consequence the coordinate $r\Delta\beta_1$ transforms in precisely the same way as does Δr_1, under all the symmetry operations of the group. Likewise $r\Delta\beta_2$ and $r\Delta\beta_3$ transform in the same way as do Δr_2 and Δr_3 respectively. The results of applying the projection operators to the coordinates of angle change will thus be analogous to those for the corresponding coordinates of bond stretch. The structure of the representation based upon the angle-change set must be the same (i.e. A_1+E) as that for the bond-stretch set. The symmetry coordinate belonging to the species A_1 will clearly be $(1/\sqrt{3})(r\Delta\beta_1+r\Delta\beta_2+r\Delta\beta_3)$. For the species E there is no unique pair of coordinates, just as there is no unique pair of basis vectors for the 2-dimensional sub-space concerned. However, if the maximum advantage is to be obtained from the use of the symmetry coordinates in dealing with the molecular vibrational problem, it is essential to make a choice for the coordinates of angle change which is strictly analogous to whatever choice has already been made for the coordinates of bond stretch. This means that having chosen as a coordinate of species E the one generated by the application of $\mathscr{P}^{(E)}$ to the particular coordinate of bond stretch Δr_1 we must, when we come to the coordinates of angle change, choose as a coordinate of species E the one generated by the application of $\mathscr{P}^{(E)}$ to that coordinate which transforms in exactly the same way as does Δr_1, namely the particular coordinate $r\Delta\beta_1$. Thus the two distinct (but analogous) sets of internal coordinates have to be treated in a symmetrically analogous way. The reason for this will become apparent when, in the next chapter, we discuss in greater detail the importance of using symmetry coordinates in the treatment of molecular vibrations.

We have now learned how a full set of $3N-6 = 6$ symmetry coordinates may be generated from the chosen six internal coordinates of the PCl_3 molecule. We can express the results in the form of the relevant U matrix, suitably constructed for use with maximum advantage in connection with the vibrational problem. It is customary to number the symmetry coordinates and arrange them in a certain order in the associated column matrix. The two which belong to the species A_1 are put first, the one generated from the coordinates of bond stretch preceding the one generated from the coordinates of angle change, and are called S_1 and S_2 respectively. The pair belonging to the species E and generated from the bond-stretch set are called S_{3a} and S_{3b}, and the corresponding pair from the angle-change set are called S_{4a} and S_{4b}. In the associated column matrix S they are written in the order S_{3a}, S_{4a}, S_{3b}, S_{4b}. In the matrix D, whose elements are the internal coordinates, the set of coordinates of bond stretch precedes the set of coordinates of angle change. In the particular example we are considering, therefore, the general matrix equation (11.24) becomes

$$
\begin{bmatrix} S_1 \\ S_2 \\ S_{3a} \\ S_{4a} \\ S_{3b} \\ S_{4b} \end{bmatrix} = \begin{bmatrix} \dfrac{1}{\sqrt{3}} & \dfrac{1}{\sqrt{3}} & \dfrac{1}{\sqrt{3}} & 0 & 0 & 0 \\ 0 & 0 & 0 & \dfrac{1}{\sqrt{3}} & \dfrac{1}{\sqrt{3}} & \dfrac{1}{\sqrt{3}} \\ \dfrac{2}{\sqrt{6}} & -\dfrac{1}{\sqrt{6}} & -\dfrac{1}{\sqrt{6}} & 0 & 0 & 0 \\ 0 & 0 & 0 & \dfrac{2}{\sqrt{6}} & -\dfrac{1}{\sqrt{6}} & -\dfrac{1}{\sqrt{6}} \\ 0 & \dfrac{1}{\sqrt{2}} & -\dfrac{1}{\sqrt{2}} & 0 & 0 & 0 \\ 0 & 0 & 0 & 0 & \dfrac{1}{\sqrt{2}} & -\dfrac{1}{\sqrt{2}} \end{bmatrix} \begin{bmatrix} \Delta r_1 \\ \Delta r_2 \\ \Delta r_3 \\ r\Delta\beta_1 \\ r\Delta\beta_2 \\ r\Delta\beta_3 \end{bmatrix}.
$$

(11.25)

The square matrix is the complete $\mathbf{U}$ matrix of the PCl_3 problem. In fact, as we shall see in the next chapter, it actually contains more information than we require for our purposes, for in order to calculate the normal vibrational frequencies of the molecule we shall not find it necessary to use the bottom two rows.

12 Internal symmetry coordinates and the factoring of the vibrational secular equation

12.1. The potential and kinetic energies in terms of internal symmetry coordinates

IN the preceding chapter we have seen how it is possible to generate, from the internal coordinates D of a molecule, a set of symmetry coordinates S. For the transformation involved we wrote the matrix equation

$$S = UD, \tag{12.1}$$

where U is an orthogonal matrix which can be determined by the projection-operator method. Differentiating eqn (12.1) with respect to time, we obtain

$$\dot{S} = U\dot{D}. \tag{12.2}$$

The internal symmetry coordinates S resemble the normal vibrational coordinates Q in that each S belongs to a symmetry species of the molecular point group. It is of course possible to envisage a transformation from the set of internal symmetry coordinates to the set of normal coordinates, as a result of which each Q would be expressed as a linear combination of S coordinates. However, for reasons of symmetry there is necessarily a very important limitation which determines which S coordinates can contribute to the expression for any particular Q. In fact, the only S coordinates that can appear are those which belong to the same symmetry species as the normal coordinate itself.

This is in harmony with the fact that both the potential energy and the kinetic energy of the system must be invariant under any symmetry operation of the molecular point group. Consider, for example, a particular normal vibrational coordinate $Q^{(\mu)}$ belonging to the non-degenerate species μ. The statement made above requires that only symmetry coordinates belonging to the same species μ can appear when $Q^{(\mu)}$ is expressed as a linear combination of the symmetry coordinates. Now let us suppose that the combination were to contain a coordinate $S^{(\nu)}$ belonging to a different non-degenerate species ν. The contribution of $Q^{(\mu)}$ to the potential energy is $\lambda(Q^{(\mu)})^2$, where λ is the corresponding root of the secular equation. This contribution, expressed in terms of the symmetry coordinates, would therefore contain a term involving the 'cross-product' $S^{(\mu)}S^{(\nu)}$. However, there will always be some symmetry operation of the group which transforms $S^{(\mu)}$ into $S^{(\mu)}$ and transforms $S^{(\nu)}$ into $-S^{(\nu)}$. This operation will therefore reverse the sign of the cross-product term, and thus alter the value of the potential energy. Such an alteration is incompatible with the necessary invariance of the potential energy. It follows, therefore, that the cross-product term must vanish, i.e. that $S^{(\nu)}$ cannot appear in the linear combination of symmetry coordinates which is equal to $Q^{(\mu)}$. The same conclusion is reached by considering the kinetic energy. We can therefore say that neither the potential energy nor the kinetic energy can contain any term involving the cross-product of symmetry coordinates belonging to different non-degenerate species. Products of different symmetry coordinates belonging to the same non-degenerate species will, however, appear. We may exemplify this by reference to the two normal vibrations of PCl_3 which belong to the totally symmetric species A_1. Let Q_1 and Q_2 be the normal coordinates. There are two symmetry coordinates of this species, S_1 and S_2, and each of the normal coordinates will be a linear combination of both S_1 and S_2. Suppose that $Q_1 = aS_1 + bS_2$ and that $Q_2 = cS_1 + dS_2$, where a, b, c, and d are numerical coefficients. Then for the kinetic energy we find

$$2T = \dot{Q}_1^2 + \dot{Q}_2^2$$
$$= (a^2+c^2)\dot{S}_1^2 + 2(ab+cd)\dot{S}_1\dot{S}_2 + (b^2+d^2)\dot{S}_2^2.$$

The kinetic energy, when expressed in terms of the symmetry coordinates, is seen to contain (in addition to terms involving the squares of $\dot{S}_1$ and $\dot{S}_2$) a term involving the product $\dot{S}_1\dot{S}_2$. The potential energy similarly contains (in addition to terms involving the squares of S_1 and S_2) a term involving the product $S_1 S_2$.

Analogous considerations apply to normal coordinates and to symmetry coordinates belonging to degenerate symmetry species. Referring again to PCl_3, we have two pairs of doubly degenerate vibrations. Let Q_{3a}, Q_{3b} and Q_{4a}, Q_{4b} be the corresponding pairs of degenerate normal coordinates. There are two pairs of symmetry coordinates of the species in question, S_{3a}, S_{3b}, and S_{4a}, S_{4b}. Since all our S coordinates belong to the same species as the normal coordinates, it might at first sight be supposed that, when any one of the Q coordinates is expressed as a linear combination of the S coordinates, the linear combination could contain all the four of them. But in fact this is not the case. If, for example, we consider Q_{3a}, the linear combination can only contain those S coordinates that transform, under all the symmetry operations of the group, in the same way as does Q_{3a} itself. This means that the linear combination can contain only S_{3a} and S_{4a}; for these do in fact transform in the same way as Q_{3a}, whereas S_{3b} and S_{4b} do not. Likewise the linear combination for Q_{4a} can only contain S_{3a} and S_{4a}, while those for Q_{3b} and Q_{4b} can only contain S_{3b} and S_{4b}. Hence it follows that, when expressed in terms of the symmetry coordinates, neither the potential energy nor the kinetic energy can contain any cross-product term involving symmetry coordinates of the same doubly degenerate species which bear different subscripts a and b.

Furthermore, using again the example of PCl_3, it is easily seen that, if the term $kS_{3a}S_{4a}$ occurs in the potential energy, it must be accompanied by a term in $S_{3b}S_{4b}$ with the same coefficient k. Only thus can the invariance of the potential energy be assured. For, supposing the coefficient of the term in $S_{3b}S_{4b}$ is k', consider a symmetry operation R such that

$$S_{3a} \xrightarrow{R} R_{11}S_{3a} + R_{12}S_{3b},$$
$$S_{4a} \longrightarrow R_{11}S_{4a} + R_{12}S_{4b},$$
$$S_{3b} \longrightarrow R_{21}S_{3a} + R_{22}S_{3b},$$
$$S_{4b} \longrightarrow R_{21}S_{4a} + R_{22}S_{4b}.$$

Then $kS_{3a} S_{4a} + k' S_{3b} S_{4b}$ will be transformed into

$$k(R_{11} S_{3a} + R_{12} S_{3b})(R_{11} S_{4a} + R_{12} S_{4b}) +$$
$$+ k'(R_{21} S_{3a} + R_{22} S_{3b})(R_{21} S_{4a} + R_{22} S_{4b})$$
$$= (kR_{11}^2 + k' R_{21}^2) S_{3a} S_{4a} + (kR_{12}^2 + k' R_{22}^2) S_{3b} S_{4b} +$$
$$+ (kR_{11} R_{12} + k' R_{21} R_{22})(S_{3a} S_{4b} + S_{3b} S_{4a}).$$

Because of the invariance of the potential energy, the last term must vanish and the coefficients of the other two terms must be k and k' respectively. But because the matrix representing R is necessarily orthogonal, we have also the relations

$$R_{11} R_{12} + R_{21} R_{22} = 0$$

and
$$R_{11}^2 + R_{12}^2 = R_{12}^2 + R_{22}^2 = 1.$$

The requirements imposed by the invariance of the potential energy can therefore only be satisfied if $k' = k$. Exactly similar conclusions apply to the products of the time derivatives $\dot{S}$ in the corresponding expression for the kinetic energy in terms of the symmetry coordinates.

Of course the expression for the potential energy will also contain terms in S_{3a}^2 and S_{3b}^2. Their coefficients must clearly be equal, for under all the group operations it is $S_{3a}^2 + S_{3b}^2$ which remains constant, and not S_{3a}^2 or S_{3b}^2 individually. The equality of the coefficients of the two terms thus follows from the invariance of the potential energy. Likewise there will be terms in S_{4a}^2 and S_{4b}^2, and they too will have equal coefficients (though not necessarily the same as for S_{3a}^2 and S_{3b}^2). Exactly similar conclusions hold for the products of the time derivatives $\dot{S}$ in the corresponding expression for the kinetic energy.

As we shall see, the above restrictions upon the quadratic terms in the expressions for $2T$ and $2V$ have very important consequences for the form of the vibrational secular equation when symmetry coordinates are used instead of ordinary internal coordinates.

12.2. Introduction of the matrices **G** and **F**

In Chapter 5 we showed that, in terms of the internal coordinates D, the kinetic energy is expressible (eqn (5.20)) in the form

$$2T = \dot{\mathbf{D}}^\dagger \mathbf{G}^{-1} \dot{\mathbf{D}} \qquad (12.3)$$

where $\mathbf{G}$ is a symmetric matrix defined in eqn (5.11). In terms of the symmetry coordinates S, let us write

$$2T = \dot{\mathbf{S}}^\dagger \mathbf{G}^{-1} \dot{\mathbf{S}}, \tag{12.4}$$

where $\mathbf{G}$ is a new matrix analogous to $\mathbf{G}$. Similarly, by analogy with

$$2V = \mathbf{D}^\dagger \mathbf{FD} \tag{12.5}$$

let us write

$$2V = \mathbf{S}^\dagger \mathbf{FS}, \tag{12.6}$$

where $\mathbf{F}$ is a new matrix corresponding to the force-constant matrix $\mathbf{F}$.

By transposing both sides of eqn (12.2) we obtain $\dot{\mathbf{S}}^\dagger = \dot{\mathbf{D}}^\dagger \mathbf{U}^\dagger$. Hence substituting in eqn (12.4) and comparing with eqn (12.3), we obtain

$$2T = \dot{\mathbf{D}}^\dagger \mathbf{U}^\dagger \mathbf{G}^{-1} \mathbf{U} \dot{\mathbf{D}} = \dot{\mathbf{D}}^\dagger \mathbf{G}^{-1} \dot{\mathbf{D}}.$$

This shows that

$$\mathbf{G}^{-1} = \mathbf{U}^\dagger \mathbf{G}^{-1} \mathbf{U}.$$

Hence on inversion we find

$$\mathbf{G} = \mathbf{U}^{-1} \mathbf{G} [\mathbf{U}^\dagger]^{-1} = \mathbf{U}^{-1} \mathbf{G} \mathbf{U}. \tag{12.7}$$

We have here made use of the fact that $\mathbf{U}$ is an orthogonal matrix, so that $\mathbf{U}^\dagger = \mathbf{U}^{-1}$. Pre-multiplying both sides of eqn (12.7) by $\mathbf{U}$ and post-multiplying by $\mathbf{U}^{-1}$, we obtain

$$\mathbf{G} = \mathbf{UGU}^\dagger. \tag{12.8}$$

Thus from a knowledge of $\mathbf{G}$ and $\mathbf{U}$, it is possible to calculate $\mathbf{G}$.

In a similar way we obtain from eqn (12.6) the relation

$$2V = \mathbf{D}^\dagger \mathbf{U}^\dagger \mathbf{FUD} = \mathbf{D}^\dagger \mathbf{FD},$$

which shows that

$$\mathbf{F} = \mathbf{U}^\dagger \mathbf{FU}.$$

Hence, as in the case of $\mathbf{G}$, we deduce that

$$\mathbf{F} = \mathbf{UFU}^\dagger. \tag{12.9}$$

This equation enables us to calculate the matrix $\mathbf{F}$ from a knowledge of $\mathbf{F}$ and $\mathbf{U}$.

In Chapter 5, when using internal coordinates D, we showed that the secular equation of the vibrational problem can be written in the determinantal form $|\mathbf{F} - \lambda \mathbf{G}^{-1}| = 0$, or in either of the equivalent forms $|\mathbf{GF} - \lambda \mathbf{E}| = 0$ and $|\mathbf{FG} - \lambda \mathbf{E}| = 0$. Now, because of the exact analogy of $\mathbf{G}$ with $\mathbf{G}$ and of $\mathbf{F}$ with $\mathbf{F}$, we can, when using symmetry coordinates, write the secular equation in the form

$$|\mathbf{F}-\lambda\mathbf{G}^{-1}| = 0, \qquad (12.10)$$

or in either of the equivalent forms

$$|\mathbf{GF}-\lambda\mathbf{E}| = 0 \qquad (12.11)$$

and

$$|\mathbf{FG}-\lambda\mathbf{E}| = 0. \qquad (12.12)$$

Evidently these forms of the secular equation involving **G** and **F** must have exactly the same roots as those involving **G** and **F**, for these roots (and the normal modes and frequencies associated with them) are properties of the molecule, and therefore in no way dependent upon the particular type of coordinate chosen for the purposes of calculating them.

12.3. Symmetry blocks in the matrices G and F

In Section 12.1 we saw that the quadratic expression for the kinetic energy in terms of symmetry coordinates must satisfy certain conditions. We will now consider how, in view of the relation $2T = \dot{\mathbf{S}}^{\dagger}\mathbf{G}^{-1}\dot{\mathbf{S}}$ expressed in eqn (12.4), these conditions affect the nature of the matrix $\mathbf{G}^{-1}$. For convenience, we recapitulate the conditions here:

(a) The expression for the kinetic energy cannot contain any term involving a cross-product of the time derivatives of symmetry coordinates which belong to different species.

(b) In the case of a degenerate pair S_a, S_b, there can be no term involving the product $\dot{S}_a\dot{S}_b$.

(c) The terms in $\dot{S}_a^2$ and $\dot{S}_b^2$ must have the same coefficient.

(d) Where there are two degenerate pairs of the same species, as in S_{3a}, S_{3b} and S_{4a}, S_{4b} for PCl_3, there can be no term involving a product of two Ss which bear different letter subscripts.

(e) In the same case as for condition (d), the terms in $\dot{S}_{3a}\dot{S}_{4a}$ and $\dot{S}_{3b}\dot{S}_{4b}$ must have identical coefficients.

Wherever the conditions forbid a term involving the product of the time derivatives of two particular S coordinates, the corresponding element of the $\mathbf{G}^{-1}$ matrix must be zero. Thus if we

188

arrange the symmetry coordinates according to symmetry species in the sort of way exemplified by the left-hand side of eqn (11.25), the only non-zero elements in $\mathbf{G}^{-1}$ will occur in square blocks, each of which concerns a single species. In the case of the PCl_3 molecule, for example, there will be one A_1 block, and two E blocks, as shown in the following diagram:

$\mathbf{G}^{-1}$	S_1	S_2	S_{3a}	S_{4a}	S_{3b}	S_{4b}
S_1						
	A_1 block					
S_2						
S_{3a}						
			E block			
S_{4a}						
S_{3b}						
					E block	
S_{4b}						

The single A_1 block will have two rows and two columns, corresponding to the fact that there are two symmetry coordinates belonging to this species, and that no quadratic term involving them is forbidden. The existence of two separate E blocks is a direct consequence of condition (*d*) above. Each of them is 2×2, corresponding to the fact that we have two pairs (subscripts 3 and 4) of degenerate symmetry coordinates. Moreover, because of conditions (*c*) and (*e*), the two E blocks must be identical in all respects.

In the inversion of $\mathbf{G}^{-1}$ to obtain $\mathbf{G}$, each of the blocks will be separately inverted (see Section 3.7), and so $\mathbf{G}$ will likewise consist of one A_1 block and two identical E blocks. Conditions exactly analogous to the five listed above apply with equal force to the expression for the potential energy in terms of the symmetry coordinates. In view of the relation $2V = \mathbf{S}^{\dagger}\mathbf{FS}$ expressed

189

in eqn (12.6), therefore, the matrix **F** must have the same block form as **G**.

12.4. Factoring of the secular equation

As pointed out at the end of Section 3.4, when two matrices with the same diagonal-block form are multiplied together, the corresponding blocks multiply independently, with the result that the product matrix is once more of the same diagonal-block form. This is true of either of the products **GF** or **FG**. It follows that the determinant of the secular equation, in either of the forms given in eqns (12.11) and (12.12), will have a symmetry-block form of exactly the same kind as that of either **G** or **F**. Now when a determinant of this form is equal to zero, it follows that each of the blocks may separately be zero. Thus, by using symmetry coordinates instead of ordinary internal coordinates, we have caused the secular equation to become factorized into a number of distinct parts, each of which corresponds to a symmetry species of the molecular group. Moreover, as we see in the example of PCl_3, the two blocks belonging to the species E must be identical with one another, so that we need only have considered one of them right from the start.

Clearly the use of internal symmetry coordinates in dealing with molecular vibrations brings important practical advantages. It leads to the maximum possible factoring of the secular equation, thereby reducing the labour involved in solving it. The necessary identity of the two blocks for a doubly degenerate species (and likewise of the three for a triply degenerate species) means that only one representative of any set of mutually degenerate symmetry coordinates need be considered, again reducing labour. It should be noted that the roots of the secular-equation factor for any symmetry block must all be distinct, for the identical roots of the whole problem, caused by degeneracies, have been separated from one another into distinct, identical blocks. In the case of the PCl_3 molecule, the advantages were referred to in advance at the end of a treatment based upon internal coordinates (in Section 6.6). It was there pointed out that, although the secular equation in the form $|\mathbf{GF}-\lambda\mathbf{E}| = 0$ amounted to a polynomial of the sixth degree in λ, it would be possible (by taking symmetry into account) to reduce it to a form amounting merely to two separate quadratics. We see now how this can be done—by transforming from internal coordinates to symmetry coordinates, and basing the treatment upon the latter. The method can be used with advantage in all molecular vibrational problems where the molecule possesses symmetry.

12.5. The matrices G and F for PCl₃

The matrix $\mathbf{G}$ for PCl_3 was derived in Chapter 6, and is presented in Table 6.7. The values of the elements (which are represented by letters in the table) are also given in full at the

TABLE 12.1

$GU^\dagger$ *matrix for* PCl_3

GU†

$\frac{1}{\sqrt{3}}(A+2B)$	$\frac{1}{\sqrt{3}}(E+2F)$	$\frac{2}{\sqrt{6}}(A-B)$	$\frac{2}{\sqrt{6}}(E-F)$	0	0
$\frac{1}{\sqrt{3}}(A+2B)$	$\frac{1}{\sqrt{3}}(E+2F)$	$\frac{1}{\sqrt{6}}(B-A)$	$\frac{1}{\sqrt{6}}(F-E)$	$\frac{1}{\sqrt{2}}(A-B)$	$\frac{1}{\sqrt{2}}(E-F)$
$\frac{1}{\sqrt{3}}(A+2B)$	$\frac{1}{\sqrt{3}}(E+2F)$	$\frac{1}{\sqrt{6}}(B-A)$	$\frac{1}{\sqrt{6}}(F-E)$	$\frac{1}{\sqrt{2}}(B-A)$	$\frac{1}{\sqrt{2}}(F-E)$
$\frac{1}{\sqrt{3}}(E+2F)$	$\frac{1}{\sqrt{3}}(C+2D)$	$\frac{2}{\sqrt{6}}(E-F)$	$\frac{2}{\sqrt{6}}(C-D)$	0	0
$\frac{1}{\sqrt{3}}(E+2F)$	$\frac{1}{\sqrt{3}}(C+2D)$	$\frac{1}{\sqrt{6}}(F-E)$	$\frac{1}{\sqrt{6}}(D-C)$	$\frac{1}{\sqrt{2}}(E-F)$	$\frac{1}{\sqrt{2}}(C-D)$
$\frac{1}{\sqrt{3}}(E+2F)$	$\frac{1}{\sqrt{3}}(C+2D)$	$\frac{1}{\sqrt{6}}(F-E)$	$\frac{1}{\sqrt{6}}(D-C)$	$\frac{1}{\sqrt{2}}(F-E)$	$\frac{1}{\sqrt{2}}(D-C)$

end of Section 6.4. The matrix $\mathbf{F}$ for a general valency force field is to be found in Table 6.8. The complete $\mathbf{U}$ matrix was derived in Chapter 11, and appears in eqn (11.25). We are thus in possession of all the information which we require in order to calculate the matrices $\mathbf{G}$ and $\mathbf{F}$ by means of eqns (12.8) and (12.9), i.e. $\mathbf{G} = \mathbf{UGU}^\dagger$ and $\mathbf{F} = \mathbf{UFU}^\dagger$. The matrix multiplications are quite straightforward. As a first stage in calculating $\mathbf{G}$, for example, we may form the product $\mathbf{GU}^\dagger$. The reader may like to work this out and so to verify that the matrix obtained is as shown in Table 12.1.

We now obtain **G** by pre-multiplying **GU**† by **U**. As may be easily verified, the result obtained is as shown in Table 12.2.

We may note that, as must be the case for all G and **G** matrices, our result turns out to be symmetrical. Also, as expected, it consists of three blocks, one A_1-block and two E-blocks, and the two E-blocks are indeed identical in all respects.

TABLE 12.2

G *matrix for* PCl_3

G	S_1	S_2	S_{3a}	S_{4a}	S_{3b}	S_{4b}
S_1	$A+2B$	$E+2F$	0	0	0	0
S_2	$E+2F$	$C+2D$	0	0	0	0
S_{3a}	0	0	$A-B$	$E-F$	0	0
S_{4a}	0	0	$E-F$	$C-D$	0	0
S_{3b}	0	0	0	0	$A-B$	$E-F$
S_{4b}	0	0	0	0	$E-F$	$C-D$

The matrix **F** corresponding to the matrix F of Table 6.8 can be obtained in exactly the same way by calculating the product **UFU**†. The reader may like to work this out, and so confirm that the result is as given in Table 12.3.

Here again the matrix turns out to be symmetrical, as indeed all F and **F** matrices must be. Also it has the same block structure as **G**. In particular the two E blocks are again identical.

The secular equation in fully factored form involves the product of **G** and **F**, the evaluation of which in algebraic form is perfectly straightforward but obviously somewhat lengthy. We shall not give the result in detail. We note, however, that it will have the same block structure as **G** and **F**, with the two E blocks identical. Clearly we need have derived only one of these two blocks. In fact we need not have evaluated S_{3b} or S_{4b} in terms of the internal coordinates. This is the reason underlying the comment in Section 11.4 that we should not find it necessary to use the two bottom rows of the complete **U** matrix of eqn (11.25). Indeed in normal practice the (incomplete) **U** matrix would have been written without these two rows. Instead of 6×6 it would thus have been 4×6, but would still have been conformable for

TABLE 12.3

F *matrix for* PCl_3

F	S_1	S_2	S_{3a}	S_{4a}	S_{3b}	S_{4b}
S_1	$f_r + 2f_{rr'}$	$f_{r\beta'} + 2f_{r\beta}$	0	0	0	0
S_2	$f_{r\beta'} + 2f_{r\beta}$	$f_\beta + 2f_{\beta\beta'}$	0	0	0	0
S_{3a}	0	0	$f_r - f_{rr'}$	$f_{r\beta'} - f_{r\beta}$	0	0
S_{4a}	0	0	$f_{r\beta'} - f_{r\beta}$	$f_\beta - f_{\beta\beta'}$	0	0
S_{3b}	0	0	0	0	$f_r - f_{rr'}$	$f_{r\beta'} - f_{r\beta}$
S_{4b}	0	0	0	0	$f_{r'\beta} - f_{r\beta}$	$f_\beta - f_{\beta\beta'}$

the formation of the products $\mathbf{UGU}^\dagger$ and $\mathbf{UFU}^\dagger$. The matrices **G** and **F** so obtained would have lacked the 5th and 6th rows and the 5th and 6th columns, as compared with the complete matrices given above. In consequence their product, and therefore also the secular equation, would have lacked the corresponding rows and columns, i.e. the secular equation would have consisted of the one A_1 block and only one of the two identical E blocks. Indeed it is always true that, in order to derive all the *distinct* blocks of the secular equation and so to be able to calculate all the distinct normal frequencies of a molecule, only one symmetry coordinate from each degenerate pair (or set of three) needs to be taken into account.

13 An example of treatment by classical mechanics: the normal vibrations of the BF_3 molecule

13.1. Introduction

UP to this point we have mostly used the PCl_3 molecule (point group C_{3v}) to illustrate the various aspects and stages of the classical treatment of molecular vibrations. Distributed piecemeal throughout Chapters 6–12, in fact, is a complete account of the application of classical mechanics to this particular case.

In the present chapter we give another example—the corresponding treatment for the BF_3 molecule (point group D_{3h}). Although in some respects similar to that for PCl_3, the treatment shows some interesting and instructive differences. In outline, it will involve the following steps: (*a*) the choice of internal coordinates and the setting up of the **G** matrix; (*b*) the choice of a type of force field and the setting up of the **F** matrix; (*c*) the setting up of the **U** matrix; (*d*) the calculation of the matrices **G** and **F**; and (*e*) the setting up of the secular equation in factored form, showing all the distinct symmetry blocks.

13.2. Choice of internal coordinates and the setting up of the G matrix

We shall use the same internal coordinates as we chose in Section 10.8—shown also in the molecular model in Fig. 10.2. First there is the complete set of three symmetrically equivalent coordinates of bond stretch Δr_1, Δr_2, and Δr_3 and secondly the analogous set of coordinates of angle change $r\Delta\beta_1$, $r\Delta\beta_2$, and $r\Delta\beta_3$, where r is the common equilibrium bond length and the angles β_1, β_2, and β_3 are the inter-bond angles respectively opposite the bonds 1, 2, and 3. All these six internal coordinates correspond to displacements in the plane of the molecule in its

194

equilibrium configuration. We shall refer to them as in-plane coordinates. As was shown in Section 10.8, they account for three out of the total of four distinct normal frequencies, i.e. for all those that belong to in-plane normal modes of vibration. The remaining frequency belongs to the sole out-of-plane vibrational mode, and in order to account for it we choose the complete set of symmetrically equivalent out-of-plane coordinates, $r\Delta\alpha_1$, $r\Delta\alpha_2$, and $r\Delta\alpha_3$. The angles α are those between the respective bonds and the plane of the molecule in its equilibrium configuration. Clearly the equilibrium values of each of these angles is zero. We are concerned, however, with small changes from zero.

We first establish that the **G** matrix must consist of two distinct blocks, one for the in-plane coordinates and the other for the out-of-plane coordinates, all elements outside these blocks being zero. To prove this, consider first the kinetic energy in terms of the chosen coordinates. According to eqn (5.20), it is given by $2T = \dot{\mathbf{D}}^\dagger\mathbf{G}^{-1}\dot{\mathbf{D}}$. This quadratic expression cannot, however, contain any term involving the 'cross-product' of the time-derivatives of an in-plane coordinate and an out-of-plane coordinate. For under the operation of reflection in the plane σ_h of the molecule in its equilibrium configuration, $\dot{D}$ for the in-plane coordinate would remain unchanged, while $\dot{D}$ for the out-of-plane coordinate would change in sign. The sign of the term involving the cross-product would thus be reversed, causing an alteration of the kinetic energy. This, however, is incompatible with the necessary invariance of $2T$ under all symmetry operations. It follows that the cross-term must vanish, and therefore that the corresponding element of the matrix $\mathbf{G}^{-1}$ must be zero. Since this is true for cross-products of every in-plane coordinate with every out-of-plane coordinate, we see that $\mathbf{G}^{-1}$ must consist of two distinct diagonally placed blocks. Inversion of $\mathbf{G}^{-1}$ merely inverts each block separately, leaving the block structure unchanged. We have thus established that the **G** matrix must consist of what we may call an in-plane block and an out-of-plane block. In setting up **G**, therefore, we may set up each block separately.

The in-plane block need not give us any trouble, as it is exactly analogous to the **G** matrix of the PCl_3 molecule, which we have already worked out in detail, in Section 6.4. The result is shown in Table 6.7, and the values of all the elements (shown as letter symbols in the table) are given at the end of Section 6.4 as functions of the inter-bond angle β. All we have to do in order to obtain the values appropriate to BF_3 is to give β the value $\frac{2}{3}\pi$ (i.e. 120°), and to write m_B and m_F in place of m_P and m_{Cl} respectively. Thus using $\sin\beta = \frac{1}{2}\sqrt{3}$ and $\cos\beta = -\frac{1}{2}$, we obtain

$$
\left.
\begin{aligned}
A &= \frac{1}{m_F} + \frac{1}{m_B} \\[6pt]
B &= -\frac{1}{2m_B} \\[6pt]
C &= \frac{2}{m_F} + \frac{3}{m_B} \\[6pt]
D &= -\frac{1}{m_F} - \frac{3}{2m_B} \\[6pt]
E &= \frac{\sqrt{3}}{m_B} \\[6pt]
F &= -\frac{\sqrt{3}}{2m_B}
\end{aligned}
\right\}. \tag{13.1}
$$

We now have to set up the matrix of the out-of-plane block of the total **G** matrix. Following the general Wilson vector method —explained in Section 6.2—we introduce three unit vectors e_1, e_2, and e_3, one for each fluorine-nucleus site. Their direction is normal to the plane of the molecule in its equilibrium configuration. We now set up a table of b vectors (similar to Table 6.1 for PCl_3). This is a simple matter, and the result is as shown in Table 13.1. The numbers 1, 2, and 3 at the heads of columns designate the respective F sites.

It is not necessary to set down a table of the dot products of the unit vectors (as we did for PCl_3 in Table 6.2), for all three have the same direction, and therefore the dot product of any two of them is simply unity. The next step is to write down, for each site in turn, the dot products of the relevant b vectors. (Compare with Tables 6.3, 6.4, 6.5, and 6.6 for PCl_3.) In the present case all the information may be presented in the single Table 13.2.

TABLE 13.1

b Vectors for out-of-plane motions of BF_3 *in terms of out-of-plane unit vectors*

	1	2	3	B
$r\Delta\alpha_1$	e_1	0	0	$-e_1$
$r\Delta\alpha_2$	0	e_2	0	$-e_2$
$r\Delta\alpha_3$	0	0	e_3	$-e_3$

TABLE 13.2

Dot products of **b** *vectors for the out-of-plane coordinates of* BCl_3

		$r\Delta\alpha_1$	$r\Delta\alpha_2$	$r\Delta\alpha_3$
F Nucleus of bond 1	$r\Delta\alpha_1$	1	0	0
	$r\Delta\alpha_2$	0	0	0
	$r\Delta\alpha_3$	0	0	0
F Nucleus of bond 2	$r\Delta\alpha_1$	0	0	0
	$r\Delta\alpha_2$	0	1	0
	$r\Delta\alpha_3$	0	0	0
F Nucleus of bond 3	$r\Delta\alpha_1$	0	0	0
	$r\Delta\alpha_2$	0	0	0
	$r\Delta\alpha_3$	0	0	1
B Nucleus	$r\Delta\alpha_1$	1	1	1
	$r\Delta\alpha_2$	1	1	1
	$r\Delta\alpha_3$	1	1	1

We are now in a position to set up the block of the **G** matrix which concerns the out-of-plane coordinates, using eqn (6.6), i.e.

$$G_{ij} = \sum_t (b_i^{(t)} \cdot b_j^{(t)}) \frac{1}{m_t}.$$

The summation extends over the nuclei t. The subscripts i and j refer to the coordinates. We thus obtain the result shown in Table 13.3.

13.3. The F matrix

By considerations analogous to those of the preceding section, but applied to the potential energy instead of the kinetic energy, it may be established that the quadratic expression for the potential energy in terms of the chosen coordinates cannot contain any term involving the cross-product of an in-plane coordinate and an out-of-plane one. In view of eqn (5.23), which

TABLE 13.3

Out-of-plane block of **G** *matrix of* BF_3

G out-of-plane	$r\Delta\alpha_1$	$r\Delta\alpha_2$	$r\Delta\alpha_3$
$r\Delta\alpha_1$	$\dfrac{1}{m_F}+\dfrac{1}{m_B}$	$\dfrac{1}{m_F}$	$\dfrac{1}{m_F}$
$r\Delta\alpha_2$		$\dfrac{1}{m_F}+\dfrac{1}{m_B}$	$\dfrac{1}{m_F}$
$r\Delta\alpha_3$			$\dfrac{1}{m_F}+\dfrac{1}{m_B}$

states that $2V = \mathbf{D}^\dagger\mathbf{F}\mathbf{D}$, it follows that the **F** matrix (like the **G** matrix) must consist of an in-plane block and an out-of-plane block, all other elements being zero. Altogether, therefore, there is no mechanical interaction at all between the in-plane motions and the out-of-plane motions. They effectively constitute two entirely distinct mechanical problems, and may be treated quite separately.

If we choose a force field of the general valency type, the matrix for the in-plane block of **F** will be exactly as for PCl_3. The force constants in question were defined in Section 6.5, and the **F** matrix is shown in Table 6.8. Entirely new force constants must be introduced as the elements of the out-of-plane block for BF_3. For the alteration of a single angle α we shall use the force constant f_α, and for the interaction between two angles of this kind, the constant $f_{\alpha\alpha'}$. The out-of-plane block of **F** will thus have the form shown in Table 13.4.

13.4. The U matrix

In setting up the matrices **G** and **F** we have used nine chosen internal coordinates, i.e. three more than the number $(3N-6 = 6)$ of normal vibrational modes. Evidently when we transform to

198

symmetry coordinates, three of these must either be redundant (in the sense of Section 10.4) or in some other way irrelevant to the vibrational problem. Having got rid of them (and we shall

TABLE 13.4

Out-of-plane block of **F** *matrix of* BF$_3$

F out-of-plane	$r\Delta\alpha_1$	$r\Delta\alpha_2$	$r\Delta\alpha_3$
$r\Delta\alpha_1$ $r\Delta\alpha_2$ $r\Delta\alpha_3$	f_α	$f_{\alpha\alpha'}$ f_α	$f_{\alpha\alpha'}$ $f_{\alpha\alpha}$ f_α

see very soon how this is done) we should be left with six symmetry coordinates which (according to the conclusion which we reached in Section 10.7 and expressed in eqn (10.15)) must include two pairs belonging to the species E'. For our purpose, however, we need consider only one from each of these pairs. Thus the number of symmetry coordinates we shall need is just the number of distinct normal frequencies of the BF$_3$ molecule, namely four. Accordingly we have to set up an incomplete **U** matrix having four rows and nine columns.

We shall use the method of projection operators explained in Chapter 11. Consider first the distinct set of symmetrically equivalent coordinates Δr_1, Δr_2, and Δr_3. We arbitrarily select Δr_1 as the one to which we shall apply in turn the projection operators $\mathscr{P}^{(\mu)}$ for the symmetry species of the molecular point group D_{3h}. According to eqn (11.1), which defines the projection operators,

$$\mathscr{P}^{(\mu)}\Delta r_1 = \sum_R \chi_R^{(\mu)} \mathscr{R}\Delta r_1$$

where $\mathscr{R}$ is the operator for the group symmetry operation R. Table 13.5 shows the values of $\mathscr{R}\Delta r_1$ for all the Rs in question. Using the characters $\chi_R^{(\mu)}$ from Table 10.2, we find that we can

TABLE 13.5

Effects of symmetry operations on Δr_1

R	I	C_3	C_3^2	$C_{2(1)}$	$C_{2(2)}$	$C_{2(3)}$	σ_h	S_3	S_3^5	σ_{v_1}	σ_{v_2}	σ_{v_3}
$\mathscr{R}\Delta r_1$	Δr_1	Δr_3	Δr_2	Δr_1	Δr_3	Δr_2	Δr_1	Δr_3	Δr_2	Δr_1	Δr_3	Δr_2

generate from Δr_1 one symmetry coordinate belonging to the totally symmetric species A_1' and one belonging to the doubly degenerate species E'. We shall call these S_1 and S_{3a} respectively. There will of course exist another symmetry coordinate of bond stretch S_{3b}, degenerate with S_{3a}, but for our purposes we may ignore it. The reader may easily verify that the values of the coordinates S_1 and S_{3a}, duly normalized to unity, are

$$S_1 = \frac{1}{\sqrt{3}}(\Delta r_1 + \Delta r_2 + \Delta r_3) \tag{13.2}$$

and

$$S_{3a} = \frac{1}{\sqrt{6}}(2\Delta r_1 - \Delta r_2 - \Delta r_3). \tag{13.3}$$

He may also confirm that the result for the in-plane species A_2' is zero. Furthermore, since Δr_1, Δr_2, and Δr_3 are in-plane coordinates, it is obvious that no linear combination of them can belong to any species which requires antisymmetry with respect to σ_h.

The distinct set of in-plane coordinates $r\Delta\beta_1$, $r\Delta\beta_2$, and $r\Delta\beta_3$ gives results exactly analogous to those for the bond stretches. It is important to remember that, if the U matrix we are setting up is to be such as to result eventually in the maximum degree of factoring of the secular equation, then in setting up the coordinate S_{4a} of species E' analogous to S_{3a}, our choice of coordinate to be subjected to $\mathscr{P}^{(E')}$ can no longer be arbitrary. Having used the particular coordinate of bond stretch Δr_1 as the one from which to generate S_{3a}, we must choose the coordinate of angle change which transforms in precisely the same way as does Δr_1 under all the symmetry operations of the group. This means that we must choose $r\Delta\beta_1$ (compare with Section 11.4, where the similar case of PCl_3 is treated). We then obtain one symmetry coordinate of species A_1' and one (S_{4a}) of species E'. Their values are analogous to those given in eqns (13.2) and (13.3). Obviously the A_1' coordinate must be redundant, since the sum of the angles β_1, β_2, and β_3 must be constant. We can therefore reject this coordinate, and are left with

$$S_{4a} = \frac{1}{\sqrt{6}}(2r\Delta\beta_1 - r\Delta\beta_2 - r\Delta\beta_3). \tag{13.4}$$

We have now obtained all the in-plane symmetry coordinates we require, for according to eqn (10.15) $\Gamma_{\text{vib}} = A_1' + 2E' + A_2''$, and we have derived the symmetry coordinate of species A_1' and one suitable representative of each of the two pairs belonging to species E'. We thus require just one more, which we shall call S_2. It must belong to the symmetry species A_2'', and must be generated by application of the appropriate projection operator to one of the set of out-of-plane coordinates $r\Delta\alpha_1$, $r\Delta\alpha_2$, and $r\Delta\alpha_3$. Let us choose $r\Delta\alpha_1$ for the purpose. The values of $\mathscr{R}(r\Delta\alpha_1)$ for all the symmetry operations of the group are shown in Table 13.6.

TABLE 13.6

Effects of symmetry operations upon $r\Delta\alpha_1$

R	I	C_3	C_3^2	$C_{2(1)}$	$C_{2(2)}$	$C_{2(3)}$	σ_h	S_3	S_3^5	σ_{v_1}	σ_{v_2}	σ_{v_3}
$\mathscr{R}(r\Delta\alpha_1)$	$r\Delta\alpha_1$	$r\Delta\alpha_3$	$r\Delta\alpha_2$	$-r\Delta\alpha_1$	$-r\Delta\alpha_3$	$-r\Delta\alpha_2$	$-r\Delta\alpha_1$	$-r\Delta\alpha_3$	$-r\Delta\alpha_2$	$r\Delta\alpha_1$	$r\Delta\alpha_3$	$r\Delta\alpha_2$

Naturally, since out-of-plane coordinates are being used, zero results are obtained when $r\Delta\alpha_1$ is subjected to the projection operators of all the symmetry species which require $S \to S$ under the operation σ_h. For the species A_1'' the result is also found to be zero; but for A_2'' we find we can generate the symmetry coordinate S_2 with the value

$$S_2 = \frac{1}{\sqrt{3}}(r\Delta\alpha_1 + r\Delta\alpha_2 + r\Delta\alpha_3). \tag{13.5}$$

This completes the requirements of the known structure of Γ_{vib}, and so must evidently complete our task of setting up the required 4-row matrix $\mathbf{U}$.

However, we find that we can also generate from $r\Delta\alpha_1$ a symmetry coordinate belonging to the doubly degenerate species E''. Its value is

$$S_{0a} = \frac{1}{\sqrt{6}}(2r\Delta\alpha_1 - r\Delta\alpha_2 - r\Delta\alpha_3). \tag{13.6}$$

Clearly S_{0a} must be irrelevant to the vibrational problem, and the same must be true of the other member S_{0b} of the degenerate pair. The value of S_{0b} is easily obtained (compare with eqn

11.19), which was derived for an exactly analogous case). It is

$$S_{0b} = \frac{1}{\sqrt{2}}(r\Delta\alpha_2 - r\Delta\alpha_3). \tag{13.7}$$

In fact, by looking at the form of S_{0a} (in Fig. 13.1 (a)) we see that it represents a pure rotation about an axis lying in the plane of the molecule and perpendicular to bond 1. The plus sign at site 1

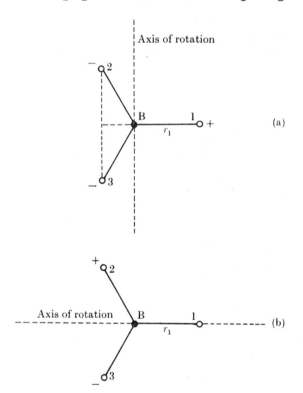

FIG. 13.1. Out-of-plane rotations of the BF_3 molecule.

indicates motion out of the plane of the figure (and of the molecule) towards the reader. From the simple geometry of the model it is clear that if the nucleus of site 1 moves by a certain small distance, then the nuclei of sites 2 and 3 will both move away from the reader (indicated by the minus sign) by a distance which is half as great. Inspection of Fig. 13.1 (b) shows likewise that S_{0b} represents the pure rotation about an axis in the plane of the figure and at right angles to the axis for S_{0a}, i.e. about bond 1. In this rotation the nucleus of site 1 remains unmoved, while those of sites 2 and 3 move by equal amounts towards and away from the reader respectively. We see that both S_{0a} and S_{0b} are indeed irrelevant in our context, since they do not represent genuine vibrations. It also becomes evident that $r\Delta\alpha_1$, $r\Delta\alpha_2$, and $r\Delta\alpha_3$ cannot properly be described as a set of internal coordinates, for they are concerned partially in 'external' types of motion.

The desired **U** matrix can now be set down in Table 13.7. It is deliberately incomplete, as it has nine columns but only four rows.

TABLE 13.7

The **U** *matrix*

U	Δr_1	Δr_2	Δr_3	$r\Delta\beta_1$	$r\Delta\beta_2$	$r\Delta\beta_3$	$r\Delta\alpha_1$	$r\Delta\alpha_2$	$r\Delta\alpha_3$
$S_1(A_1')$	$\frac{1}{\sqrt{3}}$	$\frac{1}{\sqrt{3}}$	$\frac{1}{\sqrt{3}}$	0	0	0	0	0	0
$S_2(A_2'')$	0	0	0	0	0	0	$\frac{1}{\sqrt{3}}$	$\frac{1}{\sqrt{3}}$	$\frac{1}{\sqrt{3}}$
$S_{3a}(E)$	$\frac{2}{\sqrt{6}}$	$-\frac{1}{\sqrt{6}}$	$-\frac{1}{\sqrt{6}}$	0	0	0	0	0	0
$S_{4a}(E)$	0	0	0	$\frac{2}{\sqrt{6}}$	$-\frac{1}{\sqrt{6}}$	$-\frac{1}{\sqrt{6}}$	0	0	0

13.5. The matrices **G** and **F**

The **G** matrix can now be calculated, using eqn (12.8), which states that $\mathbf{G} = \mathbf{UGU}^\dagger$. The in-plane block of **G** has the form of Table 6.7, the appropriate values of the six distinct elements being those given in eqns (13.1). In the out-of-plane block, as given in Table 13.3, all the diagonal elements have the value $(1/m_F)+(1/m_B) = X$ and all the off-diagonal elements have the value $(1/m_F) = Y$, X and Y being quite arbitrarily chosen symbols. The product $\mathbf{UGU}^\dagger$ will be found to have the form shown in Table 13.8. The reader is recommended to work through the matrix multiplications in order to see the emergence of the diagonal-block form, with 1×1 blocks for the species A_1 and A_2'' and a 2×2 block for the species E'. By inserting the values of the G-matrix elements from eqns (13.1), we finally obtain the **G** matrix shown in Table 13.9.

TABLE 13.8

Form of the **G** *matrix*

G	S_1	S_2	S_{3a}	S_{4a}
$S_1 (A_1')$	$A+2B$			
$S_2 (A_2'')$		$X+2Y$		
$S_{3a} (E')$			$A-B$	$E-F$
$S_{4a} (E')$			$E-F$	$C-D$

TABLE 13.9

The **G** *matrix*

G	S_1	S_2	S_{3a}	S_{4a}
$S_1 (A_1')$	$\dfrac{1}{m_{\mathrm{F}}}$			
$S_2 (A_2'')$		$\dfrac{3}{m_{\mathrm{F}}}+\dfrac{1}{m_{\mathrm{B}}}$		
$S_{3a} (E')$			$\dfrac{1}{m_{\mathrm{F}}}+\dfrac{3}{2m_{\mathrm{B}}}$	$\dfrac{3\sqrt{3}}{2m_{\mathrm{B}}}$
$S_{4a} (E')$			$\dfrac{3\sqrt{3}}{2m_{\mathrm{B}}}$	$\dfrac{3}{m_{\mathrm{F}}}+\dfrac{9}{2m_{\mathrm{B}}}$

The **F** matrix, see eqn (12.9), is similarly obtained by forming the product $\mathbf{UFU}^{\dagger}$. As far as the in-plane block is concerned, **F** will have a form similar to that shown in Table 6.8, in which the six distinct force-constant elements are distributed in exactly the same pattern as are the six distinct elements in the **G** matrix of Table 6.7. Accordingly the **F** matrix for BF$_3$ will have an A_1' block and an E block in which the force-constant elements are distributed in exactly the same pattern as are the corresponding **G**-matrix elements in Table 13.8. The single element of the A_2'' (out-of-plane) block is easily found to be $f_\alpha + 2f_{\alpha\alpha'}$. The **F** matrix is thus as shown in Table 13.10.

TABLE 13.10

The **F** *matrix for a general valency force-field*

F	S_1	S_2	S_{3a}	S_{4a}
$S_1(A_1')$	$f_r + 2f_{rr'}$			
$S_2(A_2'')$		$f_\alpha + 2f_{\alpha\alpha'}$		
$S_{3a}(E')$			$f_r - f_{rr'}$	$f_{r\beta'} - f_{r\beta}$
$S_{4a}(E')$			$f_{r\beta'} - f_{r\beta}$	$f_\beta - f_{\beta\beta'}$

The five distinct elements of **F** are made up of combinations of eight valency-type fs. However, we see that in three cases, i.e. $(f_\alpha + 2f_{\alpha\alpha'})$, $(f_\beta - f_{\beta\beta'})$, and $(f_{r\beta'} - f_{r\beta})$, the combination is the only one to contain the two fs which form it. There is therefore no physical meaning in keeping these combinations as such, and we accordingly replace them by the single symbols k_α, k_β, and $k_{r\beta}$ respectively. The number of distinct force constants is thus reduced to five.

13.6. The secular equation

The product **GF** involved in the secular equation will necessarily have the same diagonal-block form as each of the factor matrices. The A_1' block of the secular equation gives at once

$$\lambda_1 = \frac{1}{m_F}(f_r + 2f_{rr'}).$$

We notice that the mass of the boron nucleus is not involved. This is in harmony with the nature of the associated vibrational mode, in which the three F nuclei execute a symmetrical breathing motion while the central B nucleus remains at rest.

The root λ_2, associated with the out-of-plane mode, is at once determined by the A_2'' block as

$$\lambda_2 = \left(\frac{3}{m_F} + \frac{1}{m_B}\right)k_\alpha.$$

The remaining two distinct roots λ_3 and λ_4 are obtainable by solving the quadratic equation generated by expansion of the E'-block determinant. If we write the E' block of **GF** as

$$\begin{bmatrix} E'_{11} & E'_{12} \\ E'_{21} & E'_{22} \end{bmatrix}$$

then it follows that $\lambda_3 + \lambda_4 = E'_{11} + E'_{22}$ and that

$$\lambda_3 \lambda_4 = E'_{11} E'_{22} - E'_{21} E'_{12}.$$

Inserting the appropriate values we obtain

$$\lambda_3 + \lambda_4 = \left(\frac{1}{m_F} + \frac{3}{2m_B}\right)(f_r - f_{rr'} + 3k_\beta) + \frac{3\sqrt{3}}{m_B}\,k_{r\beta}$$

and

$$\lambda_3 \lambda_4 = \frac{3}{m_F}\left(\frac{1}{m_F} + \frac{3}{m_B}\right)\{(f_r - f_{rr'})k_\beta - k_{r\beta}^2\}.$$

Thus the vibrational problem is solved, in the sense that all the distinct roots (and hence all the distinct frequencies) can be calculated with ease, provided that the requisite force constants are available. In practice, of course, it is the frequencies that are determined experimentally and the force constants that are unknown. However, the observed frequency-values will at least serve to test suggested values of the force constants, and hopefully to give some basis for preferring one assumed type of force field over another. In Chapter 15 we shall take up the question of reversing this sort of calculation, so as to start with experimentally observed frequencies and from them obtain information about the intramolecular force field.

14 Isotopic product rules

14.1. Isotopic substitution and the vibrational secular equation

EXPERIMENTAL observations of normal frequencies provide data for calculations designed to throw light on intramolecular force fields. In a non-linear molecule containing N nuclei, the number of frequencies is limited to $3N-6$. It is therefore of importance that isotopic substitution can increase the amount of experimentally available evidence, for it leaves the force field unaffected, but gives rise to new normal frequencies because of the change of mass. However, the increase in the number of observables is not so great as might be expected at first sight, for there must exist certain relationships between the sets of frequencies of the original molecule and the isotopically substituted molecule. These are the *product rules*, which form the subject of the present chapter. They are often called the Teller–Redlich rules, after the two workers who first derived them.

We shall confine our considerations to isotopic substitutions which leave the point group of the molecule unaffected. This means that all the members of a set of symmetrically equivalent nuclei are simultaneously substituted in exactly the same way. An example would be the production from P $^{35}Cl_3$ (in which all the Cl nuclei are of the isotopic species ^{35}Cl) of the isotopically substituted molecule P $^{37}Cl_3$ (in which all the Cl nuclei are of the isotopic species ^{37}Cl).

The product rules follow from the known fact that the product $\Pi\lambda$ of all the roots λ of an equation of the form $|\mathbf{D}-\lambda\mathbf{E}| = 0$ is equal to the value of the determinant $|\mathbf{D}|$. For any particular symmetry block of the fully factored vibrational secular equation, it follows that

$$\Pi\lambda = |\mathbf{GF}| = |\mathbf{G}|\,|\mathbf{F}|. \tag{14.1}$$

The reader is asked to note that no difference of notation will be employed to distinguish between equations like eqn (14.1), which apply to one particular block, and others of the same form which will apply to the secular equation as a whole. The context will always make it clear which case is in question.

As we have already stated, isotopic substitution leaves the force field of a molecule unaffected; furthermore, substitution of a whole set of symmetrically equivalent nuclei is without effect upon the molecular point group. It is only the change of mass that changes the roots of the secular equation. In other words, the substitution leaves the **F** matrix the same as before, but alters the **G** matrix. Using primes to distinguish the isotopically substituted molecule from the original one, it therefore follows from eqn (14.1) that

$$\frac{\Pi\lambda}{\Pi\lambda'} = \frac{|\mathbf{G}|}{|\mathbf{G}'|}. \tag{14.2}$$

There will be one such relationship for each distinct symmetry block of the fully factored vibrational secular equation. For example, there will be two for the pair of molecules P ^{35}Cl$_3$ and P ^{37}Cl$_3$: one for the A_1 block and one for the E block. They are essentially the product rules, which are however usually expressed in a different form.

14.2. The Teller–Redlich product rules

The form of eqn (14.2) is not convenient for the practical application of the product rules to sets of observed frequencies. A simpler and more useful form can be derived by considering the secular equation for the complete mechanical problem, using a full set of $3N$ nuclear displacement coordinates rather than a set of $3N-6$ purely internal ones. In following this course we have to take due account of the three translational modes and the three rotational modes of the molecule. To this end we first imagine the molecule to be subject to weak extramolecular forces which are quite independent of the intramolecular field. The imagined weak forces are of such a nature that each translational displacement of the molecule, and likewise each rotational displacement, brings into play a harmonic restoring force, so that all the six otherwise non-genuine vibrations become genuine ones with very low frequencies. The vibrations arising from the rotations will be torsional in character, but this in no way affects the subsequent arguments. In due course it will be possible to imagine the weak extra-molecular forces to diminish gradually to the limit zero. In this way we shall arrive at a result which will apply to the actual free molecules.

For the time being, however, we treat the complete mechanical problem as one involving $3N$ genuine vibrations, and we express the total kinetic energy $\overline{T}$ in the form

$$2\overline{T} = \dot{\mathbf{X}}^{\dagger}\mathbf{M}\dot{\mathbf{X}}, \tag{14.3}$$

where $\dot{\mathbf{X}}$ is a single-column matrix of the time derivatives of the $3N$ nuclear displacement coordinates, and $\mathbf{M}$ is a diagonal matrix of nuclear masses, each of which occurs three times (Section 5.1). The group representation generated by the $3N$ displacement coordinates is in general reducible, and may be completely reduced by transforming to a set of $3N$ external symmetry coordinates $\overline{S}$. The reader is warned not to confuse these $\overline{S}$ coordinates with *internal* symmetry coordinates S of the kind introduced in Section 11.4 and discussed in Chapter 12. The internal symmetry coordinates ($3N-6$ in number) were produced as linear combinations of the internal coordinates D according to the matrix equation (11.24), i.e. $\mathbf{S} = \mathbf{UD}$. The external symmetry coordinates $\overline{S}$, which we are considering here, are linear combinations of the displacement coordinates x, to which they are related by the analogous equation

$$\overline{\mathbf{S}} = \overline{\mathbf{U}}\mathbf{X}. \tag{14.4}$$

The matrix $\overline{\mathbf{U}}$, which is necessarily orthogonal, can be derived in the same sort of way as was discussed for $\mathbf{U}$ in Chapter 11. Moreover the x-coordinates of each set of symmetrically equivalent nuclei can be treated separately, so as to produce a distinct square block of the $\overline{\mathbf{U}}$ matrix. Each such block must itself be orthogonal, and therefore $\sum \overline{S}^2 = \sum x^2$, where the summation extends only over the coordinates of the block, i.e. over the coordinates of the particular set of symmetrically equivalent nuclei in question. Considering the time derivatives of these coordinates instead of the coordinates themselves, we have

$$\sum \dot{\bar{S}}^2 = \sum \dot{x}^2. \tag{14.5}$$

Now all the nuclei of the set, being symmetrically equivalent, must have the same mass. Writing m for this common mass, it follows from eqn (14.5) that the contribution of the set to the total kinetic energy, i.e. $m \sum \dot{x}^2$, must be $m \sum \dot{\bar{S}}^2$. The total kinetic energy of the system is the sum of all such contributions. It is apparent that it cannot contain any cross-term involving the product of two different $\bar{S}$ coordinates, and that it is expressible in the matrix form

$$2\bar{T} = \dot{\bar{S}}^{\dagger}\mathbf{M}\dot{\bar{S}}, \tag{14.6}$$

in which $\mathbf{M}$ is the same diagonal matrix of nuclear masses as occurred in eqn (14.3). Now, by analogy with eqn (12.4) for internal symmetry coordinates S, we may write

$$2\bar{T} = \dot{\bar{S}}^{\dagger}\bar{\mathbf{G}}^{-1}\dot{\bar{S}}. \tag{14.7}$$

Comparison of eqns (14.6) and (14.7) shows that $\bar{\mathbf{G}}^{-1} = \mathbf{M}$, whence it follows that

$$\bar{\mathbf{G}} = \mathbf{M}^{-1}. \tag{14.8}$$

The matrix $\mathbf{M}^{-1}$ is of course a diagonal one, each of its elements being the reciprocal of the corresponding element of $\mathbf{M}$.

The potential energy of the system can be written in a fashion analogous to eqn (12.6) for the internal symmetry coordinates as

$$2\bar{V} = \bar{S}^{\dagger}\bar{\mathbf{F}}\bar{S}. \tag{14.9}$$

It now follows that the secular equation $|\bar{\mathbf{G}}\bar{\mathbf{F}} - \lambda \mathbf{E}_{3N}| = 0$, being based upon symmetry coordinates, must be fully factored. By virtue of eqn (14.8) it takes the form

$$|\mathbf{M}^{-1}\bar{\mathbf{F}} - \lambda \mathbf{E}_{3N}| = 0,$$

and for any one symmetry block of it we may write

$$\Pi \lambda = |\mathbf{M}^{-1}\bar{\mathbf{F}}| = \frac{|\bar{\mathbf{F}}|}{\Pi m^a}. \tag{14.10}$$

In the denominator on the right-hand side of eqn (14.10) m is the mass of a member of a set of symmetrically equivalent nuclei, and a is the number of external symmetry coordinates of the relevant symmetry species to which this set gives rise. The product Πm^a extends over all the sets of symmetrically equivalent nuclei in the molecule. The value of a for each set may be found by the general method which has been discussed in Section 10.3. It should be noted that, if the symmetry species of the block under consideration is a degenerate one, each distinct root occurs only once in the block and so only contributes a single factor to the product $\Pi\lambda$. Also the corresponding contribution to the number a of symmetry coordinates is likewise only unity.

The product rule for a particular block of the secular equation may now be written in the form, analogous to eqn (14.2),

$$\frac{\Pi\lambda}{\Pi\lambda'} = \Pi\left(\frac{m'}{m}\right)^a. \tag{14.11}$$

In applying this rule, no difficulty arises if there is no translation and no rotation belonging to the particular symmetry species of the block in question. Such non-genuine motions have zero roots, so in a block containing one or more of them both $\Pi\lambda$ and $\Pi\lambda'$ are zero. The left-hand side of eqn (14.11) thus assumes an indeterminate form. However, in our derivation of this equation we have assumed the presence of weak forces which turn the otherwise non-genuine vibrations into genuine ones. To make the rule applicable to the actual free molecule, we now imagine these forces to diminish gradually to the limit zero. Although in this process the individual values of λ and λ' for any translation or any rotation both approach zero, their ratio λ/λ' approaches a definite limiting value which is easily ascertainable. In proceeding to the limit, the forces are always the same for the unsubstituted and the isotopically substituted molecules, and therefore for a translation

$$\lim\left(\frac{\lambda}{\lambda'}\right) = \frac{M'}{M}, \tag{14.12}$$

where M and M' are the respective molecular masses. For a rotation the corresponding result is

$$\lim\left(\frac{\lambda}{\lambda'}\right) = \frac{I'}{I}, \tag{14.13}$$

211

where I and I' are the respective moments of inertia of the molecules about the axis of rotation in question. In the case of a degenerate rotational root the moments of inertia about both (or all three) axes concerned are necessarily equal, so that we may use the ratio I'/I for any one of them.

We are now in a position to express the product rule for any symmetry species in a form which is convenient for practical applications to observed frequencies. Denoting by λ_v the roots for the genuine vibrations, and using eqns (14.12) and (14.13), we may rewrite eqn (14.11) as

$$\left(\frac{M'}{M}\right)^t \left(\frac{I_x'}{I_x}\right)^{r_1} \left(\frac{I_y'}{I_y}\right)^{r_2} \left(\frac{I_z'}{I_z}\right)^{r_3} \frac{\Pi\lambda_v}{\Pi\lambda_v'} = \Pi\left(\frac{m'}{m}\right)^a. \tag{14.14}$$

Here t is the number of translations belonging to the symmetry species of the block in question, and r_1, r_2, and r_3 are each either 1 or 0 according to whether the respective rotation does or does not belong to this species. In this connection it must always be remembered that, since a distinct root only occurs once for a block, a degenerate pair (or trio) of degenerate translations or rotations is to be counted only as one. It must also be remembered that the number of factors in the product on the right-hand side of eqn (14.14) must similarly take account of non-genuine as well as genuine vibrations. Since $\lambda_v = 4\pi^2\nu^2$, where ν is the vibrational frequency, we may write the product rule in the final form

$$\frac{\Pi\nu}{\Pi\nu'} = \sqrt{\left\{\left(\frac{M}{M'}\right)^t \left(\frac{I_x}{I_x'}\right)^{r_1} \left(\frac{I_y}{I_y'}\right)^{r_2} \left(\frac{I_z}{I_z'}\right)^{r_3} \Pi\left(\frac{m'}{m}\right)^a\right\}}. \tag{14.15}$$

There will be one such rule for each symmetry species of the molecular point group.

14.3. An example of the application of the product rules

The larger the isotopic mass ratio, the greater will be the effect of isotopic substitution upon the normal frequencies of a molecule. The most striking case is that of the substitution of hydrogen atoms by deuterium atoms, and as an example we may consider the phosphine molecule PH_3 and its completely deuterated analogue PD_3. The molecular point group (C_{3v}) is the same as for PCl_3, so we have two symmetry species A_1 and E to consider. (We have shown in Section 10.2 that there is no genuine vibration belonging to the only other symmetry species A_2 of the group.)

We take first the product rule for the totally symmetric species A_1. Reference to the character table of the group (Table 8.4) shows that only one non-genuine vibration, the translation T_z, belongs to this species, so that we have $t = 1$ and $r_1 = r_2 = r_3 = 0$. As we saw in Section 9.6, the set of symmetrically equivalent nuclei consisting solely of the P nucleus contributes one symmetry coordinate of species A_1, while the other set, consisting of the three H (or D) nuclei, contributes two. Of the total of three, one is accounted for by the translation T_z: thus each molecule has two vibrational frequencies, ν_1 and ν_2, belonging to the totally symmetric species. The relevant product rule, derived from eqn (14.15), is therefore

$$\frac{(\nu_1 \nu_2)_H}{(\nu_1 \nu_2)_D} = \sqrt{\left\{ \left(\frac{m_P + 3m_H}{m_P + 3m_D} \right) \left(\frac{m_P}{m_P} \right) \left(\frac{m_D}{m_H} \right)^2 \right\}}. \tag{14.16}$$

We note that the mass m_P of the phosphorus nuclei cancels out in the factor on the right-hand side of eqn (14.16) which involves this mass alone. This is a particular instance of what will be generally true: the product rules will involve no contributions from sets of symmetrically equivalent nuclei that are not involved in the isotopic substitution. Inserting into eqn (14.16) the atomic weights ($m_H = 1$, $m_D = 2$, $m_P = 31$), we find for the right-hand side the value 1·92. As to the left-hand side, the experimentally observed values of ν_1 and ν_2 are respectively 2 327 and 991 cm^{-1} for PH_3, and 1 694 and 730 cm^{-1} for PD_3. There can be no doubt as to the correctness of the assignments, which are established by methods which we shall discuss in due course. Using these data, the left-hand side of eqn (14.16) is found to have the value 1·86. The agreement with the theoretical value (1·92) from the product rule is seen to be quite close, though not exact. The small discrepancy cannot be wholly accounted for as due to experimental error. Now while it is true that the product rules have the very great advantage that they do not involve any knowledge of force constants, it is also true that they are based upon the general assumption that the intramolecular force field is a quadratic one, i.e. that the molecular vibrations are simple-harmonic. In general this is a very good approximation, because the vibrational amplitudes are in general very small, but the presence of anharmonicity must render the rules inexact. Now it is precisely for very light nuclei, such as H and D, that the classical amplitudes will be expected to be unusually large. It is not surprising, therefore, that there should be a slight disagreement between experiment and theory in respect of the product rules for the substitution of D for H.

Coming now to the doubly degenerate species E, we have to take account of one pair of translations (T_x, T_y) and one pair of rotations (R_x, R_y). Referring to Section 9.6, we find that the P nucleus contributes one pair of symmetry coordinates; but this will not affect the product rule because the P nucleus is not involved in the isotopic substitution. We also find that the set of H (or D) nuclei contributes three pairs of coordinates of species E. Of the total of four pairs, two are accounted for as a pair of translations and a pair of rotations; each molecule therefore has two pairs of genuine vibrations belonging to this species, i.e. two distinct vibrational frequencies, ν_3 and ν_4. The product rule of eqn (14.15) thus takes the form

$$\frac{(\nu_3 \nu_4)_H}{(\nu_3 \nu_4)_D} = \sqrt{\left\{\left(\frac{m_P + 3m_H}{m_P + 3m_D}\right)\left(\frac{(I_x)_H}{(I_x)_D}\right)\left(\frac{m_D}{m_H}\right)^3\right\}}. \tag{14.17}$$

Evaluation of the right-hand side requires a knowledge of the moments of inertia of the molecules, and this in turn requires a knowledge of the angle θ between each of the bonds and the C_3 axis. Conversely, by assuming that the product rule is exactly obeyed we could use the experimentally determined value of the left-hand side of eqn (14.17) in order to calculate a value of θ, and hence a value of the bond angle in the molecules PH_3 and PD_3.

The product rules are often found to be of value for testing the admissibility of proposed assignments of observed frequencies to symmetry species. A further example of their application (to the molecules $^{10}BF_3$ and $^{11}BF_3$) is discussed in Section 24.4.

214

15 Force fields and the problem of calculating force constants from observed frequencies

15.1. The number of force constants required for a quadratic force field

THE classical treatment which has been expounded in the preceding chapters has had as its objective the derivation of the secular equation for the vibrational problem, by the solution of which all the distinct normal frequencies and their symmetry species can be obtained. The calculation of these frequencies for any particular molecule assumes a knowledge, not only of the nuclear masses and the geometry of their arrangement, but also of the intramolecular force field, assumed to be a quadratic one. In practice, of course, it is the frequencies which will be known (in as far as they are experimentally observable), and what will be desired will be to calculate from them the force constants of the field.

In fact, however, there are grave difficulties, both of principle and of practice, in attempting thus to reverse the sort of calculation we have so far envisaged. The first difficulty, which is one of principle, is that the number of parameters (i.e. force constants) required for the complete description of even a quadratic force field is in general greater than the number of observable frequencies. This means that the experimental evidence is necessarily insufficient to determine the field completely. From the secular equation in its fully factored determinantal form, it is a simple matter to deduce the minimum number of force-field parameters required for the full specification of a quadratic field. The corresponding **F** matrix will have the same diagonal-block form, with the maximum possible number of zero elements and the minimum number of force constants. Each block will necessarily be symmetric about the diagonal, and so the number of distinct elements for an $n \times n$ block will be

$$n+(n-1)+(n-2)+...+1 = \tfrac{1}{2}n(n+1).$$

If we add up these numbers for each distinct block of the **F** matrix (or of the secular equation) we shall obtain the minimum number of force constants required in order to describe the intramolecular force field completely. In general this number is greater, and often very considerably greater, than the number of distinct vibrational frequencies.

As an example we may take the PCl_3 molecule (point group C_{3v}). As has been shown in Section 10.2, $\Gamma_{vib} = 2A_1+2E$. Thus the **F** matrix has two distinct 2×2 blocks (A_1 and E). Each will require three distinct elements, making a total of six force constants. But the molecule only has four distinct normal frequencies. Here, then, is a case in which the number of observables is insufficient to determine the larger number of unknowns, i.e. is in principle insufficient to determine the quadratic field completely. This sort of situation is generally encountered with polyatomic molecules.

One way of increasing the number of observables, while leaving the force field unaltered, is by isotopic substitution. For example, if we were concerned with the PH_3 molecule, we might invoke the frequencies of the PD_3 analogue. In addition to the four frequencies of PH_3, we should then have four more, making a total of eight. However, as was discussed in Chapter 14, not all of these are independent. There will in fact be two product-rule relationships between them, one for the A_1 species and one for the E species. The number of independent pieces of experimental evidence will thus be six, instead of eight. In this particular case, as we saw above, the number of force-field parameters required is also six, so the full experimental data will here just suffice to determine the force field completely.

Of course, it must not be forgotten that a prerequisite of any calculation of a force field from observed frequencies is that these frequencies must have been correctly assigned to the symmetry species of the molecular point group. As we shall see when we come to discuss vibrational spectra, there are certain aids in making the necessary assignments; but nevertheless there is often room for doubt, and indeed the literature contains many cases of differences of opinion between different investigators. It should also be borne in mind that everything that has so far been said has been based upon the assumption that intramolecular force fields are of the quadratic type, i.e. that molecular vibrations are simple-harmonic. In fact this does not accord exactly with physical reality. The inadequacy of quadratic fields will be most evident for modes involving unusually large nuclear amplitudes, such as are to be expected for unusually light nuclei like H and D. Now it is precisely for these nuclei that the isotopic mass ratio is unusually large and that consequently the effects of isotopic substitution upon frequency-values are especially large and observable with accuracy. In the case of heavier nuclei (even for the replacement of ^{14}N by ^{15}N) the percentage change of mass on isotopic substitution is so much lower that the resulting frequency changes may be too small to measure with the requisite accuracy. For nuclei heavier than that of nitrogen the practical limitations become more and more serious.

15.2. Simplified force fields

In most cases, therefore, it becomes necessary to simplify the force field so as to reduce the number of parameters it contains. One way in which this may be effected is to start off with a field of the general valency force type (often abbreviated to G.V.F.F.), and then to set some of the interaction force constants (i.e. some of the off-diagonal elements of the corresponding **F** matrix) equal to zero. Experience has shown that interaction constants of the valency type tend to be considerably smaller than the principal constants of bond stretch and angle change which appear as diagonal elements of **F**. In this way the number of parameters may be reduced to such an extent that the number of available independent observed frequencies becomes sufficient to determine them. Of course, there is no means of knowing whether the original G.V.F.F. would have been a good representation of the actual intramolecular field; and the same is true of the simplified version. The mere fact that the parameters of the simplified field can be deduced from the experimental data does not mean that this field provides a good approximation to reality. Sometimes, for the sake of simplicity, all the off-diagonal elements of the $\mathbf{F}$ matrix of a field of the valency type are set equal to zero. The result is referred to as a simple valency force field (S.V.F.F.). By reducing the number of parameters to one or two less than the number of observed frequencies, it is possible to determine the parameters without using all the experimental data, and so to have one or two frequencies left over. The adequacy of the field can then be tested by using the parameters to calculate the frequency or frequencies which were not used in deriving them.

Unfortunately the necessity to set certain force constants equal to zero inevitably introduces a degree of arbitrariness into the whole undertaking. This circumstance has doubtless contributed to the favour in which a different type of field is held by many workers. Like the valency type it is physically plausible, but it has the practical advantage that its specification requires fewer parameters. It is generally known as the Urey–Bradley–Shimanouchi (U.B.S.) force field, after the two men (Urey and Bradley) who first suggested it and the one who has done much to develop and apply it. The U.B.S. field uses the principal bond-stretch and angle-change force constants of the ordinary S.V.F.F. but, instead of the interaction constants of the G.V.F.F., it introduces a completely new feature, namely, repulsion between non-bonded atoms. The repulsive force is such that its magnitude diminishes as the distance between the atoms increases. Its introduction represents a plausible attempt to take account of van der Waals interaction between non-bonded atoms.

Let us consider the simple case of a non-linear triatomic molecule YXZ of the kind shown in Fig. 15.1. The bond lengths are r_1 and r_2, and the bond angle is θ. The non-bonded distance between Y and Z is called ρ. The choice of Δr_1, Δr_2, $r_1 \Delta \theta$, and $\Delta \rho$ as internal coordinates obviously involves one redundancy, because necessarily

$$\rho^2 = r_1^2 + r_2^2 - 2r_1 r_2 \cos \theta. \tag{15.1}$$

It there were no repulsive force between the non-bonded atoms, the molecule would assume a certain equilibrium configuration with the potential energy at a minimum with regard to the bond stretchings Δr_1 and Δr_2 and the angle change $r_1 \Delta\theta$. We should then have a simple valency force field, and the potential energy

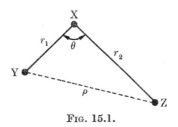

Fɪɢ. 15.1.

would necessarily contain only quadratic terms. When now we imagine the introduction of the repulsive force between the non-bonded atoms, the molecule must take up a new equilibrium configuration, in which this new force is balanced by the simple valency forces called into play by the appropriate bond stretchings and bond-angle increase. In this new configuration the molecule is in a state of internal strain. While it must be true that the potential energy is once again at a minimum with respect to small distortions from the new equilibrium configuration, it now depends not only upon Δr_1, Δr_2, and $r_1 \Delta\theta$, but also upon $\Delta\rho$, and these four internal coordinates are not independent. Consequently, if we write the potential energy as a function of all four, we have to include *linear* terms, as well as quadratic terms. In fact, because the potential energy is at a minimum, the contributions of the linear terms in Δr_1, Δr_2, and $r_1 \Delta\theta$ must be just annulled by that from the linear term in $\Delta\rho$, but the linear terms will not vanish until the redundancy has been removed.

Following a notation similar to that used by Shimanouchi, the potential energy may be expressed as

$$V = K'_{r_1} r_1 \Delta r_1 + K'_{r_2} r_2 \Delta r_2 + H' r_1 (r_1 \Delta\theta) + F' \rho \Delta\rho +$$
$$+ \tfrac{1}{2} K_{r_1} (\Delta r_1)^2 + \tfrac{1}{2} K_{r_2} (\Delta r_2)^2 + \tfrac{1}{2} H (r_1 \Delta\theta)^2 + \tfrac{1}{2} F (\Delta\rho)^2. \quad (15.2)$$

The first four terms on the right-hand side of eqn (15.2) are the linear ones. They are followed by the corresponding quadratic

219

terms. The force-constant symbols of the linear terms are the same as those of the corresponding quadratic terms, but with an added prime. Each linear term has an appropriate length as an extra factor, in order to make the force constants dimensionally equivalent to those of the quadratic terms. The lengths in question (r_1, r_2, and ρ) refer to the equilibrium configuration of the molecule. The force constants K_{r_1}, K_{r_2}, and H (equal respectively to $\partial^2 V/\partial r_1^2$, $\partial^2 V/\partial r_2^2$, and $\partial^2 V/\partial(r_1\Delta\theta)^2$) are seen to be analogous to the corresponding force constants of a simple valency force field. The expression for the potential energy in eqn (15.2) does not explicitly contain any interaction constants of the valency type. Now in order to express the potential energy in terms of the independent internal coordinates Δr_1, Δr_2, and $r_1\Delta\theta$ only, we have to remove the redundancy by substituting the appropriate value of $\Delta\rho$ as a function of these coordinates. Bearing in mind that the potential energy contains quadratic terms, it is essential to express the redundancy to the second order.

To this end we expand $\Delta\rho$ in a Taylor series, retaining only the linear and quadratic terms. We thus obtain

$$\Delta\rho = \frac{\partial\rho}{\partial r_1}\Delta r_1 + \frac{\partial\rho}{\partial r_2}\Delta r_2 + \frac{\partial\rho}{\partial\theta}\Delta\theta +$$

$$+ \frac{1}{2}\frac{\partial^2\rho}{\partial r_1^2}(\Delta r_1)^2 + \frac{1}{2}\frac{\partial^2\rho}{\partial r_2^2}(\Delta r_2)^2 + \frac{1}{2}\frac{\partial^2\rho}{\partial\theta^2}(\Delta\theta)^2 +$$

$$+ \frac{\partial^2\rho}{\partial r_1\partial r_2}\Delta r_1\Delta r_2 + \frac{\partial^2\rho}{\partial r_1\partial\theta}\Delta r_1\Delta\theta + \frac{\partial^2\rho}{\partial r_2\partial\theta}\Delta r_2\Delta\theta. \quad (15.3)$$

All the partial derivatives in eqn (15.3) refer to the equilibrium configuration. Using the value of ρ given in eqn (15.1), the values of these derivatives are easily obtained. We may conveniently express them in terms of the four quantities s_{12}, s_{21}, t_{12}, and t_{21}, the definitions of which are given in eqns (15.4), (15.5), (15.7), and (15.9) respectively. The results obtained are given below. First, the derivatives with respect to r_1 and r_2:

$$\frac{\partial\rho}{\partial r_1} = \frac{1}{\rho}(r_1 - r_2\cos\theta) = s_{12}; \quad (15.4)$$

$$\frac{\partial \rho}{\partial r_2} = \frac{1}{\rho} (r_2 - r_1 \cos \theta) = s_{21};$$ (15.5)

$$\frac{\partial^2 \rho}{\partial r_1^2} = \frac{1}{\rho^3} \{\rho^2 - (r_1 - r_2 \cos \theta)^2\},$$

whence, on substituting for ρ^2 from eqn (15.1), we obtain

$$\frac{\partial^2 \rho}{\partial r_1^2} = \frac{1}{\rho^3} (r_2^2 \sin^2 \theta) = \frac{1}{\rho} t_{12}^2,$$ (15.6)

in which

$$t_{12} = \frac{1}{\rho} r_2 \sin \theta.$$ (15.7)

Similarly, we obtain

$$\frac{\partial^2 \rho}{\partial r_2^2} = \frac{1}{\rho} t_{21}^2,$$ (15.8)

where

$$t_{21} = \frac{1}{\rho} r_1 \sin \theta.$$ (15.9)

For the derivatives with respect to θ, we obtain

$$\frac{\partial \rho}{\partial \theta} = \frac{1}{\rho} r_1 r_2 \sin \theta = (r_1 r_2 t_{12} t_{21})^{\frac{1}{2}},$$ (15.10)

and

$$\frac{\partial^2 \rho}{\partial \theta^2} = \frac{r_1 r_2}{\rho^3} (\rho^2 \cos \theta - r_1 r_2 \sin^2 \theta)$$

$$= -\frac{r_1 r_2}{\rho} s_{12} s_{21}.$$ (15.11)

Lastly, we may give the values of the mixed derivatives as

$$\frac{\partial^2 \rho}{\partial r_1 \partial \theta} = \frac{r_2 \sin \theta}{\rho^3} (\rho^2 - r_1^2 + r_1 r_2 \cos \theta)$$

$$= \frac{r_2}{\rho} t_{12} s_{12};$$ (15.12)

$$\frac{\partial^2 \rho}{\partial r_2 \partial \theta} = \frac{r_1}{\rho} t_{21} s_{12};$$ (15.13)

and finally

$$\frac{\partial^2 \rho}{\partial r_1 \partial r_2} = -\frac{1}{\rho^3} \{\rho^2 \cos \theta + (r_1 - r_2 \cos \theta)(r_2 - r_1 \cos \theta)\}$$

$$= -\frac{1}{\rho} t_{12} t_{21}.$$ (15.14)

Substituting the above values of the derivatives into the expression for $\Delta\rho$ given in eqn (15.3), and multiplying $\Delta\theta$ by r_1 in conformity with our choice of $r_1\Delta\theta$ as an internal coordinate, we obtain

$$\Delta\rho = s_{12}\,\Delta r_1 + s_{21}\,\Delta r_2 + \left(\frac{r_2}{r_1}\,t_{12}t_{21}\right)^{\frac{1}{2}} r_1\,\Delta\theta +$$

$$+ \frac{1}{2\rho}\,t_{12}^2(\Delta r_1)^2 + \frac{1}{2\rho}\,t_{21}^2(\Delta r_2)^2 - \frac{1}{2\rho}\,s_{12}s_{21}\,\frac{r_2}{r_1}\,(r_1\,\Delta\theta)^2 -$$

$$- \frac{1}{\rho}\,t_{12}t_{21}\,\Delta r_1\,\Delta r_2 + \frac{1}{\rho}\,t_{12}s_{21}\,\frac{r_2}{r_1}\,\Delta r_1(r_1\,\Delta\theta) + \frac{1}{\rho}\,t_{21}s_{12}\,r_2(r_1\,\Delta\theta)^2.$$

$$(15.15)$$

In order to remove the redundancy, we now have to substitute this value of $\Delta\rho$ in the terms $F'\rho\Delta\rho$ and $\frac{1}{2}F(\Delta\rho)^2$ of the expression for the potential energy given in eqn (15.2). The total algebra would clearly be rather extensive. However, because V must be at a minimum in the equilibrium configuration, it follows that all linear terms in Δr_1, Δr_2, and $r_1\Delta\theta$ must necessarily vanish. Furthermore, since we are interested only in quadratic terms, i.e. those involving $(\Delta r_1)^2$, $(\Delta r_2)^2$, $(r_1\Delta\theta)^2$, $\Delta r_1\Delta r_2$, $\Delta r_1(r_1\Delta\theta)$, and $\Delta r_2(r_1\Delta\theta)$, we may (in carrying out the substitutions) ignore all higher powers arising from the term in $(\Delta\rho)^2$.

As an example, let us consider the term in $(\Delta r_1)^2$. After removal of the redundancy in the manner just described, the coefficient of this term will be the sum of three parts. The first of these will be simply $\frac{1}{2}K_{r_1}$, from the original form of the potential energy given in eqn (15.2). The second will come from the term in $\Delta\rho$, and from eqn (15.3) we see that its value will be $\frac{1}{2}F'\rho(\partial^2\rho/\partial r_1^2)$, which by eqn (15.6) is equal to $\frac{1}{2}F't_{12}^2$. The third contribution to the coefficient of the term in $(\Delta r_1)^2$ will come from the term in $(\Delta\rho)^2$. Its value will be $\frac{1}{2}F(\partial\rho/\partial r_1)^2 = \frac{1}{2}Fs_{12}^2$. Thus in the expression for the potential energy as a function of the independent internal coordinates Δr_1, Δr_2, and $r_1\Delta\theta$, the complete term in $(\Delta r_1)^2$ will be $\qquad \frac{1}{2}(K_{r_1} + t_{12}^2\,F' + s_{12}^2\,F)(\Delta r_1)^2.$

The corresponding terms in $(\Delta r_2)^2$ and $(r_1\Delta\theta)^2$ can be similarly derived.

The special interest of the U.B.S. field is that the removal of the terms in $\Delta\rho$ and $(\Delta\rho)^2$ gives rise to terms in the expression for the potential energy which involve $\Delta r_1 \Delta r_2$, $\Delta r_1(r_1\Delta\theta)$, and $\Delta r_2(r_1\Delta\theta)$, i.e. terms of a 'valency-field interaction' type which are not contained at all in the original U.B.S. expression of eqn (15.2) for the potential energy. Thus, for example, a term in $\Delta r_1 \Delta r_2$ will be generated, whose coefficient will be the sum of two parts. The first part will come from the term in $\Delta\rho$ in eqn (15.2). Its value will be $F'\rho\,\partial^2\rho/(\partial r_1\,\partial r_2)$, which by eqn (15.14) is equal to $-F't_{12}t_{21}$. The second part will come from the term in $(\Delta\rho)^2$ in eqn (15.2), and will have the value

$$F(\partial\rho/\partial r_1)(\partial\rho/\partial r_2) = Fs_{12}s_{21}.$$

Thus in the expression for the potential energy as a function of the three independent internal coordinates Δr_1, Δr_2, and $r_1\Delta\theta$ there will be a term in $\Delta r_1 \Delta r_2$ with the coefficient

$$-(t_{12}t_{21}F' - s_{12}s_{21}F).$$

The other 'interaction' terms in $\Delta r_1(r_1\Delta\theta)$ and $\Delta r_2(r_1\Delta\theta)$ can be similarly derived. The complete expression for the potential energy is then

$$V = \tfrac{1}{2}(K_{r_1}+t_{12}^2\,F'+s_{12}^2\,F)(\Delta r_1)^2+\tfrac{1}{2}(K_{r_2}+t_{21}^2\,F'+s_{21}^2\,F)(\Delta r_2^2)+$$
$$+\tfrac{1}{2}\Big(H-s_{12}s_{21}\frac{r_2}{r_1}\,F'+t_{12}t_{21}\frac{r_2}{r_1}\,F\Big)(r_1\Delta\theta)^2-$$
$$-(t_{12}t_{21}\,F'-s_{12}s_{21}\,F)\Delta r_1\Delta r_2+$$
$$+(s_{21}t_{12}\frac{r_2}{r_1}\,F'+s_{12}t_{12}\,F)\Delta r_1(r_1\Delta\theta)+$$
$$+(s_{12}t_{21}\,F'+s_{21}t_{12}\,F)\Delta r_2(r_1\Delta\theta). \tag{15.16}$$

We see that the $\mathbf{F}$ matrix (based upon the three independent internal coordinates) has three off-diagonal elements in addition to the three diagonal ones. These off-diagonal elements take the place of the interaction force constants of the general valency force field, but they do not actually represent mechanical interactions, which are properties of the bonds and bond angles. Rather they have been generated in consequence of the assumption of a repulsive force between the non-bonded atoms and the redundancy thereby introduced into the problem.

In connection with the project of calculating force constants from observed frequency data, the point of special interest about the U.B.S. field is that the six distinct elements of the 3×3 symmetric $\mathbf{F}$ matrix are expressed in terms of only five force constant parameters, K_{r_1}, K_{r_2}, H, F, and F'. Moreover, these five may be reduced to four if (as is customary) it is assumed that the potential V_ρ due to the repulsion between the non-bonded atoms is inversely proportional to some plausible power of the distance ρ. Thus writing

$$V_\rho = \frac{k}{\rho^n}, \qquad (15.17)$$

where k is a constant and n is positive, we have

$$F' = \frac{1}{\rho} \frac{\partial V_\rho}{\partial \rho} = -\frac{kn}{\rho^{n+2}} \qquad (15.18)$$

and

$$F = \frac{\partial^2 V_\rho}{\partial \rho^2} = \frac{kn(n+1)}{\rho^{n+2}}. \qquad (15.19)$$

From eqns (15.18) and (15.19) we at once obtain a relation between the two parameters F' and F, namely,

$$\frac{F'}{F} = -\frac{1}{n+1}. \qquad (15.20)$$

By analogy with the repulsive van der Waals interaction between atoms of inert gases, different workers have assumed different values of n ranging from 6 to 12. In fact, within these limits, the field is not very sensitive to the exact choice. A reasonable value (and the one adopted by Shimanouchi) is $n = 9$. This gives

$$F' = -\tfrac{1}{10}F. \qquad (15.21)$$

Thus in our example of the triatomic molecule shown in Fig. 15.1, the number of U.B.S. parameters is four. The molecule has only three normal frequencies, however, so that, despite the reduction in the number of force-constant parameters (as compared with the six required by a general valency field), the number of observables still remains insufficient to determine them all.

In the case of a symmetric non-linear triatomic molecule XY_2, the general valency force field requires four force constants, whereas with $K_{r_1} = K_{r_2}$ the U.B.S. field has only three. Thus here the three normal frequencies of the molecule, though insufficient to determine the G.V.F.F., are just enough to determine the U.B.S. field. Of course, this circumstance provides no evidence at all as to which of the two types of field is the better approximation to reality.

Application of the U.B.S. field to more complicated molecules involves a summation over the inter-bond angles, each being treated (together with the two bonds defining it and the repulsion between the corresponding non-bonded atoms) in a manner similar to that outlined above for the non-linear YXZ molecule. In addition to diagonal elements, the **F** matrix obtained after redundancy removals will have a number of off-diagonal elements of the kind we have encountered in the case of non-linear YXZ. In general, all the elements will be expressible in terms of a smaller number of parameters. Certainly this number will be considerably smaller than that required by a G.V.F.F.

However, we should not lose sight of the fact that the U.B.S. field can only be an approximation to the actual one. It may be fairly claimed that in the U.B.S. field the 'interaction' constants have a clearer physical significance than those of the G.V.F.F., but it should not be forgotten that this clarity is attained at the cost of regarding these constants solely as the consequence of non-bonded repulsions—an assumption that excludes probable interactions due to the nature of the bonds themselves. In its usual form the U.B.S. field takes account only of the non-bonded repulsions between atoms belonging to neighbouring bonds. In some cases it has been deemed expedient to extend the field so as to include similar repulsions between atoms belonging to non-neighbouring bonds.

15.3. The calculation of force constants from frequency data

Even when a certain type of approximate quadratic field has been assumed, and when the number of force-constant parameters is manageable in relation to the number of observed frequencies, the procedure for deriving force constants is by no means straightforward. First of all it is essential that the observed frequencies shall have been correctly assigned to their symmetry species. In all but quite simple cases, there may well be a considerable measure of doubt about any proposed assignment. Even if we had a correct assignment, however, we should still be confronted by inevitable difficulties in obtaining the force constants from the frequencies. We have seen in the preceding chapters how the vibrational secular equation may be set up in any particular case and how, if the force constants are known, it is possible to solve this equation and so to obtain all the distinct roots and normal frequencies of the molecule. There is no straightforward way, however, to reverse this type of calculation so as to start with the roots and work back to the force constants.

The procedure used is to start with some plausible guess as to the values of the force constants, to insert them into the secular equation and to solve for the consequent roots. Of course, these calculated roots will not in general be equal to the true ones (as observed experimentally). It is then necessary to go back and alter the originally guessed force constants with the object of arriving at a new set which, when used in the secular equation, will give roots in better agreement than before. If the agreement (or as is usually said, the fit) is indeed better, we can say that we have improved the original set of force constants. By a successive repetition of this process (i.e. by iteration) we hope finally to arrive at a set of force constants which gives the best possible overall fit of calculated and observed roots. Even this best fit will not in general be a perfect one, for the assumed field can only be an approximation, not only because the number of parameters will be less than would be required for a full description of a quadratic field, but also because even the best quadratic field is itself an approximation.

In the iterative procedure outlined above, it is of course necessary to introduce some criterion by which the goodness of a fit can be quantitatively assessed. This is itself an arbitrary matter. In practice a least-squares criterion is usually adopted, and applied to the sets of calculated and observed frequencies. A detailed mathematical procedure has been developed for the successive steps in the improvement process. This takes account both of the magnitude of the difference between the calculated and observed values of every frequency, and also of the sensitivity of every frequency to variation of every force constant. The calculation is complicated and so very laborious that, except in simple cases, it is scarcely practicable without the use of a computer.

In practice, however, difficulties may arise, even with a computer to perform the very extensive calculations. What is desired is that the successive steps of 'improvement' of the initially chosen force-constant set shall cause convergence upon a 'best-fit' set representing the nearest approximation to the truth which is possible with the type of force field assumed. Things are not always as simple as that. There may be more than one set of force constants capable of giving a best fit, i.e. the problem may be undefined right from the start. In such cases it may be that the particular best-fit set upon which the iteration process converges may depend upon the nature of the guess of an initial set to be successively improved.

We hope, however, that our initial guess is such a reasonable one that the procedure will converge upon the appropriate result, i.e. upon the best approximation that the type of field and the number of parameters used will permit. In many cases this doubtless does happen. But cases are not infrequently encountered where the programme of improvement does not converge. It may start to do so and continue in a promising manner through a number of iterations, only thereafter to behave in such a manner that the further 'improvements' only cause the fit between calculated and observed frequencies to become progressively less close. A frequent reason for this is that the mathematical process of improvement is itself based upon a principle which in practice can only be followed approximately. So long as this underlying approximation remains satisfactory, the improvement procedure will in fact continue to improve the set of force constants and the closeness of the consequent fit between calculated and observed frequencies; but in cases where the approximation becomes so unsatisfactory as to do violence to the principle upon which the mathematical improvement procedure is based, the iteration will fail to converge. There are in fact certain mathematical devices which are designed to control such difficult cases, and which can be written into the computer programme. Not all of them are as unobjectionable as the one which causes the computer to undertake only very small improvements at each step and to perform a correspondingly large number of steps, rather than to attempt (as is usual) to get the whole computation over in fewer steps—and consequently in a shorter time. This device is based upon the fact that the approximation involved in the stepwise improvement becomes better as the steps become smaller and more numerous.

To go into details about such matters would take us outside the scope of this book. We may simply remark that the calculation of force fields from observed frequency data has become almost as much an art as a science. There is indeed little that is reliable to go upon in choosing the type of approximate field to be used, for approximate fields that are physically incompatible with one another may turn out to give equally good 'best fits' with the observed frequency data. Nor is there any firm principle upon which to base the guess of the initial set of force constants which are to be subjected to improvement. Lastly, in difficult cases it may be necessary to impose constraints in order to obtain the desired convergence; and there is a measure of arbitrariness as to how this shall be done. It should not be forgotten that in the end the force constants that are obtained are at best only the most satisfactory set that can be constructed with the particular type of approximate field that was arbitrarily adopted as the basis of the calculation. If the constants happen to give an excellent fit between calculated and observed frequencies, this does not necessarily mean that the field is a correspondingly good approximation to the real one.

We shall see in the next chapter how, starting with an assumed quadratic force field it is possible from the secular equation to deduce, not only the roots (i.e. the values of the normal frequencies), but also the corresponding normal coordinates (i.e. quantitative descriptions of the normal vibrational modes). Here we merely wish to note that the mathematical programme ordinarily used for the calculation of force constants from frequency data produces, for each set of force constants, not only the normal frequencies but also the normal modes associated with them. Even for the final, best-fit set of force constants these calculated modes will only be approximations, and the fact that a certain field gives an excellent fit between calculated and observed frequency values does not necessarily mean that it will give a correspondingly good account of the associated normal vibrational modes.

16 The calculation of vibrational normal coordinates

16.1. The eigenvalues and eigenvectors of a matrix

BEFORE tackling the particular problem of the vibrational normal coordinates of a molecule, we must become acquainted with some relevant general properties of square matrices. Consider a matrix $\mathbf{A}$ with n rows and n columns and a vector with n components x_i represented by a single-column matrix $\mathbf{X}$. The product $\mathbf{AX}$ will be another single-column matrix $\mathbf{X}'$, i.e. another vector with n components x_i'. Now consider a special case in which

$$\mathbf{X}' = \mathbf{AX} = \lambda\mathbf{X}, \tag{16.1}$$

λ being a scalar constant. Eqn (16.1) implies that $x_i' = \lambda x_i$ for every one of the n values of i. This is equivalent to the following set of n simultaneous equations, one for each i value:

$$\left.\begin{aligned}
(A_{11}-\lambda)x_1+ \quad A_{12}x_2+ \quad A_{13}x_3+...+ \quad A_{1n}x_n &= 0\\
A_{21}x_1+(A_{22}-\lambda)x_2+ \quad A_{23}x_3+...+ \quad A_{2n}x_n &= 0\\
A_{31}x_1+ \quad A_{32}x_2+(A_{33}-\lambda)x_3+...+ \quad A_{3n}x_n &= 0\\
\cdots \cdots \cdots \cdots \cdots \cdots \cdots \cdots \cdots\\
A_{n1}x_1+ \quad A_{n2}x_2+ \quad A_{n3}x_3+...+(A_{nn}-\lambda)x_n &= 0
\end{aligned}\right\} \tag{16.2}$$

The quantities A_{ij} are of course the elements of the matrix $\mathbf{A}$. According to the theory of equations, the condition that all the n equations (16.2) shall hold simultaneously (excluding the trivial case when all the vector components x_i are zero) is that the determinant of the coefficients shall vanish, i.e. that

$$\begin{vmatrix}
(A_{11}-\lambda) & A_{12} & A_{13} & . & . & . & A_{1n}\\
A_{21} & (A_{22}-\lambda) & A_{23} & . & . & . & A_{2n}\\
A_{31} & A_{32} & (A_{33}-\lambda) & . & . & . & A_{3n}\\
. & . & . & . & . & . & .\\
A_{1n} & A_{2n} & A_{3n} & . & . & . & (A_{nn}-\lambda)
\end{vmatrix} = 0. \tag{16.3}$$

Eqn (16.3) is known as the characteristic equation of the matrix **A**. When expanded, the determinant gives a polynomial of the nth degree in λ, the solution of which gives just n roots. These are referred to as the *latent roots* or the *eigenvalues* of the matrix.

To each eigenvalue there corresponds an *eigenvector* **X**. The eigenvectors are also referred to as the latent vectors of the matrix or (for reasons that will soon become apparent) the *modal columns*. If $\mathbf{X}_p$ is the single column representing the pth eigenvector, then

$$\mathbf{A}\mathbf{X}_p = \lambda_p \mathbf{X}_p, \tag{16.4}$$

where λ_p is the corresponding eigenvalue or root. It is evident that the eigenvector is not completely defined by eqn (16.4), for if we multiply it by any scalar factor k, the resulting vector $k\mathbf{X}_p$ is just as much an eigenvector of the root λ_p as was $\mathbf{X}_p$ itself, i.e. $\mathbf{A}[k\mathbf{X}_p] = \lambda_p[k\mathbf{X}_p]$.

16.2. The calculation of the eigenvectors of a matrix

For our purposes it will be sufficient to confine our attention to the case of a matrix whose roots are all distinct and all positive. The n components of the pth eigenvector $\mathbf{X}_p$ of the $n \times n$ matrix **A** will clearly satisfy the eqns (16.2) with the particular root λ_p in place of the general root λ. Solution of these simultaneous equations will thus give the eigenvector $\mathbf{X}_p$. The same procedure can be followed for each of the distinct roots, and so all the eigenvectors of the matrix **A** can be calculated.

As a simple numerical example, let us find the three roots and the corresponding eigenvectors of the matrix

$$\begin{bmatrix} 5 & 2 & 0 \\ -2 & 2 & -2 \\ 0 & 2 & -1 \end{bmatrix}.$$

The characteristic equation is

$$\begin{vmatrix} 5-\lambda & 2 & 0 \\ -2 & 2-\lambda & -2 \\ 0 & 2 & -1-\lambda \end{vmatrix} = 0.$$

On expansion this gives the cubic

$$6-11\lambda+6\lambda^2-\lambda^3 = 0,$$

i.e. $$(3-\lambda)(2-\lambda)(1-\lambda) = 0.$$

The desired latent roots (or eigenvalues) are thus 3, 2, and 1. Let us now calculate the eigenvector corresponding to the root $\lambda = 3$. The three equations for its components x_1, x_2, and x_3 are:

$$2x_1+2x_2 = 0,$$

$$-2x_1-x_2-2x_3 = 0,$$

$$2x_2-4x_3 = 0.$$

Hence we find $x_2 = -x_1$ and $x_3 = -\frac{1}{2}x_1$. Thus the eigenvector for the root $\lambda = 3$ has components x_1, $-x_1$, and $-\frac{1}{2}x_1$, where x_1 is an undetermined common factor. Removing this factor, we can say that this eigenvector has components in the ratios $1:-1:-\frac{1}{2}$. Exactly similar considerations show that the corresponding ratios for the roots $\lambda = 2$ and $\lambda = 1$ are respectively $1:-\frac{3}{2}:-1$ and $1:-2:-2$.

In the above very simple example the eigenvectors could be read off at once from the equations for their components. In more complicated cases it may be convenient to make use of the fact (known from the theory of equations) that the components of the pth eigenvector have values which (except for an undetermined factor) are equal to the cofactors of the successive elements of any row of the determinant in eqn (16.3) after substitution of the appropriate root λ_p.

16.3. Note on the case of coincident roots

In the discussion above we have deliberately confined our attention to matrices whose roots are all distinct, so that the number of equations for the components of any eigenvector is sufficient to determine all these components except for a common factor. When the roots of a matrix are not all distinct (i.e. when coincident roots occur), the situation is not so straightforward. In fact the pair of eigenvectors corresponding to a pair of identical roots cannot be determined uniquely. However, there is no need for us to discuss such cases here, since, in the treatment of the vibrational problem, we shall be concerned with the eigenvectors of the matrix $\mathbf{GF}$; and when pairs (or sets of three) roots of the secular equation are identical because of molecular symmetry, it is always possible (as we have seen) to obtain the equation in a fully factored form in which identical roots are separated into different blocks. The solution of the vibrational problem then breaks down into the separate solutions of the individual distinct blocks, and none of these can have coincident roots. This is one more advantage of the use of symmetry coordinates S instead of internal coordinates D, for only in this way can the secular equation be obtained in fully factored form.

16.4. The relation of internal symmetry coordinates to vibrational normal coordinates

The vibrational normal coordinates Q are characterized by the fact that when the kinetic and potential energies are expressed in terms of them they both assume simple forms without cross-product terms,

$$2T = \dot{\mathbf{Q}}^\dagger \dot{\mathbf{Q}}, \tag{16.5}$$

$$2V = \mathbf{Q}^\dagger \mathbf{\Lambda} \mathbf{Q}. \tag{16.6}$$

In the expression for $2V$, $\mathbf{\Lambda}$ is a diagonal matrix whose elements are the roots λ of the problem. Let $\mathbf{L}$ be the square matrix which effects the transformation of the normal coordinates into the symmetry coordinates S. Then we may write

$$\mathbf{S} = \mathbf{LQ} \tag{16.7}$$

and

$$\dot{\mathbf{S}} = \mathbf{L}\dot{\mathbf{Q}}. \tag{16.8}$$

The matrix $\mathbf{G}^{-1}$ is defined by eqn 12.4 as

$$2T = \dot{\mathbf{S}}^\dagger \mathbf{G}^{-1} \dot{\mathbf{S}}.$$

Substituting for $\dot{\mathbf{S}}$ from eqn (16.8), we obtain

$$2T = \dot{\mathbf{Q}}^\dagger \mathbf{L}^\dagger \mathbf{G}^{-1} \mathbf{L} \dot{\mathbf{Q}}. \tag{16.9}$$

Comparing eqn (16.9) with eqn (16.5), we see that

$$\mathbf{L}^\dagger \mathbf{G}^{-1} \mathbf{L} = \mathbf{E},$$

233

whence $\qquad\qquad \mathbf{G}^{-1} = [\mathbf{L}^\dagger]^{-1}\mathbf{L}^{-1},$ $\qquad\qquad$ (16.10)

and finally $\qquad\qquad \mathbf{G} = \mathbf{LL}^\dagger.$ $\qquad\qquad$ (16.11)

We may proceed in an analogous fashion in the case of the potential energy. The matrix $\mathbf{F}$ is defined in eqn (12.6) as

$$2V = \mathbf{S}^\dagger\mathbf{FS}.$$

Substituting for $\mathbf{S}$ from eqn (16.7) we obtain

$$2V = \mathbf{Q}^\dagger\mathbf{L}^\dagger\mathbf{FLQ}. \qquad (16.12)$$

Comparing eqn (16.12) with eqn (16.6), we see that

$$\mathbf{L}^\dagger\mathbf{FL} = \mathbf{\Lambda}. \qquad (16.13)$$

Hence we obtain the following value of $\mathbf{F}$:

$$\mathbf{F} = [\mathbf{L}^\dagger]^{-1}\mathbf{\Lambda L}^{-1} = [\mathbf{L}^{-1}]^\dagger\mathbf{\Lambda L}^{-1}. \qquad (16.14)$$

16.5. The eigenvectors of the matrices GF and FG

Consider the product of the matrix $\mathbf{GF}$ and the matrix $\mathbf{L}$. Substituting the expressions for $\mathbf{G}$ and $\mathbf{F}$ from eqns (16.11) and (16.14) respectively, we find

$$\mathbf{GFL} = \mathbf{LL}^\dagger[\mathbf{L}^\dagger]^{-1}\mathbf{\Lambda L}^{-1}\mathbf{L},$$

$$= \mathbf{L\Lambda}. \qquad (16.15)$$

Now if we consider the square matrix $\mathbf{L}$ as made up of the successive columns $\mathbf{L}_1$, $\mathbf{L}_2$, $\mathbf{L}_3$, etc., it follows that the product $\mathbf{GFL}$ will likewise be made up of the successive columns $\mathbf{GFL}_1$, $\mathbf{GFL}_2$, $\mathbf{GFL}_3$, etc., as shown in eqn (16.16):

$$\mathbf{GFL} = \begin{bmatrix} \mathbf{GFL}_1 & \mathbf{GFL}_2 & \mathbf{GFL}_3 & \cdots \end{bmatrix}. \qquad (16.16)$$

According to eqn (16.15) this must be identical with $\mathbf{L\Lambda}$ which has the form

$$\mathbf{L\Lambda} = \left[\; \mathbf{L}_1\lambda_1 \;\middle|\; \mathbf{L}_2\lambda_2 \;\middle|\; \mathbf{L}_3\lambda_3 \;\middle|\; \cdots \;\right], \tag{16.17}$$

where $\lambda_1, \lambda_2, \lambda_3$, etc. are the roots which form the elements of the diagonal matrix $\mathbf{\Lambda}$. Comparing eqns (16.16) and (16.17) we see that

$$\mathbf{GFL}_1 = \lambda_1\,\mathbf{L}_1,$$
$$\mathbf{GFL}_2 = \lambda_2\,\mathbf{L}_2,$$
$$\mathbf{GFL}_3 = \lambda_3\,\mathbf{L}_3, \quad \text{etc.}$$

The columns $\mathbf{L}_1, \mathbf{L}_2, \mathbf{L}_3$, etc. of $\mathbf{L}$ are therefore eigenvectors of the matrix $\mathbf{GF}$ corresponding to the roots (or eigenvalues) $\lambda_1, \lambda_2, \lambda_3$, etc.

For the sake of completeness we may mention another approach to the matrix $\mathbf{L}$, which is however really equivalent to the above. Consider the product $\mathbf{FG}[\mathbf{L}^{-1}]^\dagger$, instead of the product $\mathbf{GFL}$. Inserting the values of $\mathbf{G}$ and $\mathbf{F}$ from eqns (16.11) and (16.14) respectively, we find

$$\mathbf{FG}[\mathbf{L}^{-1}]^\dagger = [\mathbf{L}^{-1}]^\dagger\mathbf{\Lambda}\mathbf{L}^{-1}\mathbf{L}\mathbf{L}^\dagger[\mathbf{L}^{-1}]^\dagger$$
$$= [\mathbf{L}^{-1}]^\dagger\mathbf{\Lambda}. \tag{16.18}$$

Eqn (16.18) is seen to have a form analogous to that of eqn (16.15) but with $\mathbf{FG}$ in place of $\mathbf{GF}$ and $[\mathbf{L}^{-1}]^\dagger$ in place of $\mathbf{L}$. It therefore follows that the columns of $[\mathbf{L}^{-1}]^\dagger$ (or the rows of $\mathbf{L}^{-1}$, which is the same thing) are eigenvectors of the matrix $\mathbf{FG}$. Of course the eigenvalues (roots) of $\mathbf{FG}$ are exactly the same as those of $\mathbf{GF}$, for the vibrational secular equation can be written in either of the two equivalent forms $|\mathbf{GF}-\lambda\mathbf{E}| = 0$ or $|\mathbf{FG}-\lambda\mathbf{E}| = 0$, and the roots (being properties of the molecule) are obviously independent of the particular form used.

We may note here that the computerized iteration programme commonly used for the calculation of force constants from frequency data is such that, for every set of force constants, it produces not only the roots but also the corresponding eigenvectors.

16.6. The normalization of the **L** matrix

We have seen that the columns of **L** are eigenvectors of the matrix **GF**, and that the rows of L^{-1} are eigenvectors of the matrix **FG**. When these eigenvectors are derived by the methods described in Section 16.2, they contain undetermined multiplicative factors. These factors, one for each eigenvector, have to be determined appropriately. This process is called the *normalization* of the **L** matrix (or the L^{-1} matrix).

A convenient method makes use of eqn (16.13), i.e. $L^{\dagger}FL = \Lambda$. Let L_1, L_2, etc. be a set of known, but unnormalized columns of **L**, and let the desired normalizing factors be N_1, N_2, etc. Then the normalizing condition takes the form

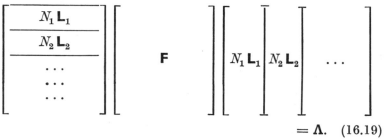

$$= \Lambda. \quad (16.19)$$

By multiplying out the left-hand side of eqn (16.19) and equating each of the resulting diagonal elements to the corresponding element (i.e. λ value) of the diagonal matrix Λ, we obtain a number of equations equal to the number of unknown Ns. Solution of these equations thus gives all the desired normalizing constants for the columns of the **L** matrix.

When **L** has been duly normalized in this way it will necessarily satisfy, not only the equation $L^{\dagger}FL = \Lambda$, but also eqn (16.11), i.e. $LL^{\dagger} = G$. This may be proved as follows. From eqn (16.15) we know that

$$GFL = L\Lambda. \quad (16.20)$$

236

Now since **L** has been normalized so as to satisfy the relation $\mathbf{L^\dagger FL} = \mathbf{\Lambda}$, it follows that

$$\mathbf{FL} = [\mathbf{L^\dagger}]^{-1}\mathbf{\Lambda}. \qquad (16.21)$$

Substitution of eqn (16.21) into eqn (16.20) gives

$$\mathbf{G}[\mathbf{L^\dagger}]^{-1}\mathbf{\Lambda} = \mathbf{L\Lambda},$$

from which we see that

$$\mathbf{G}[\mathbf{L^\dagger}]^{-1} = \mathbf{L},$$

and consequently $\qquad \mathbf{G} = \mathbf{LL^\dagger}.$

For the sake of completeness we may call attention to the alternative procedure of normalizing $\mathbf{L}^{-1}$ instead of **L**. Here the appropriate normalizing condition, obtained by inverting both sides of eqn (16.13), is

$$\mathbf{L}^{-1}\mathbf{F}^{-1}[\mathbf{L}^{-1}]^\dagger = \mathbf{\Lambda}^{-1}. \qquad (16.22)$$

The matrix $\mathbf{\Lambda}^{-1}$ is like $\mathbf{\Lambda}$ in being diagonal, but its elements are the reciprocals of the roots instead of the roots themselves. The unnormalized columns of $[\mathbf{L}^{-1}]^\dagger$ are obtained as eigenvectors of the matrix **FG** instead of the matrix **GF** (discussed in Section 16.5). Then proceeding in the same way as for **L**, we may use the normalizing condition of eqn (16.22) to determine all the factors required to convert the matrix $\mathbf{L}^{-1}$ to the appropriately normalized form.

16.7. Vibrational normal coordinates in terms of internal symmetry coordinates and internal coordinates

Having duly normalized the **L** matrix we may use it to transform from the set of internal symmetry coordinates S to the vibrational normal coordinates Q, using the inverse of the relation expressed in eqn (16.7), i.e.

$$\mathbf{Q} = \mathbf{L}^{-1}\mathbf{S}. \qquad (16.23)$$

It should be noted that the **L** matrix is not in general an orthogonal one, so that it is not possible simply to write $\mathbf{L}^\dagger$ for $\mathbf{L}^{-1}$. The normalized **L** matrix must be inverted, if eqn (16.23) is to be used. For this reason it might appear more convenient to calculate and normalize the $\mathbf{L}^{-1}$ matrix (as in the preceding section) and not to handle the **L** matrix as such at all. However, reference to eqn (16.22) shows that the normalizing condition for **L**$^{-1}$ contains **F**$^{-1}$. Thus the trouble of inverting **L** can be avoided only at the cost of the trouble of inverting **F**.

We may remark in passing that, because the kinetic energy (which involves masses) can be expressed in the form $2T = \dot{\mathbf{Q}}^\dagger \dot{\mathbf{Q}}$, it follows that the normal coordinates must be suitably mass-reduced. (In this connection we recall that, in the simple example treated in Chapter 1, the normal coordinates were expressed as linear combinations of *mass-reduced* Cartesian coordinates with pure numerical coefficients.) Now the internal symmetry coordinates S, being linear combinations of internal displacement coordinates with pure numerical coefficients, are not mass-reduced. It follows therefore that the matrix **L**$^{-1}$ which transforms the coordinates S into the normal coordinates Q must involve the masses of the vibrating nuclei. This fact is expressed quantitatively in eqn (16.10).

It may well be desired to have the normal coordinates Q expressed in terms of ordinary internal coordinates D (such as bond stretches and changes of bond angle) rather than in terms of internal symmetry coordinates S. To this end we merely have to make use of the relation between the D coordinates and the S coordinates, as expressed in eqn (11.24), i.e. **S** = **UD**. Substituting this into eqn (16.23) ,we at once find that

$$\mathbf{Q} = \mathbf{L}^{-1}\mathbf{UD}. \tag{16.24}$$

This is the desired relationship. Thus knowing the matrices **L** and **U**, we can use them to express the normal coordinates as linear combinations of the internal coordinates, so obtaining a clear picture of the normal modes of vibration.

By analogy with eqn (16.23), the relationship expressed in eqn (16.24) is often written in the form

$$\mathbf{Q} = \mathbf{L}^{-1}\mathbf{D}. \tag{16.25}$$

The matrices **L** and **L** are related by the equation **L**$^{-1}$ = **L**$^{-1}$**U**, from which it follows that **L** = **UL**.

Part II

The quantum-mechanical treatment of molecular vibrational spectroscopy

17 The vibrational wave equation and its solutions

17.1. Introduction

IN Part I of this book we have outlined the treatment of the problem of molecular vibrations by classical mechanics. Assuming the vibrations to be simple-harmonic, we have seen how the classical secular equation can be derived, and how the solution of it gives on the one hand the roots λ which are directly related to the classical mechanical frequencies ν_{cl} by the equation $\lambda = 4\pi^2\nu_{\mathrm{cl}}^2$, and on the other hand information from which the normal coordinates Q can be found in terms of internal symmetry coordinates S or internal coordinates D. We have seen that each normal coordinate Q_k has its own root λ_k and vibrates in accordance with the equation $Q_k = A_k \cos(\lambda_k^{\frac{1}{2}} t + \epsilon_k)$, where A_k is an amplitude (determined by the initial conditions at time $t = 0$) and ϵ_k is a phase angle. We have become acquainted with the very important property which characterizes the normal coordinates, namely that in terms of them the expressions for the kinetic and potential energies assume the very simple forms

$$2T = \sum_k \dot{Q}_k^2$$

and

$$2V = \sum_k \lambda_k Q_k^2.$$

Thus in terms of the normal coordinates the Hamiltonian H of the system, i.e. the total energy, can be written as

$$H = T + V = \tfrac{1}{2} \sum_k \dot{Q}_k^2 + \tfrac{1}{2} \sum_k \lambda_k Q_k^2. \tag{17.1}$$

Now, in Part II, our object is to consider molecular vibrational states and vibrational spectroscopy by the methods of quantum mechanics, and in particular Schrödinger wavemechanics. It will be assumed that the reader is already acquainted with this form of quantum theory, and no general exposition of it will be given. From time to time, however, attention will be drawn to aspects which are of particular interest in the treatment of vibrations.

Before beginning to discuss this treatment in detail, however, the reader should realize that, in wave mechanics as in classical mechanics, it is not possible to consider nuclear motions in isolation without having first invoked the Born–Oppenheimer approximation (discussed in Section 2.1) in order to separate them from electronic motions. Moreover, just as in the classical treatment, the use of a purely vibrational wave equation must be based upon a previous separation from rotations, a step that cannot be taken without making the approximate assumption that rotation–vibration interactions are negligible. In the first instance we shall also be assuming that the vibrations are simple-harmonic, but, as we shall find later, it will be possible to reach certain important conclusions that transcend this special limitation.

17.2. The vibrational wave equation in terms of normal coordinates

The Schrödinger amplitude equation, like Newton's laws of motion, is essentially a discovered expression of how natural systems are observed to behave. As such, it cannot be derived or deduced from anything more fundamental than itself. It has close analogies with the classical Hamiltonian H, from which indeed it may be constructed according to certain postulatory rules. From the Hamiltonian

$$H = T + V$$

we form the so-called Hamiltonian operator $\mathscr{H}$ defined as

$$\mathscr{H} = \mathscr{T} + \mathscr{V},$$

where $\mathscr{T}$ and $\mathscr{V}$ are respectively the operators corresponding to T and V. The wave equation is then obtained by allowing $\mathscr{H}$ to operate upon the so-called wave function (or more correctly the amplitude wave function) ψ, and setting the result equal to

242

$W\psi$ where W is the total energy. Thus the wave equation is simply

$$\mathscr{H}\psi = \mathscr{T}\psi + \mathscr{V}\psi = W\psi.$$

The wave function ψ is a function of the positional coordinates of the problem—in our case the vibrational coordinates. The operators $\mathscr{T}$ and $\mathscr{V}$ (and hence the Hamiltonian operator $\mathscr{H}$) are constructed by rules which are essentially part of the postulatory basis of the Schrödinger form of quantum mechanics. The potential energy V is of course a function of the vibrational coordinates. The operator $\mathscr{V}$ is simply a multiplying one, i.e. $\mathscr{V}\psi = V\psi$. The case of the operator $\mathscr{T}$ is more complicated. In the Hamiltonian H the kinetic energy is expressed in terms of the momental coordinates p which are conjugate to the positional ones ξ. In forming the operator $\mathscr{T}$ the rule is that each momental coordinate p is replaced by the differential operator

$$\frac{h}{2\pi i}\frac{\partial}{\partial\xi},$$

where h is Planck's constant and $i = \sqrt{(-1)}$. The square of p is replaced by the operator representing two successive operations of the above kind, i.e. by

$$-\frac{h^2}{4\pi^2}\left(\frac{\partial^2}{\partial\xi^2}\right).$$

When the vibrational kinetic energy is expressed in terms of the normal coordinates Q, so that $T = \frac{1}{2}\sum_k \dot{Q}_k^2$, the operator $\mathscr{T}$ becomes

$$\mathscr{T} = -\frac{h^2}{8\pi^2}\sum_k \frac{\partial^2}{\partial Q_k^2}.$$

Thus, following the above rules and writing ψ for the total vibrational wave-function of the molecule, we obtain the wave equation in the following form:

$$-\frac{h^2}{8\pi^2}\sum_k \frac{\partial^2\psi}{\partial Q_k^2} + \frac{1}{2}\sum_k \lambda_k Q_k^2 \psi = W\psi. \tag{17.2}$$

Slight rearrangement gives the more customary form

$$\sum_k \frac{\partial^2\psi}{\partial Q_k^2} + \frac{8\pi^2}{h^2}\left(W - \frac{1}{2}\sum_k \lambda_k Q_k^2\right)\psi = 0. \tag{17.3}$$

From a purely mathematical point of view, a differential

243

equation of this kind permits a continuous range of solutions ψ, associated with a continuous range of values W of the total energy. However, it is an essential feature of wave mechanics that we attribute a physical significance to ψ. In consequence, only physically 'well-behaved' functions can be accepted as solutions—for example, functions that remain finite at every point in configurational space. These acceptable functions are the *eigenfunctions* of the system (sometimes known as characteristic or proper functions), and each of them is associated with a discrete *eigenvalue* W (sometimes known as a characteristic or proper value) which is in fact the energy of the quantum state characterized by the wave function ψ. The criteria which a wave function must satisfy in order to be physically acceptable (i.e. that it must be continuous, single-valued, and finite throughout the whole configuration space of the system) constitute or imply the *boundary conditions*. For example, in the case of a simple-harmonic vibrator the wave function must tend to zero in the limit where the vibrational coordinate becomes infinite.

17.3. Separability of the vibrational wave equation with respect to normal coordinates

Because of the absence of cross-terms in the expressions for the kinetic and potential energies in terms of normal coordinates, the simple-harmonic vibrational problem in classical mechanics could be separated into individual parts, one for each Q. We shall now show that the wave equation (17.2) is likewise separable. This is the great practical advantage of using normal coordinates. Let us try a solution of the form

$$\psi = \psi_1(Q_1)\psi_2(Q_2)\psi_3(Q_3)...\psi_k(Q_k)... \qquad (17.4)$$

in which the total vibrational function ψ (which is a function of all the Q's) is written as the product of individual functions ψ_k, each of which is a function only of the single normal coordinate Q_k. Differentiating with respect to Q_k we have

$$\frac{\partial^2\psi}{\partial Q_k^2} = \psi_1(Q_1)\psi_2(Q_2)...\frac{\mathrm{d}^2\psi_k}{\mathrm{d}Q_k^2}...$$

$$= \frac{\psi}{\psi_k}\frac{\mathrm{d}^2\psi_k}{\mathrm{d}Q_k^2}. \qquad (17.5)$$

Substitution of eqn (17.5) into the wave equation (17.2) gives

$$\left(-\frac{h^2}{8\pi^2}\sum_k \frac{1}{\psi_k}\frac{\mathrm{d}^2\psi_k}{\mathrm{d}Q_k^2}+\tfrac{1}{2}\sum_k \lambda_k Q_k^2\right)\psi = W\psi.$$

Dividing through by ψ, we obtain

$$\sum_k \left(-\frac{h^2}{8\pi^2}\frac{1}{\psi_k}\frac{\mathrm{d}^2\psi_k}{\mathrm{d}Q_k^2}+\tfrac{1}{2}\lambda_k Q_k^2\right) = W. \tag{17.6}$$

Now each of the summands on the left-hand side of eqn (17.6) is a function of one normal coordinate only. Since the sum of them is equal to W, which is a constant, it follows that each separate summand must itself be equal to a constant, say W_k for the kth summand. Thus for each value of k we may write

$$-\frac{h^2}{8\pi^2}\frac{\mathrm{d}^2\psi_k}{\mathrm{d}Q_k^2}+(\tfrac{1}{2}\lambda_k Q_k^2)\psi_k = W_k\psi_k. \tag{17.7}$$

This is a wave equation corresponding to the expressions $2T_k = \dot{Q}_k^2$ and $2V_k = \lambda_k Q_k$ for the kinetic and potential energies respectively. We see that eqn (17.7) is in fact the wave equation for the kth simple-harmonic mode with the normal coordinate Q_k and the classical frequency $(\nu_{\mathrm{cl}})_k$ given by $\lambda_k = 4\pi^2(\nu_{\mathrm{cl}})_k^2$. Thus in wave mechanics, as in classical mechanics, the use of normal coordinates causes the total problem of the vibrations of a molecule to separate into individual problems, one for each normal mode. In the wave-mechanical treatment, each of the individual wave equations can be separately solved to give the eigenfunction ψ_k and the corresponding eigenvalue W_k. The total eigenfunction ψ is then given by

$$\psi = \prod_k \psi_k, \tag{17.8}$$

and the total eigenvalue W by

$$W = \sum_k W_k. \tag{17.9}$$

17.4. The mutual orthogonality of eigenfunctions

Because of its importance to us in our later discussions, we here draw attention to the following general property of the eigenfunctions of any system. Let $\psi^{(i)}$ and $\psi^{(j)}$ be two such eigenfunctions, and let the corresponding eigenvalues be $W^{(i)}$

and $W^{(j)}$. The property in question is that, if $W^{(i)} \neq W^{(j)}$, then

$$\int \psi^{(i)}\psi^{(j)} \, d\tau = 0. \tag{17.10}$$

Here $d\tau$ is a volume element in the configurational space of the system, and the integration (usually multiple) extends over the whole of this space. This remarkable property is conventionally expressed by saying that eigenfunctions with different eigenvalues are mutually *orthogonal*. Of course the word 'orthogonal' as so used, does not have the same meaning as when we say that two vectors are mutually orthogonal (as in Section 4.3). Nevertheless it is deliberately adopted in order to express a formal analogy (however far-fetched this may appear at first sight) between the dot product of two vectors, on the one hand, and the integral of eqn (17.10) on the other. When the dot product of two vectors vanishes, we say that the vectors are mutually orthogonal; we use the same words to describe two eigenfunctions when the integral in question vanishes. We shall see later that it is reasonable to generalize the term 'vector' so as to include eigenfunctions.

We shall now prove that the eigenfunctions of a single vibrator (not necessarily simple-harmonic) obey eqn (17.10), i.e. are mutually orthogonal. If V is the potential energy as a function of the single mass-reduced vibrational coordinate Q, we have

$$\frac{d^2\psi^{(i)}}{dQ^2} + \frac{8\pi^2}{h^2}(W^{(i)}-V)\psi^{(i)} = 0 \tag{17.11}$$

and

$$\frac{d^2\psi^{(j)}}{dQ^2} + \frac{8\pi^2}{h^2}(W^{(j)}-V)\psi^{(j)} = 0. \tag{17.12}$$

We may eliminate V by multiplying eqn (17.11) by $\psi^{(j)}$, multiplying eqn (17.12) by $\psi^{(i)}$, and then subtracting. We thus obtain

$$\psi^{(j)}\frac{d^2\psi^{(i)}}{dQ^2} - \psi^{(i)}\frac{d^2\psi^{(j)}}{dQ^2} + \frac{8\pi^2}{h^2}(W^{(i)}-W^{(j)})\psi^{(i)}\psi^{(j)} = 0. \tag{17.13}$$

In order to proceed further, we note that

$$\frac{d}{dQ}\left(\psi^{(j)}\frac{d\psi^{(i)}}{dQ} - \psi^{(i)}\frac{d\psi^{(j)}}{dQ}\right)$$
$$= \frac{d\psi^{(j)}}{dQ}\frac{d\psi^{(i)}}{dQ} + \psi^{(j)}\frac{d^2\psi^{(i)}}{dQ^2} - \frac{d\psi^{(i)}}{dQ}\frac{d\psi^{(j)}}{dQ} - \psi^{(i)}\frac{d^2\psi^{(j)}}{dQ^2}. \tag{17.14}$$

The first and third terms on the right-hand side of eqn (17.14) cancel out, and the remaining two are identical with the first two on the left-hand side of eqn (17.13). Thus eqn (17.13) becomes

$$\frac{d}{dQ}\left(\psi^{(j)}\frac{d\psi^{(i)}}{dQ}-\psi^{(i)}\frac{d\psi^{(j)}}{dQ}\right)+\frac{8\pi^2}{h^2}(W^{(i)}-W^{(j)})\psi^{(i)}\psi^{(j)}=0. \quad (17.15)$$

Integration over the whole range of Q (i.e. between the limits $-\infty$ and $+\infty$) now gives

$$\left(\psi^{(j)}\frac{d\psi^{(i)}}{dQ}-\psi^{(i)}\frac{d\psi^{(j)}}{dQ}\right)_{-\infty}^{+\infty}+\frac{8\pi}{h^2}(W^{(i)}-W^{(j)})\int_{-\infty}^{+\infty}\psi^{(i)}\psi^{(j)}\,dQ=0. \quad (17.16)$$

Because of the boundary conditions which both $\psi^{(i)}$ and $\psi^{(j)}$ must obey, the first of the two parts of the left-hand side of eqn (17.16) must vanish. The second part must therefore vanish also. We have thus proved that, provided $W^{(i)}\neq W^{(j)}$, the integral must vanish. This is the condition shown in eqn (17.10) for the mutual orthogonality of the two eigenfunctions $\psi^{(i)}$ and $\psi^{(j)}$.

Because of the elimination of the potential energy V in the above argument, the orthogonal property of the eigenfunctions is independent of whether the vibrator in question is simple-harmonic or not. Indeed, as may be proved by an extension of the reasoning used above, the mutual orthogonality of eigen-functions with different eigenvalues is a very general property of the solutions of all Schrödinger wave equations.

We shall not give the general proof here, but shall merely note that the argument for a single vibrator can be very easily extended to the normal vibrations of a polyatomic molecule, provided that these are regarded as being simple-harmonic. This follows at once from the separability of the wave equation when expressed in terms of the normal coordinates Q_k. Any total eigenfunction $\psi^{(i)}$ of the system can be written as in eqn (17.4):

$$\psi^{(i)}=\psi_1^{(i)}(Q_1)\psi_2^{(i)}(Q_2)\psi_3^{(i)}(Q_3)\dots\psi_k^{(i)}(Q_k)\dots.$$

Consequently, if $\psi^{(j)}$ is another total eigenfunction, we have

$$\int_{-\infty}^{+\infty} \psi^{(i)}\psi^{(j)}\,\mathrm{d}\tau$$

$$= \int_{-\infty}^{+\infty} \psi_1^{(i)}\psi_1^{(j)}\,\mathrm{d}Q_1 \int_{-\infty}^{+\infty} \psi_2^{(i)}\psi_2^{(j)}\,\mathrm{d}Q_2 \int_{-\infty}^{+\infty} \psi_3^{(i)}\psi_3^{(j)}\,\mathrm{d}Q_3 \dots \int_{-\infty}^{+\infty} \psi_k^{(i)}\psi_k^{(j)}\,\mathrm{d}Q_k \dots$$

$$(17.17)$$

Now if $\psi^{(i)}$ and $\psi^{(j)}$ have eigenvalues which are not equal, it follows that for at least one normal coordinate (say Q_k) the eigenfunctions $\psi_k^{(i)}$ and $\psi_k^{(j)}$ must have eigenvalues $W_k^{(i)}$ and $W_k^{(j)}$ which are not equal. This being so, the right-hand side of eqn (17.17) must vanish. We have thus proved that the total vibrational eigenfunctions $\psi^{(i)}$ and $\psi^{(j)}$ are mutually orthogonal.

17.5. The normalization of eigenfunctions

If ψ is an eigenfunction of a wave equation and k is a numerical factor, then it is at once apparent from the form of the wave equation, as given in eqn (17.3), that $k\psi$ is also an eigenfunction with the same eigenvalue as ψ. Thus all eigenfunctions are indeterminate as regards a multiplicative factor. The removal of this indeterminacy is called *normalization*.

Just as the dot product of an ordinary vector with itself is non-zero, so also the integral $\int \psi^2\,\mathrm{d}\tau$ (analogous to the integral in eqn (17.10)) will have some value, w say, which is different from zero. It follows that by multiplying ψ by the so-called normalizing factor $1/\sqrt{w}$ we obtain a new eigenfunction with the same eigenvalue as ψ but with the special property

$$\int \left(\frac{\psi}{\sqrt{w}}\right)^2 \mathrm{d}\tau = 1. \tag{17.18}$$

The new function is said to have been normalized to unity, or simply to have been normalized.

We note here the analogy with the case of a unit vector e_i, for which $e_i . e_i = 1$; whereas for two different unit vectors which are mutually orthogonal, $e_i . e_j = 0$. Unless otherwise stated, all eigenfunctions will be assumed to have been normalized, i.e. to include the appropriate multiplicative factor ensuring that $\int \psi^2 \, d\tau = 1$.

17.6. Degenerate eigenfunctions

Just as we refer to two ordinary vectors as being linearly independent when neither is expressible as the other multiplied by a numerical factor, so also we may describe as linearly independent two eigenfunctions (both with the same eigenvalue) which have this same property. Such eigenfunctions are said to be degenerate. An example is the pair $\psi^{(a)}$, $\psi^{(b)}$ corresponding to a doubly degenerate pair of normal coordinates $Q_k^{(a)}$, $Q_k^{(b)}$.

It is an easy matter to verify, by substitution into the wave equation, that any linear combination of degenerate eigenfunctions is itself an eigenfunction (though not a normalized one) with the same eigenvalue. Now since $W^{(a)} = W^{(b)}$, the members $\psi^{(a)}$, $\psi^{(b)}$ of a degenerate pair are not necessarily orthogonal (see Section 17.4); but if they are not, it is always possible to construct a pair of linear combinations of them which are. Indeed it is always possible to do this in an unlimited number of ways. Thus let us suppose that both $\psi^{(a)}$ and $\psi^{(b)}$ are normalized, and consider the pair of linear combinations $(\alpha\psi^{(a)}+\beta\psi^{(b)})$ and $(\gamma\psi^{(a)}+\delta\psi^{(b)})$. The condition that these shall be mutually orthogonal is

$$\int (\alpha\psi^{(a)}+\beta\psi^{(b)})(\gamma\psi^{(a)}+\delta\psi^{(b)}) \, d\tau = 0.$$

This means that

$$\alpha\gamma \int (\psi^{(a)})^2 \, d\tau + (\alpha\delta+\beta\gamma) \int \psi^{(a)}\psi^{(b)} \, d\tau + \beta\delta \int (\psi^{(b)})^2 \, d\tau = 0. \tag{17.19}$$

By virtue of the assumed normalized character of $\psi^{(a)}$ and $\psi^{(b)}$, eqn (17.19) reduces to

$$\alpha\gamma + w(\alpha\delta+\beta\gamma) + \beta\delta = 0, \tag{17.20}$$

where w is the value of $\int \psi^{(a)}\psi^{(b)} \, d\tau$. If each of the linear combinations is to be normalized, we have the further conditions

$$\left.\begin{aligned} \alpha^2 + 2w(\alpha\beta) + \beta^2 &= 1 \\ \gamma^2 + 2w(\gamma\delta) + \delta^2 &= 1 \end{aligned}\right\}. \tag{17.21}$$

The value of w being known, it is clearly possible to select the coefficients α, β, γ, and δ in an unlimited number of ways.

It is worth remarking here on the analogy with ordinary vectors, which runs as follows. Two linearly independent vectors in a two-dimensional vector space are not necessarily orthogonal to one another; but when they are not, it is always possible to construct a pair of linear combinations of them which are. Indeed it is always possible to do this in an unlimited number of ways, depending upon the orientation chosen for the first member of a pair. Of course all linear combinations of the two vectors (including those combinations which are mutually orthogonal) will belong to the same vector space as the original vectors themselves.

Exactly similar considerations can be extended to sets of three linearly independent vectors in a three-dimensional space, and analogously also to sets of three linearly independent eigenfunctions all having the same eigenvalue.

17.7. The simple-harmonic wave functions

It will be assumed that the reader is already acquainted with the method of solving the wave equation for a simple-harmonic vibrator,

$$\frac{d^2\psi}{dQ^2} + \frac{8\pi^2}{h^2}(W - \tfrac{1}{2}\lambda Q^2)\psi = 0,$$

and with the solutions obtained. (Accounts will be found in books on wave mechanics: for example, in that authoritative and admirably lucid work, *Introduction to Quantum Mechanics* by L. Pauling and E. B. Wilson Jr. which, although first published as long ago as 1935, can still be very highly recommended.)

Each eigenfunction is associated with a quantum number v, which takes the values 0, 1, 2,..., etc. The solution $\psi^{(v)}$ for the quantum number v is given by

$$(\psi^{(v)})_k = N_v \exp(-\tfrac{1}{2}\gamma_k Q_k^2) H_v(\gamma_k^{\frac{1}{2}} Q_k) \qquad (17.22)$$

in which N_v is a numerical normalizing factor, γ_k is a constant of the system, namely,

$$\gamma_k = \frac{4\pi^2}{h}(\nu_{\text{cl}})_k, \qquad (17.23)$$

and H_v is a so-called Hermite polynomial. We need not consider the details of the mathematical properties of these polynomials. For our purposes it will suffice to list the values for the
first few values of the vibrational quantum number v:

$$\left.\begin{aligned}
H_0(x) &= 1 \\
H_1(x) &= 2x \\
H_2(x) &= 4x^2 - 2 \\
H_3(x) &= 8x^3 - 12x \\
H_4(x) &= 16x^4 - 48x^2 + 12 \\
H_5(x) &= 32x^5 - 160x^3 + 120x
\end{aligned}\right\} . \tag{17.24}$$

In the eigenfunctions it is merely necessary to substitute
$x = \gamma_k^{\frac{1}{2}} Q_k$. We see that when the vibrational quantum number
v is even, the polynomials contain only even powers of Q_k,
whereas when v is odd, they contain only odd powers.

It is an easy matter to pass from a single simple-harmonic
oscillator to a vibrating system (such as a polyatomic molecule)
which has a larger number of simple-harmonic normal modes.
In such a case (as we saw in Section 17.3) the total wave equation
is separable into parts (one for each mode), which are of exactly
the same type as that for a single simple-harmonic oscillator.
The total eigenfunction ψ is thus simply the product of the
individual ones ψ_k for the different modes. Naturally each mode
has its own value of ν_{cl} and hence its own value of γ_k. The state
of the whole system is characterized by a set of vibrational
quantum numbers v_k.

17.8. The eigenvalues of the simple-harmonic vibrator

Associated with the wave function $\psi^{(v)}$ and the vibrational
quantum number v is the eigenvalue W_v, which is given by

$$W_v = (v + \tfrac{1}{2})h\nu_{cl}. \tag{17.25}$$

We see that even in the lowest vibrational state (i.e. the so-called
ground vibrational state), for which $v = 0$ and the vibrational
energy is the minimum possible, the vibrator must still possess
the so-called residual or zero-point vibrational energy $\tfrac{1}{2}h\nu_{cl}$.

Radiation is emitted or absorbed by transitions between an upper state with the eigenvalue (i.e. the energy) W' and a lower state with the energy W''. The frequency ν_{qu} of the radiation is given by the well-known Bohr relation

$$\nu_{qu} = \frac{W' - W''}{h}. \tag{17.26}$$

In the case of a simple-harmonic vibrator, the nature of the wave functions is such that transitions can only occur between states for which the vibrational quantum numbers differ by unity, i.e. between adjacent energy levels. Writing $W' = (n' + \frac{1}{2})h\nu_{cl}$ and $W'' = (n'' + \frac{1}{2})h\nu_{cl}$, as in eqn (17.25), we find from eqn (17.26) that

$$\nu_{qu} = (n' - n'')\nu_{cl}.$$

Since only transitions for which $n' - n'' = 1$ can occur, we reach the very simple conclusion

$$\nu_{qu} = \nu_{cl}. \tag{17.27}$$

This is the result which was quoted at the end of Section 1.1, where it was described as 'conveniently uncomplicated'. The convenience is, of course, that radiation frequencies, as observed in spectroscopic experiments, can be identified with classical mechanical frequencies and used directly for such purposes as the calculation of force fields.

We may note here a specially simple feature of the simple-harmonic vibrator. It is that, in accordance with eqn (17.25), its eigenvalues (or energy levels) are equally spaced, i.e. the difference in energy between levels with the quantum numbers v and $v+1$ is independent of the magnitude of v. Thus all transitions for which $\Delta v = \pm 1$ must absorb or emit one and the same frequency ν_{qu}, irrespective of the value of v. Vibrational frequencies are usually observed in absorption ($\Delta v = +1$), and in fact usually involve the transition $v = 0 \rightarrow v = 1$. This particular transition is known as the *fundamental transition*. By way of contrast, the energy levels of an anharmonic vibrator are not quite equally spaced, so that the frequency for the transition $v = 0 \rightarrow v = 1$ (i.e. the fundamental frequency) is not exactly equal to that for such a transition as $v = 1 \rightarrow v = 2$.

18 Infra-red absorption and Raman scattering: simple theory and restricted selection rules

18.1. Classical theory of absorption and emission in the infra-red

ACCORDING to classical electrodynamics a system emits radiation by virtue of periodic changes in its electric dipole moment, the frequency of the emitted radiation being the same as that of the dipole oscillations. Absorption is the inverse of emission, and so the system is able to absorb radiation of the same frequency (or frequencies) as it is able to emit. Infra-red spectroscopy is generally concerned with the absorption of radiation incident upon a sample. We shall therefore refer mainly to absorption, but the reader should bear in mind throughout that all considerations will apply with equal force to the inverse process of emission.

The electric dipole $\boldsymbol{\mu}$ of a molecule is a vector, and so has three components μ_x, μ_y, and μ_z in a Cartesian system. On the basis of classical electrodynamics we can say that the molecule will only be able to absorb (or emit) radiation of the frequency ν, provided that $\boldsymbol{\mu}$ (or in greater detail at least one of its three components) can oscillate with this frequency. Now the dipole moment is a function of the nuclear configuration and so, when the molecule vibrates, it varies correspondingly. In the simple-harmonic approximation, all the molecular vibrations can be regarded as superpositions of a limited number of normal modes, each with its own normal frequency ν_k. It follows therefore that the electric dipole moment can only oscillate with these normal frequencies, and that only radiation of these frequencies can be absorbed. Such radiation lies in the infra-red region of the spectrum. We shall see later that, in the case of certain normal

253

modes, the amplitude of vibration of μ may necessarily be zero. The intensity of absorption of radiation of the corresponding normal frequency will then also be zero. We say that the absorption is forbidden by a selection rule.

In general the magnitudes of the components of the molecular dipole moment will be functions of all the vibrational coordinates Q, and thus capable of expansion as a Taylor series:

$$\left.\begin{aligned}
\mu_x &= (\mu_x)_0 + \sum_k \left\{\left(\frac{\partial \mu_x}{\partial Q_k}\right)_0 Q_k\right\} + \text{higher terms} \\
\mu_y &= (\mu_y)_0 + \sum_k \left\{\left(\frac{\partial \mu_y}{\partial Q_k}\right)_0 Q_k\right\} + \text{higher terms} \\
\mu_z &= (\mu_z)_0 + \sum_k \left\{\left(\frac{\partial \mu_z}{\partial Q_k}\right)_0 Q_k\right\} + \text{higher terms}
\end{aligned}\right\} \quad (18.1)$$

The zero subscripts in eqns (18.1) indicate values at the equilibrium configuration of the molecule. We may conveniently adopt the convention whereby the three separate expressions for μ_x, μ_y, and μ_z are all implied by the single condensed form

$$\mu = \mu_0 + \sum_k \left\{\left(\frac{\partial \mu}{\partial Q_k}\right)_0 Q_k\right\} + \text{higher terms.} \quad (18.2)$$

Since all the amplitudes of the normal vibrations are very small, it is a good approximation to neglect terms of higher than the first degree in the Qs. We may therefore write

$$\mu = \mu_0 + \sum_k \left\{\left(\frac{\partial \mu}{\partial Q_k}\right)_0 Q_k\right\}. \quad (18.3)$$

Evidently the condition that the molecular dipole moment shall be able to oscillate with the frequency ν_k, i.e. the condition that this normal frequency shall be capable of being absorbed (or, as we say, shall be active in absorption), is that $(\partial \mu/\partial Q_k)_0$ shall differ from zero. This implies that

$$\left(\frac{\partial \mu_i}{\partial Q_k}\right)_0 \neq 0 \quad (18.4)$$

for at least one of the components ($i = x$, y, or z).

254

This is a general statement of the *selection rule* for infra-red absorption. It belongs to the kind of selection rule which we shall describe as *restricted*, in order to emphasize that its derivation is dependent upon two special approximate assumptions. The first is that the molecular vibrations are simple-harmonic, for otherwise the normal modes would not be separable and the meaning of the individual normal coordinates Q_k would be lost. The second assumption is that in the Taylor expansion of the electric dipole moment represented by eqn (18.2) all the higher terms are negligible. We shall later see how the restrictions imposed by these two approximate assumptions may be relaxed, so as to allow the derivation of what we shall call the *general* (as opposed to the restricted) selection rule.

18.2. Quantum theory of absorption and emission in the infra-red

It will be assumed that the reader is acquainted with the wave-mechanical quantity known as the *transition moment*. For the transition between the two states characterized by the wave functions $\psi^{(n)}$ and $\psi^{(m)}$ we shall denote it by $\boldsymbol{\mu}_{nm}$. Like the ordinary molecular dipole moment $\boldsymbol{\mu}$, it is a vector, and so has three Cartesian components with the magnitudes $(\mu_x)_{nm}$, $(\mu_y)_{nm}$, and $(\mu_z)_{nm}$. The transition moment is defined by the three following equations:

$$(\mu_x)_{nm} = \int \psi^{(n)} \mu_x \psi^{(m)} \, d\tau,$$

$$(\mu_y)_{nm} = \int \psi^{(n)} \mu_y \psi^{(m)} \, d\tau,$$

$$(\mu_z)_{nm} = \int \psi^{(n)} \mu_z \psi^{(m)} \, d\tau,$$

in which μ_x, μ_y, and μ_z are the magnitudes of the components of $\boldsymbol{\mu}$, $d\tau$ is a volume element in configurational space, and the integrals (in general multiple) are to be extended over the whole of this space. As with eqns (18.1), we may conveniently condense the three separate equations for the components $(\mu_x)_{nm}$, $(\mu_y)_{nm}$, and $(\mu_z)_{nm}$ into the single conventional equation

$$\boldsymbol{\mu}_{nm} = \int \psi^{(n)} \boldsymbol{\mu} \psi^{(m)} \, d\tau. \tag{18.5}$$

The importance of the transition moment is that it determines the intensity of the absorption (or emission) of radiation by the transition in question. In fact the relation of this intensity to the magnitude μ_{nm} of the transition moment is similar to that of the intensity of classical absorption (or emission) to the amplitude of oscillation of an ordinary dipole moment. Thus the total intensity is proportional to the square of μ_{nm}, i.e. to the sum of the squares of $(\mu_x)_{nm}$, $(\mu_y)_{nm}$, and $(\mu_z)_{nm}$. Clearly a very general statement of the selection rule (indeed one that is too general to be immediately useful to the spectroscopist) is that a transition is forbidden in absorption (or emission) if $\boldsymbol{\mu}_{nm} = 0$, i.e. if $(\mu_x)_{nm} = (\mu_y)_{nm} = (\mu_z)_{nm} = 0$.

A more particular, and therefore more useful, statement of the selection rule can be derived by substituting the Taylor-expansion value of $\boldsymbol{\mu}$ from eqn (18.2) into the integral defining the transition moment in eqn (18.5). This gives

$$\boldsymbol{\mu}_{nm} = \boldsymbol{\mu}_0 \int \psi^{(n)}\psi^{(m)} \,\mathrm{d}\tau + \sum_k \left\{ \left(\frac{\partial \boldsymbol{\mu}}{\partial Q_k}\right)_0 \int \psi^{(n)} Q_k \psi^{(m)} \,\mathrm{d}\tau \right\}. \quad (18.6)$$

Because of the mutual orthogonality of the wave functions, the first integral on the right-hand side of eqn (18.6) vanishes unless the two states n and m are the same. This, however, corresponds to no transition. The first term may therefore be disregarded as far as absorption (or emission) is concerned. Consider now the kth term of the remaining summation. Each of the total wave functions is the product of the respective individual wave functions for the normal modes, i.e.

$$\psi^{(n)} = \prod_k \psi_k^{(n)}$$

and
$$\psi^{(m)} = \prod_k \psi_k^{(m)}.$$

It follows that the integral in the second term of the right-hand side of eqn (18.6) may be written

$$\int \left(\prod_k \psi_k^{(n)} \right) Q_k \left(\prod_k \psi_k^{(m)} \right) \mathrm{d}\tau$$
$$= \int \psi_1^{(n)} \psi_1^{(m)} \,\mathrm{d}Q_1 \int \psi_2^{(n)} \psi_2^{(m)} \,\mathrm{d}Q_2 \dots$$
$$\dots \int \psi_k^{(n)} Q_k \psi_k^{(m)} \,\mathrm{d}Q_k \int \psi_{k+1}^{(n)} \psi_{k+1}^{(m)} \,\mathrm{d}Q_{k+1} \dots . \quad (18.7)$$

From eqn (18.7) we see that the right-hand side will only be non-vanishing provided that every factor is non-vanishing, and that the conditions for this are the following: (*a*) for all modes except the kth the two states must be identical; and (*b*) for the kth mode the vibrational quantum number must change only by unity. Condition (*a*) is a direct consequence of the mutual orthogonality of wave functions of states with different eigenvalues. Condition (*b*) is a consequence of the known particular properties of the Hermite polynomials in the expressions (see eqn (17.22)) for the simple-harmonic eigenfunctions $\psi_k^{(n)}$ and $\psi_k^{(m)}$.

Conclusions of exactly the same kind apply to each of the normal vibrational modes and to each of the three components of the transition moment. Thus in general we can say that, in any act of absorption, the vibrational quantum numbers of all but one of the normal modes must remain unchanged, and that there must be a change of unity in the quantum number of the single remaining mode. We have already seen in Section 17.8 that the transition with $\Delta v_k = 1$ is associated with absorption of radiation of the frequency ν_k of the normal mode in question. Thus the wave-mechanical treatment agrees with the classical treatment in showing that the only frequencies that can be active in absorption are the individual normal frequencies.

However, it does not follow that all such frequencies will in fact be active, for the kth term of the summation in the second term on the right-hand side of eqn (18.6) contains not only the integral which we have been considering, but also the factor $(\partial\boldsymbol{\mu}/\partial Q_k)_0$. Only if this is non-vanishing (or in more detail only if $(\partial\mu_i/\partial Q_k)_0$ is non-vanishing for at least one of the components of $\boldsymbol{\mu}$, i.e. for $i = x, y,$ or z) will the frequency ν_k in fact be active in absorption (or emission). This is the same selection rule as we derived, in eqn (18.4), on classical grounds.

It is still a restricted selection rule, in the sense that its wave-mechanical derivation involves the same two approximate assumptions as did the classical derivation. We recall that these two restrictive assumptions are that the molecular vibrations are simple-harmonic, and that in the Taylor expansion of the electric

S

dipole moment as a function of the normal coordinates, all terms higher than the linear ones are negligible.

18.3. Classical theory of Raman scattering

In the two preceding sections of this chapter we have discussed infra-red absorption, which is the basis of one of the two methods of molecular vibrational spectroscopy. The other makes use of a physical phenomenon of a quite different kind—the Raman effect. This is a molecular light-scattering phenomenon in which a change of frequency occurs. It does not involve absorption at all. In fact it uses light in the visible region of the spectrum, the frequency of which is deliberately chosen so that it is not absorbed by the system under investigation. What is observed is the spectrum of the light after it has been scattered by the molecules of the sample. Frequency shifts occur which, in the vibrational Raman effect, are found to be equal to normal vibrational frequencies of the scattering species. Because the mechanism of Raman scattering is essentially different from that of absorption, the selection rules for the two kinds of molecular vibrational spectroscopy are in general different. Vibrational transitions that are forbidden in infra-red absorption may be permitted in the Raman effect, and vice versa. The two experimental methods are thus essentially complementary in character.

The fundamental difference between absorption on the one hand and Raman scattering on the other is that in absorption we are concerned with the intrinsic electric dipole moment μ of the molecule, whereas in the Raman effect this dipole moment is of no account at all, and we are concerned instead with the dipole moment π which is *induced* in the molecule by the electric field of the incident light.

Consisting as it does of positively charged nuclei in a 'cloud' of negative electrons, a molecule is electrically polarizable. An applied electric field E will therefore induce in it a dipole moment π which will be quite distinct from any moment μ which the molecule may chance to possess in the absence of the applied field. For field strengths of the magnitude ordinarily used in

Raman spectroscopy, the relation between $\boldsymbol{\pi}$ and $\boldsymbol{E}$ may be written

$$\boldsymbol{\pi} = \alpha\boldsymbol{E}, \tag{18.8}$$

where α is the electric polarizability of the molecule. In general the vector $\boldsymbol{\pi}$ will have a different direction from that of the vector $\boldsymbol{E}$, and therefore α is not a simple scalar quantity.

Eqn (18.8) must be understood to imply that the magnitudes of the components of $\boldsymbol{\pi}$ are related to the magnitudes of the components of the electric field $\boldsymbol{E}$ by the following three relations:

$$\left.\begin{aligned}
\pi_x &= \alpha_{xx}E_x + \alpha_{xy}E_y + \alpha_{xz}E_z \\
\pi_y &= \alpha_{yx}E_x + \alpha_{yy}E_y + \alpha_{yz}E_z \\
\pi_z &= \alpha_{zx}E_x + \alpha_{zy}E_y + \alpha_{zz}E_z
\end{aligned}\right\}. \tag{18.9}$$

Equations (18.9) express the fact that all three components of $\boldsymbol{E}$ make contributions to each of the three components of $\boldsymbol{\pi}$. The nine coefficients α_{ij} involved are called the components of the polarizability α. In consequence of eqns (18.9), α is said to be a *tensor*. In fact it is an example of what is known as a symmetric tensor, which means that $\alpha_{ij} = \alpha_{ji}$. (The proof of this property will be found in Section 19.5.) This symmetric nature provides three relations ($\alpha_{xy} = \alpha_{yx}$, $\alpha_{yz} = \alpha_{zy}$, $\alpha_{zx} = \alpha_{xz}$) between the nine components, so that in fact only six are distinct.

If we form single-column matrices from the magnitudes of the components of $\boldsymbol{\pi}$ and $\boldsymbol{E}$ respectively, the eqns (18.9) can be re-expressed in matrix form as

$$\begin{bmatrix} \pi_x \\ \pi_y \\ \pi_z \end{bmatrix} = \begin{bmatrix} \alpha_{xx} & \alpha_{xy} & \alpha_{xz} \\ \alpha_{yx} & \alpha_{yy} & \alpha_{yz} \\ \alpha_{zx} & \alpha_{zy} & \alpha_{zz} \end{bmatrix} \begin{bmatrix} E_x \\ E_y \\ E_z \end{bmatrix}. \tag{18.10}$$

Here the components of the tensor $\overset{\bullet}{\alpha}$ are used as elements of a square matrix, which we may represent by the symbol $[\alpha]$. The fact that α is a symmetric tensor means that $[\alpha]$ is a symmetric matrix. Of course, the reader will appreciate that the actual values of the individual elements of $[\alpha]$, or of the components of α

259

(which are the same), will depend upon the particular choice of the basis Cartesian coordinate system.

To avoid possible misunderstanding, the reader should be careful to distinguish between the polarizability tensor α itself and the square array in eqn (18.10). The latter is not a tensor, but a matrix whose elements are the components of the tensor α, and it is introduced here simply in order to enable us to re-express the three equations (18.9) in convenient matrix form. The transformation properties of tensor components (discussed in Section 19.5) are not the same as those of matrix elements. Tensors and matrices are thus essentially different in kind.

Now, like the dipole moment, the electric polarizability of a molecule will in general be a function of all the normal vibrational coordinates. We may therefore expand α as a Taylor series with respect to these coordinates and neglect powers higher than the first. We thus obtain

$$\alpha = \alpha_0 + \sum_k \left\{ \left(\frac{\partial \alpha}{\partial Q_k} \right)_0 Q_k \right\}. \qquad (18.11)$$

In this equation, which is analogous to eqn (18.2), α_0 is the polarizability tensor in the equilibrium configuration of the molecule, and $(\partial \alpha / \partial Q_k)_0$ is the so-called derived polarizability (also at the equilibrium configuration) for the kth normal mode. Like α itself, $(\partial \alpha / \partial Q_k)_0$ is a symmetric tensor, each of its components being the value, at the equilibrium configuration, of the derivative $\partial \alpha_{ij} / \partial Q_k$ of the corresponding component of α. The derived tensor is often denoted by α', and a component of it by α'_{ij}. If it is necessary to specify that the kth normal mode is concerned, an extra subscript k may be added to these symbols.

In Raman spectroscopy the electric field is applied by irradiating the molecule with monochromatic light of a frequency ν_0 which usually lies in the visible region and is chosen so as to ensure that no absorption by the molecule can occur. Using eqn (18.11) we may write

$$\boldsymbol{\pi} = \alpha \boldsymbol{E} = \alpha_0 \boldsymbol{E} + \sum_k \left\{ \left(\frac{\partial \alpha}{\partial Q_k} \right)_0 Q_k \right\} \boldsymbol{E}. \qquad (18.12)$$

This conveniently condensed form is to be understood as implying, for each of the components of the induced dipole $\boldsymbol{\pi}$, an
equation which takes into account all the components of the
applied electric field $\boldsymbol{E}$. For instance,

$$\pi_x = (\alpha_{xx})_0 E_x + (\alpha_{xy})_0 E_y + (\alpha_{xz})_0 E_z +$$
$$+ \sum_k \left[\left\{ \left(\frac{\partial \alpha_{xx}}{\partial Q_k}\right)_0 E_x + \left(\frac{\partial \alpha_{xy}}{\partial Q_k}\right)_0 E_y + \left(\frac{\partial \alpha_{xz}}{\partial Q_k}\right)_0 E_z \right\} Q_k \right]. \quad (18.13)$$

Consider the first term $\alpha_0 E$ in the expression for $\boldsymbol{\pi}$ on the
right-hand side of eqn (18.12). Since every component of α_0 is
simply a molecular constant and every component of E oscillates
with the incident light frequency ν_0, it follows that the corresponding part of every component of $\boldsymbol{\pi}$ must oscillate with this
same frequency. Thus light of the incident frequency ν_0 will be
emitted and will be observable in directions which differ from
that of the incident light. This is the phenomenon known as
classical or Rayleigh scattering. It is of no interest to us at
present, as it in no way involves the vibrations of the scattering
molecule. We can therefore ignore the first term on the right-
hand side of eqn (18.12).

Considering now the second term, let us fix our attention on the
contribution from the particular vibrational mode with the
normal coordinate Q_k. Every component of the derived tensor
$(\partial \alpha / \partial Q_k)_0$ is simply a constant. The time-dependent factors are
Q_k, which oscillates with the normal frequency ν_k, and all the
components of E, which oscillate with the incident frequency ν_0.
These time dependences could be expressed by including the
respective factors $\cos(2\pi \nu_k t)$ and $\cos(2\pi \nu_0 t)$. In view of the
identity $\cos \theta \cos \phi = \frac{1}{2}\{\cos(\theta + \phi) + \cos(\theta - \phi)\}$, we see that all
the corresponding contributions to all the components of the
induced dipole moment $\boldsymbol{\pi}$ are characterized by the two new
frequencies $\nu_0 + \nu_k$ and $\nu_0 - \nu_k$. This is sometimes referred to as the
effect of 'optical beating'. Light of the two 'beat frequencies'
will be emitted by the molecule, and will be observable in directions different from that of the incident light. They constitute
the contribution of the kth normal mode to the Raman spectrum
of the scattering molecule.

The frequency *shifts* (relative to the incident frequency) are known as Raman frequencies. They are clearly equal to the normal vibrational frequencies of the molecule in question, such as might (if active in absorption) have been observed directly in the infra-red.

Evidently the condition that a particular normal frequency ν_k shall be active in Raman scattering is that the factor $(\partial\alpha/\partial Q_k)_0$ in eqn (18.12) shall be different from zero. This condensed form of the selection rule implies that

$$\left(\frac{\partial\alpha_{ij}}{\partial Q_k}\right)_0 \neq 0 \qquad (18.14)$$

for at least one of the components (i or $j = x$, y, or z) of the polarizability tensor α. This condition should be compared with the corresponding one for infra-red absorption, as given in eqn (18.4).

As in the case of infra-red absorption, we again remark that this is a *restricted* selection rule, since its derivation is based upon corresponding approximate assumptions.

18.4. Quantum theory of Raman scattering

The translation of the above classical treatment into wave-mechanical language follows lines which will be already familiar from the case of infra-red absorption (discussed in Sections 18.1 and 18.2). The transition moment arising from the induced dipole moment π is given by

$$\int \psi^{(n)}\pi\psi^{(m)}\,\mathrm{d}\tau = E\int \psi^{(n)}\alpha\psi^{(m)}\,\mathrm{d}\tau$$
$$= E\alpha_0\int \psi^{(n)}\psi^{(m)}\,\mathrm{d}\tau + E\sum_k\left\{\left(\frac{\partial\alpha}{\partial Q_k}\right)_0\int \psi^{(n)}Q_k\psi^{(m)}\,\mathrm{d}\tau\right\}.$$
$$(18.15)$$

Because of the mutual orthogonality of the eigenfunctions, the integral in the first term on the right-hand side of eqn (18.15) must vanish unless $\psi^{(n)} = \psi^{(m)}$, in which case its value is unity. Thus this first term accounts for the Rayleigh scattering without change of frequency. (We may note incidentally that the components of α_0 must obviously be non-zero for all molecules, and even for single atoms. It follows, therefore, that Rayleigh scattering is never forbidden.) Coming now to the second term, the kth summand represents the contribution of the kth normal mode to the Raman spectrum. The integral in this summand is identical with the corresponding one in the case of infra-red absorption, see eqn (18.6), our discussion of which showed that it vanishes except when $\Delta v_k = \pm 1$, i.e. except for transitions associated with the frequency v_k. Thus the incidence of the light of frequency v_0 induces an interaction with the molecule in which the latter either gains or loses vibrational energy by transitions involving the quantum hv_k. In the consequent scattering act this amount of energy is either taken from or added to the incident quantum hv_0, and quanta $h(v_0-v_k)$ and $h(v_0+v_k)$ respectively are scattered. It is important to realize that at no stage is the incident quantum absorbed by the scattering molecule.

As in the classical treatment of the preceding section, the restricted selection rule which governs whether a particular normal vibrational frequency shall be permitted or forbidden in the Raman effect depends upon the factor $(\partial\alpha/\partial Q_k)_0$ in the second term on the right-hand side of eqn (18.15). If this factor is different from zero, the frequency will be permitted; if this factor vanishes, the frequency will be forbidden. In greater detail the rule states that only those normal modes can be active in Raman scattering which satisfy the condition $(\partial\alpha_{ij}/\partial Q_k)_0 \neq 0$ for at least one of the components of the molecular polarizability (i or $j = x$, y, or z).

18.5. Inadequacy of restricted selection rules when applied to real molecules

The restricted selection rules will only be valid in so far as the two simplifying assumptions made in their derivation are justified. Relaxation of the first of these assumptions (namely, that the vibrations are simple-harmonic) means a change in the nature of the potential energy and hence also in the wave equation and its solutions. For small amounts of mechanical anharmonicity, such as are usually encountered in practice, the energy levels are not very different from those of a simple-harmonic oscillator, but are no longer exactly equally spaced. Moreover, the associated eigenfunctions no longer possess those special properties of the Hermite polynomials which form the basis of the restriction of transitions to $\Delta v = \pm 1$ in infra-red absorption or in the Raman effect.

Even if the assumption of simple-harmonic vibrations were fully justified, the rule $\Delta v = \pm 1$ would be rendered invalid if the second assumption (namely that the Taylor expansions of μ and α contain only linear terms in the normal coordinates) were not justified. Failure of this second assumption is sometimes referred to as being due to 'electrical anharmonicity', as opposed to the mechanical anharmonicity which would render the first assumption invalid. If we had included a term in Q_k^2 in either of the expansions given in eqns (18.2) and (18.11), we should have had a contribution to the relevant transition moment involving the integral $\int \psi_k^{(n)} Q_k^2 \psi_k^{(m)} \, d\tau$. Now it is a known special property of the Hermite polynomials occurring in the simple-harmonic eigenfunctions that this integral vanishes, except when $\Delta v_k = \pm 2$. Thus transitions involving the frequency $2\nu_k$ (the *first overtones*) could become permitted. Higher overtones, $|\Delta v_k| > 2$, could become permitted by the inclusion of higher powers of Q_k in the Taylor expansions. Furthermore, higher terms involving more than one normal coordinate simultaneously could permit transitions (*combination tones*) involving simultaneous changes in the vibrational quantum numbers of more than one normal mode, the resulting frequencies being corresponding sums or differences of the respective normal vibrational frequencies.

Thus overtones and combination tones may become active (both in infra-red absorption and in the Raman effect) in consequence of either mechanical or electrical anharmonicity. In practice doubtless both factors operate, but observation shows that their combined effects are still only weak. Generally speaking, the restricted selection rules give a fairly good account of the observed spectra, so that to a fairly good approximation both kinds of anharmonicity can indeed be neglected, or at most treated as small perturbations. It is a matter of experience that the restricted selection rules are more nearly obeyed in the Raman effect than in infra-red absorption. Overtones and combination tones are generally so weak in the Raman effect as only rarely to be observed, whereas they are relatively common features (though generally rather weak ones) in the infra-red.

That the vibrations of real molecules are in fact mechanically anharmonic is evident from the experimentally established fact that the spacing between successive vibrational energy levels of a mode is not quite constant, but decreases appreciably with increase of the vibrational quantum number v. Observation shows that not all transitions for which $\Delta v = 1$ have exactly the same frequency. The one in which v changes from 0 to 1 is generally referred to as the fundamental. Similarly not all transitions for which $\Delta v = 2$ have exactly the same frequency. The one in which v changes from 0 to 2 is called the first overtone. Its frequency is not exactly twice that of the fundamental, but slightly less than this. Nevertheless, if the fundamental frequency is v_k, that of the first overtone is formally written as $2v_k$. Similarly the frequency of the second overtone ($v = 0$ to $v = 3$) is formally written as $3v_k$, although its actual value is somewhat less.

18.6. Relative intensities of corresponding Stokes and anti-Stokes lines in Raman spectra

As stated in Section 18.2, the intensity of emission or absorption of radiation in a transition between quantized energy levels is proportional to the square of the value μ_{nm} of the transition moment. The selection rules which permit or forbid emission or absorption accordingly depend upon the non-vanishing of μ_{nm}. Although it is certain (within the approximations being used) that the intensity will be zero when μ_{nm} is zero, the rules tell us nothing about how large (or how small) the intensity will be when μ_{nm} is different from zero. The magnitude of the intensity of a permitted transition will be determined by the magnitude of μ_{nm}^2, but beyond this the rules are silent. In practice, cases may occur when a frequency, although permitted by a selection rule, has such a low intensity as to be very difficult (and even, with the available techniques, impossible) to observe. It will thus seem to the observer to be forbidden—a false impression that may mislead him in the interpretation of his results. We shall have more to say of such cases in the last chapter (Chapter 24) of this book.

There is one case, however, where the relative intensities of different lines in the Raman spectrum can be foreseen theoretically. In Section 18.4 we showed that, for an active vibrational mode, Raman scattering involves transitions in which Δv is either $+1$ or -1. These result respectively in the scattering of light with the frequency $v_0 - \Delta v$ or $v_0 + \Delta v$, where Δv is the magnitude of the Raman frequency shift, which is equal to the vibrational frequency of the mode in question. The Raman line of the diminished frequency $v_0 - \Delta v$ is referred to as the *Stokes line*, and the one with the enhanced frequency $v_0 + \Delta v$ as the corresponding *anti-Stokes line*. Now for any pair of energy levels the definition in eqn (18.5) shows that the transition moment for the upward transition between them (in our case the vibrational transition with $\Delta v = +1$) is exactly the same as for the downward transition (in our case $\Delta v = -1$). As far as their transition moments are concerned, therefore, we should expect the anti-Stokes and Stokes lines to have equal intensities. A small alteration to this expectation arises from the fact that the intensity of emission is not only proportional to the square of the transition moment, but also to the fourth power of the actual frequency emitted. Since the anti-Stokes frequency $v_0 + \Delta v$ is necessarily greater than the Stokes frequency $v_0 - \Delta v$, we should expect all anti-Stokes lines to be more intense than their Stokes counterparts. Incident frequencies v_0 are generally around $20\,000$ cm^{-1}, and vibrational shifts in the range 1 to 10 per cent of v_0. We should accordingly expect the intensity ratio $I_{\text{anti-Stokes}}/I_{\text{Stokes}}$ to be only slightly above unity for the lower vibrational frequencies and to increase up to about 2 for the higher ones. In fact, however, observation reveals a very different state of affairs. All anti-Stokes lines are found to be *less intense* than the corresponding Stokes lines, and moreover the intensity ratio *decreases* very rapidly with increasing frequency shift. Indeed it is very difficult in practice to observe any anti-Stokes lines at all for vibrational shifts of about 1000 cm^{-1} and higher, although the Stokes lines are of normal intensity.

The reason for this striking behaviour lies in the fact that the experimentally observed intensities are summations over the scattering acts of the very large number of identical molecules which the sample under investigation inevitably contains. Each intensity is therefore proportional to the molecular population of the initial energy levels of the transitions involved. Now, as we have seen, the Raman effect involves pairs of vibrational levels whose vibrational quantum numbers differ by unity, the initial level for the emission of an anti-Stokes line being the upper one of the pair, and the initial level for the corresponding Stokes line being the lower. According to the Boltzmann distribution, the ratio of the corresponding molecular populations is $\exp\{-(W'-W'')/kT\}$, where W' and W'' are the energies of the upper and lower levels respectively, k is the Boltzmann constant, and T is the temperature. At ordinary temperatures kT/h is equivalent to about 200 cm^{-1}, while vibrational shifts $\Delta\nu$, i.e. values of $(W'-W'')/h$, range from about this value up to about 10 times greater (or more for bond-stretching modes involving hydrogen atoms). Thus with increase of $\Delta\nu$ the Boltzmann factor will decrease exponentially from a little less than unity to the order of 10^{-4} or less. It will thus be dominant in determining the anti-Stokes/Stokes intensity ratio. In fact experimental measurements are in good agreement with the theoretical expression for this ratio, i.e.

$$\frac{I_{\text{anti-Stokes}}}{I_{\text{Stokes}}} = \left(\frac{\nu_{\text{anti-Stokes}}}{\nu_{\text{Stokes}}}\right)^4 \exp\left(-\frac{h\Delta\nu}{kT}\right)$$

$$= \left(\frac{\nu_0+\Delta\nu}{\nu_0-\Delta\nu}\right)^4 \exp\left(-\frac{h\Delta\nu}{kT}\right), \qquad (18.16)$$

in which ν_0 is the frequency of the incident light and $\Delta\nu$ the magnitude of the Raman frequency shift.

18.7. Elementary symmetry considerations in relation to the restricted selection rules

As we shall see in detail later, it is molecular symmetry that determines whether or not $(\partial\mu/\partial Q_k)_0 \neq 0$ for a particular mode, i.e. whether the mode shall be active in infra-red absorption. We may illustrate this by the very simple example of a linear Y–X–Y molecule in which the two bonds are identical. The mechanical system treated in Chapter 1 was exactly of this type, and its two modes of vibration along the line of the masses are shown in Fig. 1.2. Remembering that the molecule is composed of positively charged nuclei and negatively charged electrons, we see that in its equilibrium configuration its structural symmetry will carry with it a corresponding electrical symmetry, and will result in a zero molecular dipole moment. In the totally symmetric vibrational mode a (Fig. 1.2) this symmetry is conserved throughout. The dipole moment therefore remains unchanged. This means that $(\partial\mu/\partial Q_a)_0 = 0$, i.e. that the mode must be forbidden in infra-red absorption.

In the antisymmetric mode b, on the other hand, the molecule does not conserve its equilibrium symmetry. The structural asymmetry through which it passes in the course of a vibration causes a corresponding electrical asymmetry, i.e. produces a non-zero dipole moment which oscillates synchronously with the mechanical vibration, passing through values of equal magnitude but opposite sign at corresponding values of the normal coordinate Q_b. This is shown diagrammatically in Fig. 18.1, where the abscissae represent the normal coordinates and the ordinates the magnitude of the dipole moment. (In this simple case the moment is always directed along the line of nuclei, which we may call the x-direction. Thus we need consider the component μ_x only, the other two (μ_y and μ_z) being always zero.) From Fig. 18.1 we see that $(\partial\mu/\partial Q_b)_0 \neq 0$. The mode is therefore active in infra-red absorption. Incidentally this is a very simple instance of the fact that, quite generally, infra-red activity of certain modes of a molecule is in no way incompatible with the possession by that molecule of a zero dipole moment in its equilibrium configuration.

Similarly it is molecular symmetry that determines whether or not $(\partial\alpha/\partial Q_k)_0 \neq 0$, i.e. whether a normal vibrational mode shall be active or inactive in Raman scattering. Let us consider again the modes a and b of a linear symmetric Y–X–Y molecule, as depicted in Fig. 1.2. In the totally symmetric mode a, both

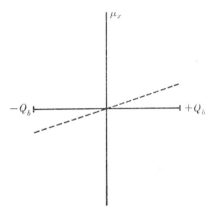

Fig. 18.1. Variation of an electric dipole moment component in a mode that is active in infra-red absorption.

bonds are stretched in one phase and both are compressed in the other. Although this leaves the molecular dipole moment unaltered, it is clear that the electrical polarizability will be different in the two phases. In general therefore $(\partial\alpha/\partial Q_a)_0$ will differ from zero. In greater detail we may say that $(\partial\alpha_{ij}/\partial Q_a)_0 \neq 0$ for at least one of the components α_{ij} of the polarizability tensor. The diagrammatic representation of this would give a figure resembling Fig. 18.1, with values of $\alpha_{ij}-(\alpha_{ij})_0$ instead of μ_x as ordinates, and values of Q_a instead of Q_b as abscissae.

In the mode b, on the other hand, the structural situation is the same in both phases, in that one bond is stretched and the other is compressed. The end-for-end interchange caused a reversal in the sign of the non-zero electric dipole moment vector, but there is no such reversal for the components of the electrical polarizability tensor. Thus the value of this tensor (and in more detail the value of each of its components) is the same for $+Q_b$ as for $-Q_b$. This is shown diagrammatically in Fig. 18.2. It is true

269

that the value of any polarizability component at each extreme value of the normal coordinate may be different from its value at $Q_b = 0$, but from Fig. 18.2 it is clear that the variation is such that the slope at the equilibrium configuration, i.e. $(\partial \alpha_{ij}/\partial Q_b)_0$, is zero. The mode b is thus forbidden in the Raman effect. In this

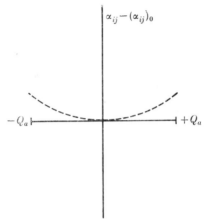

Fig. 18.2. Variation of a polarizability component in a mode that is forbidden in the Raman effect.

very simple but very instructive example it happens that the mode which is forbidden in infra-red absorption is permitted in Raman scattering, and *vice versa*. This, however, is a special circumstance, for in general many molecular vibrational modes are permitted in both kinds of spectrum, and some are forbidden in both.

The example illustrates strikingly the importance of symmetry in the context of the restricted selection rules. Generalization of the symmetry considerations in the form of group theory makes it possible to predict the spectroscopic consequences of the application of the rules to molecules containing any number of nuclei and belonging to any point group.

But symmetry theory is more powerful than this. Being independent of the particular physical nature of the systems to which it is applied, the theory enables us to escape from the special restrictions to which the selection rules so far considered have been subject. The relevant symmetry properties of the potential energy are quite independent of whether the intra-molecular force-field is simple-harmonic or anharmonic, and the relevant symmetry properties of molecular dipole moments and electrical polarizabilities are likewise independent of any special assumptions about their respective Taylor expansions in terms of normal vibrational coordinates. Therefore, by taking account of the general symmetry properties of the vibrational eigen-functions and electrical properties, rather than their particular mathematical formulations, we shall be enabled to derive selection rules whose validity will extend to the spectra of molecules with unknown amounts of mechanical and electrical anharmonicities.

However, before continuing with the discussion of selection rules (in Chapter 20 ff.), we must first become acquainted with an important observable property of light that has been scattered by molecules—its state of polarization.

19 States of polarization of light scattered by molecules

19.1. Directional effects in molecular vibrational spectroscopy

In the preceding chapter we considered the processes of infra-red absorption and Raman scattering by a single molecule. In doing this we introduced a Cartesian coordinate system fixed in physical space and, assuming that the molecule in question was fixed in a certain spatial orientation, we referred its properties (such as the components of its dipole moment vector and its polarizability tensor) to this same coordinate system. It will be realized, of course, that the values of these components must change as we go from one fixed orientation of the molecule to another. We therefore anticipate that associated directional effects will arise.

In practice a spectroscopically observable signal is the sum of contributions from each of the very large number of molecules present in even the smallest of practicable samples. If these molecules are free (or effectively so) they will certainly not all be oriented in the same way relative to a space-fixed direction such as that of the incident light or (in the Raman effect) that in which the scattered light is observed. Since, in this book, we are concerned only with free molecules, we have to consider whether any observable directional effects will remain in view of the averaging over all molecular orientations which the observation necessarily entails. Freedom of the molecules in a sample is realized by using gases at sufficiently low pressures. It is a matter of experience that the vibrational spectra of pure liquids and of solutes in solution generally approximate quite closely to expectations for the gaseous state.

19.2. Directional properties of the radiation emitted by an oscillating electric dipole

According to classical electrodynamics, the intensity of emission of radiation from a small oscillating electric dipole is by no means the same in all directions. Fig. 19.1 shows its direc-

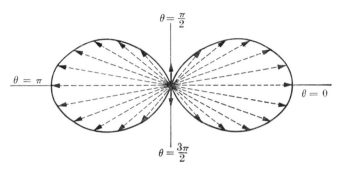

Fig. 19.1. Directional properties of the emission of radiation from an oscillating electric dipole.

tional distribution in any plane through the dipole. The direction of oscillation of the dipole is indicated by the short vertical double-headed arrow at the centre of the figure. The lengths of the dotted arrows are proportional to the intensities of emission in the different directions in the plane of the figure. We see that the dipole does not radiate at all along its own direction, and that the maximum intensity of emission is at right angles to this. This maximum value I_{max} is in fact equal to $K\mu^2$, where μ is the magnitude of the oscillating dipole moment (i.e. the amplitude of its oscillation) and K is a constant. As is easily appreciated, the intensity I_θ in a direction making the angle θ with that of maximum intensity is K times the square of the resolved part of μ in the direction making the angle θ with μ itself, i.e. the intensity is $K(\mu \cos \theta)^2$. We thus have

$$I_\theta = I_{\text{max}} \cos^2\theta. \tag{19.1}$$

This is the function shown in Fig. 19.1. Of course the complete three-dimensional directional distribution is obtained by rotation of Fig. 19.1 about the line of μ as axis.

In addition to its intensity, there is one other important

directional property of the radiation emitted by an oscillating dipole, namely, its polarization. As is well known, electromagnetic radiation is associated with a transversely oscillating electric field. In natural (unpolarized) light all transverse directions for this field are equally represented, but with plane-polarized light, one particular direction alone is present. We shall define the plane of polarization as the plane to which the electric vector is confined. Referring now to Fig. 19.1, we can say that all the radiation emitted by the oscillating dipole is plane-polarized and that, for all the rays shown by the dotted lines in Fig. 19.1, the plane of polarization is the plane of the figure, i.e. the plane containing the dipole and the line of propagation of the emitted ray.

It is important to realize that the above directional properties of intensity and polarization apply not only to the intrinsic and induced dipole moments of a molecule, but also to the corresponding wave-mechanical transition moments.

19.3. Directional effects in infra-red absorption

The oscillating electric dipole associated with a particular normal vibrational mode of a molecule has a definite direction in a coordinate system which is fixed in the molecule; therefore, relative to this system, the radiation emitted (or absorbed) by the molecule will have corresponding directional properties as regards both intensity and polarization. For example, consider either of the totally symmetric modes of the PCl_3 molecule. Obviously, because the equilibrium-configuration symmetry is conserved throughout, the molecular dipole moment will oscillate along the direction of the C_3 axis. It follows that, despite the fact that both of the totally symmetric modes are permitted by the infra-red selection rule $(\partial\mu/\partial Q)_0 \neq 0$, they can give rise to no emission (and therefore no absorption) along the molecular axis. Infra-red radiation of the appropriate frequency will be absorbed if it is incident at an angle to this axis, and the intensity of absorption will have a distribution like that in Fig. 19.1, but a beam incident along the axis will not be absorbed at all. Moreover, in the case of beams which are absorbed, it will only be the appropriately polarized part that will be affected. No incident beam (no matter what its direction) will be absorbed at all if it is plane-polarized with its electric vector at right angles to the axis of the molecule.

The discussion of all the above directional effects for PCl_3 has assumed that the molecule had a fixed orientation in space, so that there was a unique relation between the directions of its axis and of the incident infra-red beam. In a sample containing free molecules, this is certainly not the case. Instead, the individual molecules are randomly oriented relative to space-fixed axes, and all the directional effects must disappear in the actually observed absorptions. Of course the selection rule holds good.

Only if, in some way, the molecules of the sample could all be held in the same spatial orientation, would the directional effects become apparent. Such a regularity may be realized in single crystals, and when studying these the investigator has the power of changing at will the relative directions of molecular dipoles and incident radiation. But in crystals the constituent molecules are necessarily far from free, and their motions can no longer be dealt with in terms of an intramolecular force field alone. They therefore lie outside the scope of this book. We may remark, however, that the modes of vibration of single crystals do indeed give rise to electric dipoles which oscillate in characteristic directions, and so directional effects are in fact observed in infra-red absorption. In certain cases it may be that, in some of the crystal modes, the constituent molecular units vibrate in ways that approximate more or less closely to normal modes which they would possess if free. Directional effects then become observable that are absent when gaseous or liquid samples are used.

19.4. Polarization of Rayleigh scattering from a single molecule with a fixed orientation in space

In view of what has been said above about infra-red absorption, it might be anticipated that, on account of the random spatial orientation of the molecules, all directional properties would disappear in light-scattering phenomena using samples containing free molecules. In fact, however, this is not the case.

Before considering such samples and the averaging over all molecular orientations which is then involved, we must look a little more closely than in Sections 18.3 and 18.4 at the process of the scattering of light by a single molecule with a fixed orientation in space, i.e. a fixed orientation relative both to the direction of the incident light and also to that in which the scattered light is observed. For the time being it will be convenient to restrict our considerations to Rayleigh scattering, i.e. to that part of the scattered light which has suffered no frequency shift in the scattering act.

As in Section 18.3, we introduce a space-fixed Cartesian coordinate system, to which we relate not only the directions of the incident and scattered beams, but also the polarizability components of the molecule in its assumed fixed orientation. Let the molecule be situated at the origin of the coordinate system and irradiated along the positive y-direction by natural light of frequency ν_0. Unpolarized light of this kind may be regarded as composed of two parts with equal intensities and with no phase relationship, one polarized in the xy plane and the other polarized in the yz plane. The equal intensities of the two parts means that the amplitudes of the corresponding electric vectors (E_x and E_z respectively) are equal (see Fig. 19.2).

In Sections 18.3 and 18.4 we saw that Rayleigh scattering is accounted for by the first term on the right-hand side of either eqn (18.12) or eqn (18.15), i.e. by that part of the induced dipole moment (or of the corresponding induced transition moment) which is determined solely by the polarizability tensor α_0 of the molecule in its equilibrium configuration. In the present discussion, we shall confine our attention to Rayleigh scattering in the positive x-direction (shown in Fig. 19.2), i.e. at right angles to the direction of the incident beam. We shall therefore be concerned only with the y- and z-components of the induced moment; for, in accordance with Section 19.2 and Fig. 19.1, the x-component will make no contribution to the emission in the

x-direction. On the understanding that throughout this discussion of Rayleigh scattering we shall be dealing only with the tensor α_0 and with that part of the induced moment which oscillates with frequency ν_0, we may (without risk of confusion)

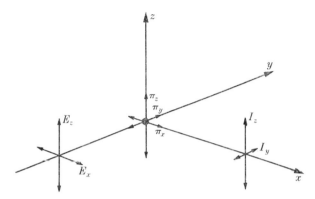

Fig. 19.2. Rayleigh scattering at right angles to the direction of the incident beam, showing polarized components.

omit zero subscripts from the symbols for the respective components α_{ij} and π_i. Since $E_y = 0$, we may write

$$\left.\begin{aligned}
\pi_y &= \alpha_{yx} E_x + \alpha_{yz} E_z \\
\pi_z &= \alpha_{zx} E_x + \alpha_{zz} E_z
\end{aligned}\right\}. \tag{19.2}$$

In general, therefore, π_y will not be equal to π_z, and consequently the intensities of the two corresponding plane-polarized components of the light scattered along the x-direction will not be equal. These intensities, which we shall call I_y and I_z respectively, could be measured separately by interposing in the scattered beam an optical device which, according to its orientation, would transmit first the one and then the other of the polarized components.

The observable ratio I_y/I_z is called the *degree of depolarization* of the Rayleigh scattering and is denoted by the symbol ρ_n. Two points should be noted concerning this symbol: first, the subscript n is used to signify that the incident light in question is natural; and secondly, all ρ-values conventionally refer to scattering at right angles to the direction of the incident light.

Bearing in mind the absence of any phase relationship between E_x and E_z, and taking account of their equality by writing simply E for both, we find from eqn (19.2) that the expressions for the intensities of the plane-polarized components of the scattered light take the form

$$I_y = KE^2(\alpha_{yx}^2 + \alpha_{yz}^2) \atop I_z = KE^2(\alpha_{zx}^2 + \alpha_{zz}^2) \Bigg\}, \qquad (19.3)$$

where K is the same proportionality factor in both cases. Hence it follows that the value of the degree of depolarization must be

$$\rho_n = \frac{I_y}{I_z} = \frac{\alpha_{yx}^2 + \alpha_{yz}^2}{\alpha_{zx}^2 + \alpha_{zz}^2}. \qquad (19.4)$$

Instead of using natural incident light, it is possible experimentally to use plane-polarized incident light. Moreover, there is a choice of planes of polarization, including two very important cases. In the first, the incident light is polarized in the zy plane, i.e. E_z alone is present, both E_x and E_y being zero, so that the electric vector is *perpendicular* to the direction of scattering (Fig. 19.2). In the second case the incident light is polarized in the xy plane, i.e. E_x alone is present, both E_y and E_z being zero, so that the electric vector is *parallel* to the direction of scattering. The corresponding degrees of depolarization of the Rayleigh scattering are denoted by the common symbol ρ_p, and the two cases are distinguished as $\rho_{p\perp}$ and $\rho_{p\parallel}$ respectively. From what has already been said in connection with ρ_n, it follows at once that

$$\rho_{p\perp} = \frac{\alpha_{yz}^2}{\alpha_{zz}^2} \qquad (19.5)$$

and

$$\rho_{p\parallel} = \frac{\alpha_{yx}^2}{\alpha_{zx}^2}. \qquad (19.6)$$

The expressions for the degrees of depolarization in eqns (19.4), (19.5), and (19.6) are valid for scattering by a single molecule with a fixed orientation in space. In order to obtain corresponding expressions for the practically important cases of fluid samples containing very large numbers of effectively free molecules, it is clearly necessary to undertake an averaging over

278

all possible molecular orientations. In particular what we require are the average values $\overline{\alpha_{ij}^2}$ of the squares of the components (i or $j = x$, y, or z) of the polarizability tensor α_0 of the scattering molecule in its equilibrium configuration. In order to find these values we must look more closely into the nature of the polarizability and the way in which its components transform with change of the orientation of the molecule.

19.5. The transformation properties of the components of the polarizability tensor

In any electrically polarizable system (for example a molecule) there are three mutually perpendicular directions, such that the application of an electric field in any one of them results in the production of an electric dipole moment which is in the same direction. The directions are those of the *principal axes*, and may be denoted by X, Y, and Z. The relations between the induced dipole moments in these directions (π_X, π_Y, and π_Z) and the corresponding electric field strengths (E_X, E_Y, and E_Z) are accordingly

$$\left. \begin{aligned} \pi_X &= \alpha_X E_X \\ \pi_Y &= \alpha_Y E_Y \\ \pi_Z &= \alpha_Z E_Z \end{aligned} \right\}, \qquad (19.7)$$

where α_X, α_Y, and α_Z are the *principal values* of the polarizability. The simple form of eqns (19.7), obtained when the tensor is referred to the principal axes, is to be contrasted with the form of eqns (18.9) which is required when the tensor is referred to general Cartesian axes. If, as in Section 18.3, we form single-column matrices from the components of π and E respectively, the equations (19.7) can be re-expressed in the matrix form

$$\begin{bmatrix} \pi_X \\ \pi_Y \\ \pi_Z \end{bmatrix} = \begin{bmatrix} \alpha_X & 0 & 0 \\ 0 & \alpha_Y & 0 \\ 0 & 0 & \alpha_Z \end{bmatrix} \begin{bmatrix} E_X \\ E_Y \\ E_Z \end{bmatrix}. \qquad (19.8)$$

Thus in the basis whose coordinate axes are the principal polarizability axes, the matrix corresponding to α_0 is a diagonal matrix. By analogy we may say that the tensor is a 'diagonal' one or that its 'off-diagonal' components are all zero.

What, then, is the relationship between the tensor components in a general Cartesian system x, y, z, and in the special Cartesian system X, Y, Z? Let us introduce into the discussion the cosines of the angles between the i (or j) axes and the principal axes of X, Y, and Z, and write as $\cos(x, Y)$, for example, the cosine of the angle between the x-axis and the Y-axis. There will be nine such cosines with the general designation $\cos(i, p)$ or $\cos(j, p)$, where i or $j = x$, y, or z and $p = X$, Y, or Z. Consider the polarizability component α_{ji}, which is the one which determines the magnitude of that part π_{ji} of the induced dipole component π_j which is due to the component E_i of the applied electric field. First resolve E_i along the three principal axes p. The resolved parts are $E_i \cos(i, p)$. They induce the respective dipole moments $\alpha_p E_i \cos(i, p)$, each directed along the respective p-axis. Now resolve each of these moments along the j-axis and form the sum of the resolved parts, thus obtaining the contribution π_{ji} to the induced dipole component π_j due to the component E_i of the applied field. This gives

$$\pi_{ji} = \sum_p \alpha_p E_i \cos(i, p)\cos(j, p). \qquad (19.9)$$

We know, however, that

$$\pi_{ji} = \alpha_{ji} E_i. \qquad (19.10)$$

Comparison of eqns (19.9) and (19.10) shows that

$$\alpha_{ji} = \sum_p \alpha_p \cos(i, p)\cos(j, p). \qquad (19.11)$$

This is the required relationship between the polarizability tensor components referred respectively to the general and principal coordinate systems.

Inspection of eqn (19.11) shows at once that interchange of j and i leaves the right-hand side unchanged, thus providing the proof of the fact (mentioned in Section 18.3) that $\alpha_{ij} = \alpha_{ji}$, i.e. that the polarizability is a symmetric tensor.

In the particular case where $j = i$, eqn (19.11) gives the value for a 'diagonal' component as

$$\alpha_{ii} = \sum_p \alpha_p \cos^2(i, p).$$ (19.12)

In eqns (19.11) and (19.12) the summations extend over the three values of p, i.e. X, Y, and Z.

The nine cosines $\cos(i, p)$ are by no means independent. In fact they constitute the elements of the square matrix in the transformation

$$\begin{bmatrix} x \\ y \\ z \end{bmatrix} = \begin{bmatrix} \cos(x, X) & \cos(x, Y) & \cos(x, Z) \\ \cos(y, X) & \cos(y, Y) & \cos(y, Z) \\ \cos(z, X) & \cos(z, Y) & \cos(z, Z) \end{bmatrix} \begin{bmatrix} X \\ Y \\ Z \end{bmatrix}.$$

Since this is an orthogonal transformation, the square matrix must be an orthogonal one and consequently (as shown in Section 3.8) the elements must satisfy relations of the kinds

$$\cos^2(x, X) + \cos^2(y, X) + \cos^2(z, X) = 1$$ (19.13)

and

$$\cos(x, X)\cos(x, Y) + \cos(y, X)\cos(y, Y) + \cos(z, X)\cos(z, Y) = 0.$$
(19.14)

In consequence of these relations, certain combinations of the components of the tensor α_0 remain unchanged when the orientation of the principal axes relative to the space-fixed axes is varied, i.e. as the molecule in question rotates. These combinations are the *invariants* of the tensor. They are two in number, the so-called mean-value invariant $\overline{\alpha_0}$, and the so called aniso-tropy invariant γ_0. Their definitions are as follows:

$$\overline{\alpha} = \tfrac{1}{3}(\alpha_{xx} + \alpha_{yy} + \alpha_{zz})$$

and

$$\gamma_0^2 = \tfrac{1}{2}\{(\alpha_{xx} - \alpha_{yy})^2 + (\alpha_{yy} - \alpha_{zz})^2 + (\alpha_{zz} - \alpha_{xx})^2 + 6(\alpha_{xy}^2 + \alpha_{yz}^2 + \alpha_{zx}^2)\}.$$
(19.15)

That they are indeed invariants may be shown by substituting the values of the tensor components in terms of the principal values, as given in eqn (19.11), and bearing in mind the necessary relations between the cosines, of the kinds given in eqns (19.13) and (19.14). The case of $\overline{\alpha_0}$ is very simple. We have

$$
\begin{aligned}
\overline{\alpha_0} &= \tfrac{1}{3}(\alpha_{xx}+\alpha_{yy}+\alpha_{zz}) \\
&= \tfrac{1}{3}\sum_p \alpha_p \cos^2(x,p)+\tfrac{1}{3}\sum_p \alpha_p \cos^2(y,p)+\tfrac{1}{3}\sum_p \alpha_p \cos^2(z,p) \\
&= \tfrac{1}{3}\sum_p \alpha_p\{\cos^2(x,p)+\cos^2(y,p)+\cos^2(z,p)\}.
\end{aligned}
$$

But according to relations of the type of eqn (19.13) the sum of the squares of the cosines is equal to unity for each value of p. It follows, therefore, that

$$
\overline{\alpha_0} = \tfrac{1}{3}\sum_p \alpha_p = \tfrac{1}{3}(\alpha_X+\alpha_Y+\alpha_Z). \tag{19.16}
$$

Thus the relative orientation of the two coordinate systems has vanished, and we are left with the one third of the sum of the principal values, which is of course a constant of the molecule.

The case of the anisotropy invariant γ_0 is rather more complicated. From the definition in eqn (19.15) we have

$$
\gamma_0^2 = \alpha_{xx}^2+\alpha_{yy}^2+\alpha_{zz}^2-\alpha_{xx}\alpha_{yy}-\alpha_{yy}\alpha_{zz}-\alpha_{zz}\alpha_{xx}+ \\
+3(\alpha_{xy}^2+\alpha_{yz}^2+\alpha_{zx}^2). \tag{19.17}
$$

Into this we have to substitute the values of the components from eqn (19.11). Instead of going to the trouble of setting out the whole rather cumbersome result, let us first pick out the coefficient in it of α_X^2. Each term of the right-hand side of eqn (19.17) makes a contribution, and writing these contributions in order we obtain

$$
\cos^4(x,X)+\cos^4(y,X)+\cos^4(z,X)-\cos^2(x,X)\cos^2(y,X)- \\
-\cos^2(y,X)\cos^2(z,X)-\cos^2(z,X)\cos^2(x,X)+ \\
+3\{\cos^2(x,X)\cos^2(y,X)+\cos^2(y,X)\cos^2(z,X)+ \\
+\cos^2(z,X)\cos^2(x,X)\}. \tag{19.18}
$$

This is easily seen to be

$$
\{\cos^2(x,X)+\cos^2(y,X)+\cos^2(z,X)\}^2,
$$

which by eqn (19.13) must have the value unity. We have thus shown that the coefficient of α_X^2 in γ_0^2 is unity. Clearly the coefficients of α_Y^2 and α_Z^2 must be the same as that of α_X^2, i.e. unity also. Now let us pick out similarly the coefficient of $\alpha_X\alpha_Y$.

Again using eqn (19.11) and writing in order the contributions from the successive terms of the right-hand side of eqn (19.17), we find

$$2\cos^2(x, X)\cos^2(x, Y) + 2\cos^2(y, X)\cos^2(y, Y) +$$
$$+ 2\cos^2(z, X)\cos^2(z, Y) -$$
$$- \{\cos^2(x, X)\cos^2(y, Y) + \cos^2(x, Y)\cos^2(y, X)\} -$$
$$- \{\cos^2(y, X)\cos^2(z, Y) + \cos^2(y, Y)\cos^2(z, X)\} -$$
$$- \{\cos^2(z, X)\cos^2(x, Y) + \cos^2(z, Y)\cos^2(x, X)\} +$$
$$+ 6\{\cos(x, X)\cos(y, X)\cos(x, Y)\cos(y, Y) +$$
$$+ \cos(y, X)\cos(z, X)\cos(y, Y)\cos(z, Y) +$$
$$+ \cos(z, X)\cos(x, X)\cos(z, Y)\cos(x, Y)\}.$$

As the reader may easily verify, this coefficient is of the form $3C^2 - AB$, where

$$A = \cos^2(x, X) + \cos^2(y, X) + \cos^2(z, X),$$
$$B = \cos^2(x, Y) + \cos^2(y, Y) + \cos^2(z, Y),$$

and

$$C = \cos(x, X)\cos(x, Y) + \cos(y, X)\cos(y, Y) + \cos(z, X)\cos(z, Y).$$

Now by eqn (19.13) we have $A = 1$. For similar reasons $B = 1$ also. By eqn (19.14) $C = 0$. Thus the value of the coefficient of $\alpha_X \alpha_Y$ in γ_0^2 reduces to minus unity. This must also be the coefficient of $\alpha_Y \alpha_Z$ and of $\alpha_Z \alpha_X$. Altogether, therefore, we have proved that

$$\gamma_0^2 = \alpha_X^2 + \alpha_Y^2 + \alpha_Z^2 - \alpha_X \alpha_Y - \alpha_Y \alpha_Z - \alpha_Z \alpha_X$$
$$= \tfrac{1}{2}\{(\alpha_X - \alpha_Y)^2 + (\alpha_Y - \alpha_Z)^2 + (\alpha_Z - \alpha_X)^2\}. \quad (19.19)$$

This is, of course, a constant of the molecule, and independent of its orientation relative to space-fixed axes.

19.6. Averages of the squares of the polarizability tensor components over all molecular orientations

Eqns (19.4), (19.5), and (19.6) give the values of ρ_n, $\rho_{p\perp}$, and $\rho_{p\parallel}$ for scattering by a single molecule with a fixed orientation in space. In order to adapt them for the practically important case of a gaseous sample, we have to find the averages of the squares

283

of the tensor components over all orientations of the principal axes (and hence of the scattering molecule) relative to the space-fixed axes of x, y, and z. Quantities averaged thus will be distinguished by a bar over their symbols. Using eqn (19.11) and supposing the averaging process to have been carried out, we obtain

$$\overline{\alpha_{ji}^2} = \sum_p \alpha_p^2 \overline{\cos^2(i,p)\cos^2(j,p)} +$$
$$+ \sum_{p \neq q} \sum \alpha_p \alpha_q \overline{\cos(i,p)\cos(j,p)\cos(i,q)\cos(j,q)}. \quad (19.20)$$

As before, i or $j = x$, y, or z, and p or $q = X$, Y, or Z. The principal values of α_p (or α_q) are of course constants. Therefore, in order to be able to evaluate $\overline{\alpha_{ji}^2}$, we need to find the average values $\overline{\cos^2(i,p)\cos^2(j,p)}$ and $\overline{\cos(i,p)\cos(j,p)\cos(i,q)\cos(j,q)}$, the latter in the case when $p \neq q$.

We start with the necessary relation between the cosines which is expressed in eqn (19.13). Squaring both sides of this equation, we have

$$\sum_i \cos^4(i,X) + \sum_{i \neq j} \sum \cos^2(i,X)\cos^2(j,X) = 1. \quad (19.21)$$

We now average over all orientations and make use of the facts that $\overline{\cos^4(i,X)}$ must clearly be the same for all i-values, i.e. for all three summands in the first summation, and that all six summands of the second summation must likewise be equal. We thus obtain

$$3\,\overline{\cos^4(i,X)} + 6\,\overline{\cos^2(i,X)\cos^2(j,X)} = 1. \quad (19.22)$$

Now the first term on the left-hand side of eqn (19.22) is easy to evaluate:

$$\overline{\cos^4(i,p)} = \frac{\int_0^\pi \int_0^{2\pi} \cos^4\theta \sin\theta \, d\theta d\phi}{\int_0^\pi \int_0^{2\pi} \sin\theta \, d\theta d\phi} = \frac{1}{5}. \quad (19.23)$$

Here p may have any of the values X, Y, or Z. Substituting from eqn (19.23) into eqn (19.22) we find

$$\overline{\cos^2(i,X)\cos^2(j,X)} = \tfrac{1}{6}(1-\tfrac{3}{5}) = \tfrac{1}{15}.$$

Clearly the same average would have been obtained with either Y or Z in place of X. The general result for $i \neq j$ is

$$\overline{\cos^2(i, p)\cos^2(j, p)} = \tfrac{1}{15}. \tag{19.24}$$

This is one of the results required for the evaluation of $\overline{\alpha_{ji}^2}$ by means of eqn (19.20).

To obtain the second average in eqn (19.20), we start with the necessary relation between the cosines which is expressed in eqn (19.14). Squaring both sides of this equation, we have

$$\sum_i \cos^2(i, X)\cos^2(i, Y) +$$
$$+ \sum_i \sum_j \cos(i, X)\cos(i, Y)\cos(j, X)\cos(j, Y) = 0. \tag{19.25}$$

We now average over all orientations. As far as the first term is concerned, arguments exactly analogous to those used above in evaluating $\overline{\cos^2(i, p)\cos^2(j, p)}$ show that the average value of $\cos^2(i, X)\cos^2(i, Y)$ is likewise 1/15. Since all six summands in the second term must be equal, we have

$$6\overline{\cos(i, X)\cos(i, Y)\cos(j, X)\cos(j, Y)} = -\tfrac{3}{15}.$$

Hence the average involved is equal to $-1/30$. It is clear that the same value would have been obtained if we had had any pair p, q (with $p \neq q$) instead of the pair X, Y. So the required result is

$$\overline{\cos(i, p)\cos(i, q)\cos(j, p)\cos(j, q)} = -\tfrac{1}{30}, \tag{19.26}$$

with $p \neq q$ and $i \neq j$.

We are now in a position to evaluate the averages of the squares of the components of the polarizability tensor. Let us take first the case of a 'diagonal' component α_{ii}. We use eqn (19.20), setting $j = i$, and obtain

$$\overline{\alpha_{ii}^2} = \sum_p \alpha_p^2 \,\overline{\cos^4(i, p)} + \sum_p \sum_q \alpha_p \alpha_q \,\overline{\cos^2(i, p)\cos^2(j, p)}.$$

Inserting the values of the averages (see eqns (19.23) and (19.24) respectively) we obtain

$$\overline{\alpha_{ii}^2} = \tfrac{1}{5}(\alpha_X^2 + \alpha_Y^2 + \alpha_Z^2) + \tfrac{1}{15}(2\alpha_X \alpha_Y + 2\alpha_Y \alpha_Z + 2\alpha_Z \alpha_X).$$

It is usual to express this average value in terms of the two invariants $\overline{\alpha_0}$ and γ_0 whose values are given in eqns (19.16) and (19.19) respectively. The reader may easily confirm that

$$\overline{\alpha_{ii}^2} = \tfrac{1}{45}\{45(\overline{\alpha_0})^2 + 4\gamma_0^2\}. \tag{19.27}$$

The value is obviously the same for all three values of i, and so eqn (19.27) holds for $\overline{\alpha_{xx}^2}$, $\overline{\alpha_{yy}^2}$, and $\overline{\alpha_{zz}^2}$.

It remains only to evaluate the corresponding value for an 'off-diagonal' component. Again using eqn (19.20), and inserting the values of the averages from eqns (19.24) and (19.26), we obtain

$$\overline{\alpha_{ji}^2} = \tfrac{1}{15}(\alpha_X^2 + \alpha_Y^2 + \alpha_Z^2) - \tfrac{1}{30}(2\alpha_X\,\alpha_Y + 2\alpha_Y\,\alpha_Z + 2\alpha_Z\,\alpha_X). \tag{19.28}$$

The reader will see, by comparing it with eqn (19.19), that the right-hand side of eqn (19.28) is simply equal to $(1/15)\gamma_0^2$. Thus we have found that

$$\overline{\alpha_{ji}^2} = \tfrac{1}{15}\gamma_0^2. \tag{19.29}$$

Of course, all 'off-diagonal' elements give the same average.

19.7. Degrees of depolarization of Rayleigh scattering by free molecules

Eqns (19.4), (19.5), and (19.6) give the values of the degrees of depolarization of Rayleigh scattering by a single molecule with a fixed orientation in space. In order to make them applicable to the experimentally observable scattering from a sample containing a large number of free molecules with random orientations, we have to replace the squares of the tensor components by the corresponding averages. These are given, in terms of the invariants $\overline{\alpha_0}$ and γ_0, in eqns (19.27) and (19.29). The total intensity of Rayleigh scattering, as well as the intensities I_y and I_z of the polarized components involved in the degrees of depolarization (Fig. 19.2 and eqn (19.4)), will be proportional to the number of scattering molecules in the sample, but this number will cancel out as far as the ρ values are concerned.

These values are easily found to be

$$\rho_n = \frac{6\gamma_0^2}{45(\overline{\alpha_0})^2 + 7\gamma_0^2}, \tag{19.30}$$

$$\rho_{p\perp} = \frac{3\gamma_0^2}{45(\overline{\alpha_0})^2 + 4\gamma_0^2}, \tag{19.31}$$

$$\rho_{p\parallel} = 1. \tag{19.32}$$

The mean-value invariant $\overline{\alpha_0}$ of the polarizability tensor α_0 can never vanish, for all molecules are electrically polarizable to some extent. On the other hand, the anisotropy invariant γ_0 may be zero. This is the case for a molecule that is electrically isotropic, i.e. for which the three principal values of the polarizability, α_X, α_Y, and α_Z, are equal—as will be seen from the definition of γ_0 in eqn (19.19). From eqn (19.30) we see that $\rho_n = 0$ for the Rayleigh scattering from isotropic molecules. From eqn (19.31) we see that $\rho_{p\perp} = 0$ also. Indeed these results would be equally true if the isotropic molecules had any arbitrary fixed orientation in space, for the directions of the principal axes of such a molecule are not defined, and all orientations in space are equivalent with respect to polarizability. Equation (19.32) is not applicable to isotropic species. The value of unity which it gives for $\rho_{p\parallel}$ is the ratio of two intensities, each of which is proportional to γ_0^2 and therefore zero when γ_0 is zero. Thus no Rayleigh scattering occurs under the conditions normally used for the observation of $\rho_{p\parallel}$. For anisotropic scattering species eqns (19.30) and (19.31) show that, since $\overline{\alpha_0}$ can never be zero, ρ_n must always be less than 6/7 and $\rho_{p\perp}$ always less than 3/4. In general, therefore, we conclude that

$$\tfrac{6}{7} > \rho_n \geqslant 0 \tag{19.33}$$

and $\qquad \tfrac{3}{4} > \rho_{p\perp} \geqslant 0. \tag{19.34}$

When the incident light is natural (i.e. unpolarized), the total intensity of Rayleigh scattering at 90° to the incident direction (i.e. the sum of the two polarized components) is proportional to the number of scattering molecules in the sample and—as shown by eqn (19.30)—to $45(\overline{\alpha_0})^2 + 13\gamma_0^2$. When the incident light is plane-polarized and the observation is in the direction perpendicular to the incident electric vector, the total intensity—as shown by eqn (19.31)—is proportional to $45(\overline{\alpha_0})^2 + 7\gamma_0^2$. Finally when the incident light is plane-polarized and the scattering is observed in the direction parallel to the incident electric vector, the total intensity (with the same proportionality factor as before) is proportional to $6\gamma_0^2$. In this case, since the degree of depolarization $\rho_{p\parallel}$ is unity, the Rayleigh-scattered light is unpolarized.

19.8. Degrees of depolarization of Raman scattering by free molecules

In Sections 18.3 and 18.4 we saw that (except for the presence of a proportionality factor) the Raman scattering for a particular vibrational mode of normal coordinate Q_k is determined by the derived polarizability tensor $(\partial\alpha/\partial Q_k)_0$ in precisely the same way as the Rayleigh scattering is determined by the polarizability tensor α_0. Throughout the present discussion we shall be concerned only with a single normal mode, and so we may for simplicity omit the subscript k which differentiates it from the other modes. We may accordingly write the derived polarizability tensor as $(\partial\alpha/\partial Q)_0$ or simply as α'.

Like the polarizability itself, α' is a symmetric tensor and has a mean-value invariant $\overline{\alpha'}$ and an anisotropy invariant γ'. Their definitions are analogous to those of the invariants of α_0 in eqns (19.16) and (19.19). Moreover, the averages of the squares of the components α'_{ij} of α' are precisely analogous to those of α_0 in eqns (19.27) and (19.29).

There is, however, one very important difference between α_0 and α'. Since all molecules are polarizable, all the 'diagonal' components of α_0 must be positive, and so $\overline{\alpha_0}$ (which is equal to one third of the sum of these components) must also be positive. For $(\partial\alpha/\partial Q)_0$, however, all the components are rates of change with respect to Q, and so may be either positive or negative or zero. It follows that in certain cases $\overline{\alpha'}$ may actually vanish. This is a notable difference as compared with $\overline{\alpha_0}$, for in view of

288

the fact that γ' (like γ_0) may also vanish, it means that in certain circumstances both $\overline{\alpha'}$ and γ' may be simultaneously zero. In this case the Raman scattering will have no intensity or, in other words, will be forbidden. Rayleigh scattering, in contrast, is never forbidden.

For Raman scattering (always at 90° to the direction of the incident light) by a single molecule with a fixed orientation in space, the expressions for the degrees of depolarization are exactly similar to eqns (19.4), (19.5), and (19.6), but with α'_{ij} in place of α_{ij}. When modified so as to be applicable to scattering by a sample containing a large number of free molecules with random orientations, they become analogous to eqns (19.30), (19.31), and (19.32). Thus in place of eqn (19.30), which refers to Rayleigh scattering, we have for Raman scattering

$$\rho_n = \frac{6(\gamma')^2}{45(\overline{\alpha'})^2 + 7(\gamma')^2}. \tag{19.35}$$

The corresponding total intensity (i.e. the sum of the intensities of the two plane-polarized components) is proportional to $45(\overline{\alpha'})^2 + 13(\gamma')^2$. Here the constant of proportionality includes not only the number of scattering molecules, but also a suitable factor which takes account of their Boltzmann distribution amongst the initial levels involved and the magnitudes of the transition moments from these levels. However, the whole factor of proportionality cancels out in the expression for the degree of depolarization, which is essentially a ratio of two intensities. Because $\overline{\alpha'}$ may vanish without γ' vanishing, the value of ρ_n may actually reach $6/7$. For the Raman scattering, therefore, we can write

$$\tfrac{6}{7} \geqslant \rho_n \geqslant 0. \tag{19.36}$$

This should be contrasted with (19.33) for Rayleigh scattering. When $\rho_n = 6/7$ for a Raman line, the line is conventionally described as *depolarized*. Any line for which $\rho_n < 6/7$ is said to be *polarized*. In particular, when $\rho_n = 0$ the line is said to be completely polarized.

As we shall see in detail later, it is molecular symmetry which determines, not only whether a certain normal mode shall be permitted or forbidden in the Raman effect, but also whether (if it is permitted) its Raman line shall be depolarized or polarized. In fact the value of ρ_n is related to the symmetry species of the mode, and is therefore a very important piece of experimental evidence in connection with frequency assignments.

For Raman scattering the analogue of eqn (19.31) is

$$\rho_{p\perp} = \frac{3(\gamma')^2}{45(\overline{\alpha'})^2 + 4(\gamma')^2}. \tag{19.37}$$

We may accordingly write

$$\tfrac{3}{4} \geqslant \rho_{p\perp} \geqslant 0. \tag{19.38}$$

This should be contrasted with (19.34) for Rayleigh scattering. The total intensity corresponding to eqn (19.37) is proportional to $45(\overline{\alpha'})^2 + 7(\gamma')^2$.

For $\rho_{p\parallel}$ the result for Raman scattering is identical with that for Rayleigh scattering, as given in eqn (19.32), i.e. $\rho_{p\parallel} = 1$. The Raman-scattered light, like the Rayleigh-scattered, is unpolarized and its total intensity (with the same proportionality factor as before) is proportional to $6(\gamma')^2$.

19.9. Alternative experimental methods for measuring ρ_n

The results of the preceding section provide the basis for two alternative (but essentially equivalent) experimental methods for determining ρ_n values in Raman spectroscopy. The first of these uses natural incident light and makes separate intensity measurements on the two plane-polarized components of the scattered light, after successively isolating each of them by interposing a suitable optical device in the scattered beam. This gives the individual values of I_y and I_z (in Fig. 19.2) on the same arbitrary scale. Then

$$\rho_n = \frac{I_y}{I_z} = \frac{6(\gamma')^2}{45(\overline{\alpha'})^2 + 7(\gamma')^2}.$$

The second method uses plane-polarized incident light, and measures successively the total intensities scattered in directions respectively perpendicular and parallel to that of the incident electric vector. In the preceding section we saw that these total intensities are respectively proportional to $45(\overline{\alpha'})^2+7(\gamma')^2$ and $6(\gamma')^2$. Thus

$$\frac{(I_{\text{total}})_{\parallel}}{(I_{\text{total}})_{\perp}} = \frac{6(\gamma')^2}{45(\overline{\alpha'})^2+7(\gamma')^2} = \rho_n.$$

For practical reasons the second (slightly less direct) method is often preferred. When it is used, the directions of incidence and scattering are kept the same throughout, and for the measurements of $(I_{\text{total}})_{\perp}$ and $(I_{\text{total}})_{\parallel}$ the incident light is polarized in the yz plane and in the xy plane respectively.

20 Symmetry properties of wave functions

20.1. The effect of a symmetry operation upon a function

W E shall be concerned in this chapter with eigenfunctions ψ of the molecular vibrational wave equation, which (in the simple-harmonic case) are functions of the normal coordinates Q. (The considerations of the present section, however, are not confined to this particular kind of function, but apply generally to any function of a set of configurational coordinates.) An orthonormal basis having been chosen for configuration space, each point P in it is associated with a particular set of values of the co-ordinates Q, and therefore with a particular value $\psi(P)$ of the function ψ. We may thus think of the function as attaching to each point of configuration space (or every infinitesimal volume element thereof) a particular number, which is the value of the function at that point (or in that volume element). According to this view the function is to be thought of as constituting a 'cloud' or continuum of values, extending throughout the whole of the configurational space. A different function of the coordinates Q would also be thought of as such a continuum, but with different values in each volume element. A number of different functions could thus be likened, in certain respects, to a number of different gases contained in the same vessel, though of course, like any other simile, this one must not be pushed too far.

Consider now the consequence of performing a symmetry operation R of the molecular point group. As we have discussed at length earlier, the effect upon the point P is to move it to another point P' in the same space:

$$P \xrightarrow{R} P' = \mathscr{R}P. \qquad (20.1)$$

If the sets of coordinates associated with the points P and P' are written as single-column matrices (i.e. vector columns) then the operator $\mathscr{R}$ takes the form of premultiplication by an orthogonal matrix. By analogy with eqn (20.1), we may express the effect of R upon the function ψ by writing

$$\psi \xrightarrow{\ R\ } \psi' = \mathscr{R}\psi, \qquad (20.2)$$

where $\mathscr{R}$ is the appropriate operator. It is necessary, however, to define exactly what we mean by ψ'. Now as the operation R moves P to P', it is natural to regard it as also moving to P' the value of ψ at P. On this view, the effect of R is to attach to every point (or volume element) of the space a different number from the one which was previously attached to it by the function ψ. The corresponding definition of ψ' follows at once, namely

$$\psi'(P') = \psi(P). \qquad (20.3)$$

This very important definition is the one which is almost universally adopted as the foundation for the application of symmetry theory to functions, and we shall adopt it in all future discussions.

If in any particular case we know in detail how a symmetry operation affects the points of configuration space, we shall be able by using eqn (20.3) to find the form of the function ψ' into which the operation transforms a function ψ of known form. We shall be dealing later with examples of this.

20.2. Invariance of the Hamiltonian operator under symmetry operations

In the classical treatment of molecular vibrations we took due account of the important fact that the potential energy V and the kinetic energy T are unaffected by the performance upon the distorted molecule of a symmetry operation characteristic of its equilibrium configuration. Another way of stating this is to say that the classical Hamiltonian H is invariant with respect to such symmetry operations. Now in wave mechanics the Hamiltonian operator $\mathscr{H}$ (like H) is invariant under any symmetry operation of the molecular point group. This may be seen as follows.

In Section 17.2 it was seen that analogous to $H = T + V$, we have $\mathscr{H} = \mathscr{T} + \mathscr{V}$, where $\mathscr{V}$ signifies simply multiplication by V and

$$\mathscr{T} = -\frac{h^2}{8\pi^2} \sum_k \frac{\partial^2}{\partial Q_k^2}.$$

Clearly the invariance of $\mathscr{V}$ under any symmetry operation follows at once from the invariance of V. In order to establish the invariance of $\mathscr{T}$, it is sufficient to prove that $\sum_k (\partial^2/\partial Q_k^2)$ is invariant. A simple example will suffice to indicate the nature of the general proof. We shall take the case of just two normal coordinates Q_a and Q_b and a symmetry operation R which transforms them according to the matrix equation

$$\begin{bmatrix} Q_a' \\ Q_b' \end{bmatrix} = \begin{bmatrix} R_{11} & R_{12} \\ R_{21} & R_{22} \end{bmatrix} \begin{bmatrix} Q_a \\ Q_b \end{bmatrix}$$

so that we may write

$$Q_a' = R_{11} Q_a + R_{12} Q_b, \tag{20.4}$$

$$Q_b' = R_{21} Q_a + R_{22} Q_b. \tag{20.5}$$

From eqns (20.4) and (20.5) we find that

$$\frac{\partial}{\partial Q_a} = \frac{\partial Q_a'}{\partial Q_a} \frac{\partial}{\partial Q_a'} + \frac{\partial Q_b'}{\partial Q_a} \frac{\partial}{\partial Q_b'}$$

$$= R_{11} \frac{\partial}{\partial Q_a'} + R_{21} \frac{\partial}{\partial Q_b'}$$

$$= D \text{ (an arbitrarily chosen symbol)}.$$

Hence

$$\frac{\partial^2}{\partial Q_a^2} = R_{11} \frac{\partial D}{\partial Q_a'} + R_{21} \frac{\partial D}{\partial Q_b'}$$

$$= R_{11}^2 \frac{\partial^2}{(\partial Q_a')^2} + 2 R_{11} R_{21} \frac{\partial^2}{\partial Q_a' \partial Q_b'} + R_{21}^2 \frac{\partial^2}{(\partial Q_b')^2}. \tag{20.6}$$

In exactly the same way we find that

$$\frac{\partial^2}{\partial Q_b^2} = R_{12}^2 \frac{\partial^2}{(\partial Q_a')^2} + 2 R_{12} R_{22} \frac{\partial^2}{\partial Q_a' \partial Q_b'} + R_{22}^2 \frac{\partial^2}{(\partial Q_b')^2}. \tag{20.7}$$

Adding eqns (20.6) and (20.7), we obtain

$$\frac{\partial^2}{\partial Q_a^2}+\frac{\partial^2}{\partial Q_b^2} = (R_{11}^2+R_{12}^2)\frac{\partial^2}{(\partial Q_a')^2}+2(R_{11}\,R_{21}+R_{12}\,R_{22})\frac{\partial^2}{\partial Q_a'\,\partial Q_b'}+$$

$$+(R_{21}^2+R_{22}^2)\frac{\partial^2}{(\partial Q_b')^2}. \quad (20.8)$$

Now the matrix R of the symmetry operation is necessarily orthogonal, and therefore $R_{11}^2+R_{12}^2 = R_{21}^2+R_{22}^2 = 1$ and $R_{11}\,R_{21}+R_{12}\,R_{22} = 0$. Thus eqn (20.8) reduces to

$$\frac{\partial^2}{\partial Q_a^2}+\frac{\partial^2}{\partial Q_b^2} = \frac{\partial^2}{(\partial Q_a')^2}+\frac{\partial^2}{(\partial Q_b')^2},$$

whence it follows, in this simple case, that $\mathscr{T}$ is invariant under the operation R. Extension of the argument provides a general proof of the invariance of $\mathscr{T}$. Since $\mathscr{V}$ is also invariant, so also must be the Hamiltonian operator $\mathscr{H}$.

20.3. The concept of function space

Consider now the effect of a symmetry operation R upon the Schrödinger wave equation which, as we saw in Section 17.2, may be written in the form $\mathscr{H}\psi = W\psi$. Because of the invariance of the operator $\mathscr{H}$ and the factor W, we may write

$$\mathscr{H}\psi' = W\psi', \quad (20.9)$$

where (as before) ψ' is the function into which ψ is transformed by R. From eqn (20.9) it follows that the transformed function ψ' is an eigenfunction of the wave equation with the same eigenvalue (i.e. the same energy) as ψ.

This important result makes it advantageous to invoke again the analogy between eigenfunctions and ordinary vectors which we first encountered in Section 17.6 in connection with the idea of the mutual orthogonality of eigenfunctions. We saw in Section 17.6 that all wave functions having a particular eigenvalue are expressible as linear combinations of a certain number of linearly independent (mutually orthogonal) wave functions. This number is the degeneracy of the energy level in question. We can accordingly extend the analogy by regarding eigenfunctions with the same eigenvalue as belonging to a purely conceptual *function space*, just as we regarded ordinary vectors as belonging to a vector space.

Of course this function space is an entirely different concept from that of configuration space, and the view of a function as the analogue of an ordinary vector is entirely different from the view of it as a continuum of values in configuration space. It is perhaps not altogether easy at first to keep both pictures before the mind's eye at the same time, but any initial difficulty of this kind soon vanishes with increasing familiarity and with increasing realization of the respective advantages of the two views and the close relationship between them.

As an orthonormal basis of the conceptual function space we take a set of mutually orthogonal normalized functions (analogous to a set of mutually orthogonal unit vectors in the case of an ordinary vector space). Then the 'line' drawn from the origin of the associated coordinate system to any point F in the function space represents a particular function ψ, and the coordinates of F are the magnitudes of the components of ψ relative to the chosen basis. The possibility of thus extending the analogy between eigenfunctions and ordinary vectors arises, of course, from the analogy (see Section 17.4) between the integral $\int \psi^{(i)}\psi^{(j)} \, d\tau$ and the dot product of two ordinary vectors; for upon this is based the idea of the analogues for eigenfunctions of mutual orthogonality and magnitude, as understood for ordinary vectors.

We now see how the 'vector view' of an eigenfunction is linked to the 'continuum view'. In thinking of the integral $\int \psi^{(i)}\psi^{(j)} \, d\tau$, one has in mind the continuum view of each of the functions, i.e. their values at each volume element throughout configuration space, and the integration over all such elements. Thereafter one goes over to the vector picture, taking with one the results of the previous continuum integrations, and now interpreting them as the analogues of dot products, i.e. as supplying information on the mutual orthogonality and the magnitudes of the eigenfunctions.

As a simple illustration of a vector space, take the level with $v = 1$ for a doubly degenerate vibrational mode. Here the space is two-dimensional. If ψ_a and ψ_b are the orthonormal basis functions, any eigenfunction of the level must be of the form $A\psi_a + B\psi_b$, where A and B are numerical factors. Such a function is not in general normalized. It is associated with the point F in the function space with the coordinates A and B; or it may be thought of as represented by the directed 'line' (analogous to an ordinary vector) from the origin to this point. Now consider the effect upon this function ψ of a symmetry operation R. We have already proved that ψ is transformed into a new function ψ' which has the same eigenvalue as ψ. In terms of the function space, this means that the point F is moved by R to a new point F' *in the same space*; or, otherwise expressed, that R transforms the vector-like line OF into another vector-like line OF'. If we write the coordinates of F and F' as single-column matrices $\mathbf{F}$ and $\mathbf{F}'$ respectively, we may express the effect of R in the matrix form

$$\mathbf{F}' = \mathbf{RF}. \tag{20.10}$$

In order to complete the analogy with ordinary vectors, it remains to show that $\mathbf{R}$ is necessarily an orthogonal matrix. This will be true if the 'magnitude' of ψ' is the same as that of ψ, i.e. if

$$\int (\psi')^2 \, d\tau = \int \psi^2 \, d\tau. \tag{20.11}$$

Now, reverting for the moment to the 'continuum view', and bearing in mind eqn (20.3), we see that in the transformation of ψ into ψ' the value of ψ in any volume element $d\tau$ of configuration space is simply transferred to some other volume element; consequently, since the integrations in eqn (20.11) are both extended over all such volume elements, they must obviously be equal to one another. We have thus proved that the matrix $\mathbf{R}$ in eqn (20.10) is orthogonal, and we see that there is a complete analogy between the transformation of ψ and the transformation of an ordinary vector under the symmetry operation R. For this reason the meaning of the term 'vector' is sometimes extended so as to include functions as well as ordinary vectors, and the term 'vector space' is taken to include what we have called function space. This is entirely justified, of course, in the interests of mathematical generalization. In this book, however, we shall find it more convenient to continue to use the term 'vector' in its restricted sense.

20.4. Wave functions as bases for matrix representations of point groups

Consider the function space to which belong all the vibrational eigenfunctions of a system which have a particular eigenvalue, i.e. which correspond to the same energy level. In the preceding section we have recognized the complete analogy with a vector space of the same dimensions. A wave function of the function space is transformed just like an ordinary vector, i.e. each symmetry operation is represented by an orthogonal matrix. Also if $\mathbf{R}$, $\mathbf{S}$, and $\mathbf{T}$ are the matrices representing the operations R, S, and T respectively, and if $SR = T$, then $\mathbf{SR} = \mathbf{T}$. It follows that a set of mutually orthogonal normalized wave functions, which spans the function space, will form an orthonormal basis for a matrix representation of the molecular point group. The dimension of the representation (or of the function space) will be the degeneracy of the energy level concerned. In general the representation will be an irreducible one, its dimension (1, 2, or 3) being determined entirely by the molecular symmetry. However, as we shall see, special circumstances may give rise to additional degeneracy of the level. The representation based upon an orthonormal set of wave functions will then be of correspondingly higher dimension, and so reducible. We shall discuss examples of this kind in Chapter 22 when we come to consider levels with $v > 1$ for degenerate simple-harmonic modes.

20.5. Symmetry species of simple-harmonic wave functions for levels with $v = 0$

As very simple examples of the use of the fundamental eqn (20.3), we shall first consider the symmetry species of simple-harmonic wave functions for levels with $v = 0$, where v is the vibrational quantum number. Substituting this value of v into the general expression given in eqn (17.22), and using the value of H_0 from eqn (17.24), we obtain

$$\psi_0(Q) = N_0 \exp(-\tfrac{1}{2}\gamma Q^2). \tag{20.12}$$

From eqn (20.12) we note that

$$\psi_0(-Q) = \psi_0(Q). \tag{20.13}$$

If Q is non-degenerate, then

either $\qquad\qquad Q \xrightarrow{\ R\ } Q' = Q \qquad\qquad (20.14)$

or $\qquad\qquad Q \xrightarrow{\ R\ } Q' = -Q. \qquad\qquad (20.15)$

In the case of eqn (20.14), the fundamental equation (20.3) tells us at once that $\psi_0' = \psi_0$. In the case of eqn (20.15), the relation expressed in eqn (20.13) gives

$$\psi_0(Q) = \psi_0(-Q) = \psi_0(Q'). \tag{20.16}$$

But by the fundamental equation (20.3) we have

$$\psi_0'(Q') = \psi_0(Q). \tag{20.17}$$

Combining eqns (20.16) and (20.17) we conclude that

$$\psi_0'(Q') = \psi_0(Q'),$$

i.e. that $\psi_0' = \psi_0$. Thus in both the cases represented by eqns (20.14) and (20.15), the function ψ_0 remains unchanged under all the operations of the group. In other words, if Q is non-degenerate, ψ_0 belongs to the totally symmetric species.

It remains to consider degenerate vibrational modes. Take for example the case of double degeneracy with the pair of normal coordinates Q_a and Q_b. The wave function of the level with $v = 0$ has the form

$$\psi_{00} = \psi_0(Q_a)\psi_0(Q_b) = N_{00} \exp\{-\tfrac{1}{2}\gamma(Q_a^2+Q_b^2)\}. \tag{20.18}$$

Every symmetry operation R leaves unchanged the value of $Q_a^2+Q_b^2$ and hence also the value of ψ_{00}. This means that, if R moves the point P of coordinate space (with coordinates Q_a, Q_b) to the point P' (with coordinates Q_a', Q_b'), then

$$\psi_{00}(P') = \psi_{00}(P). \tag{20.19}$$

The fundamental equation (20.3) states that

$$\psi_{00}'(P') = \psi_{00}(P).$$

Thus from eqn (20.3) and eqn (20.19) we conclude that

$$\psi'_{00}(P') = \psi_{00}(P'),$$

i.e. that $\psi'_{00} = \psi_{00}$. As with non-degenerate vibrations, the wave function with $v = 0$ remains unaffected by all the operations of the group. Similar arguments lead to the same result in the case of triply degenerate modes.

The general conclusion is therefore as follows. Simple-harmonic wave functions of so-called ground levels (i.e. those with $v = 0$) are always totally symmetric, no matter what the symmetry species of the vibrations themselves.

20.6. Symmetry species of simple-harmonic wave functions for levels with $v = 1$

We consider first the case where the vibrational mode is non-degenerate. From the general expression in eqn (17.22), and the value of H_1 in eqn (17.24), we obtain

$$\psi_1(Q) = N_1 \exp(-\tfrac{1}{2}\gamma Q^2).2\gamma^{\frac{1}{2}}Q. \tag{20.20}$$

We note that

$$\psi_1(-Q) = -\psi_1(Q). \tag{20.21}$$

The right-hand side of eqn (20.20) is totally symmetric, with the possible exception of the final factor Q. If Q is totally symmetric, then ψ_1 will be totally symmetric. If Q belongs to some other non-degenerate species, then there will be some symmetry operation R for which $Q \xrightarrow{R} Q' = -Q$. In a case of this kind, eqn (20.21) gives

$$\psi_1(Q') = \psi_1(-Q) = -\psi_1(Q), \tag{20.22}$$

while according to the fundamental eqn (20.3)

$$\psi'_1(Q') = \psi_1(Q).$$

Combining eqns (20.22) and eqn (20.3), we conclude that $\psi'_1(Q') = -\psi_1(Q')$, i.e. that $\psi'_1 = -\psi_1$. Thus whenever an operation reverses the sign of Q it also reverses the sign of ψ_1. In other words, ψ_1 belongs to the same symmetry species as does Q.

As an instance of a degenerate vibration, consider the doubly degenerate case with normal coordinates Q_a and Q_b. Here the energy level with $v = 1$ is realizable in two ways: first, by having

$v_a = 1\,$and $v_b = 0$ and secondly by having $v_a = 0$ and $v_b = 1$. As orthonormal basis functions we choose the wave functions $\psi_{10} = N_{10}(\psi_1)_a(\psi_0)_b$ and $\psi_{01} = N_{01}(\psi_0)_a(\psi_1)_b$. From eqns (17.22) and (17.24) we have

$$\psi_{10} = N_{10}\exp\{-\tfrac{1}{2}\gamma(Q_a^2+Q_b^2)\}.2\gamma^{\frac{1}{2}}Q_a = KQ_a, \quad (20.23)$$

$$\psi_{01} = N_{01}\exp\{-\tfrac{1}{2}\gamma(Q_a^2+Q_b^2)\}.2\gamma^{\frac{1}{2}}Q_b = KQ_b, \quad (20.24)$$

where K is a factor which is totally symmetric. Under the group operation R let the coordinates Q_a and Q_b be transformed according to the matrix equation $\mathbf{Q}' = \mathbf{RQ}$, where $\mathbf{R}$ is the matrix representing the operation R. We will designate the elements of $\mathbf{R}$ in the usual way:

$$\mathbf{R} = \begin{bmatrix} R_{11} & R_{12} \\ R_{21} & R_{22} \end{bmatrix}.$$

Thus if $\boldsymbol{e}_a$ and $\boldsymbol{e}_b$ are the mutually orthogonal unit vectors forming the basis of the Q-space, we may write

$$\boldsymbol{e}_a' = R_{11}\,\boldsymbol{e}_a + R_{21}\,\boldsymbol{e}_b \quad (20.25)$$

and

$$\boldsymbol{e}_b' = R_{12}\,\boldsymbol{e}_a + R_{22}\,\boldsymbol{e}_b. \quad (20.26)$$

From the fundamental eqn (20.3) and eqn (20.23) we have

$$\psi_{10}'(Q_a', Q_b') = \psi_{10}(Q_a, Q_b) = KQ_a. \quad (20.27)$$

Now from $\mathbf{Q}' = \mathbf{RQ}$, we have $\mathbf{Q} = \mathbf{R}^{-1}\mathbf{Q}' = \mathbf{R}^{\dagger}\mathbf{Q}'$, whence

$$Q_a = R_{11}Q_a' + R_{21}Q_b'. \quad (20.28)$$

Substitution of the value of Q_a from eqn (20.28) into eqn (20.27) gives

$$\psi_{10}'(Q_a', Q_b') = KR_{11}Q_a' + KR_{21}Q_b' = R_{11}\psi_{10}(Q_a') + R_{21}\psi_{01}(Q_b'). \quad (20.29)$$

In an exactly similar way we obtain

$$\psi_{01}'(Q_a', Q_b') = R_{12}\psi_{10}(Q_a') + R_{22}\psi_{01}(Q_b'). \quad (20.30)$$

Comparing eqn (20.29) with eqn (20.25), and eqn (20.30) with eqn (20.26), we see that the basis wave functions of the function space transform in precisely the same way as do the basis vectors of the Q-space.

The general conclusion is therefore as follows. Simple-harmonic wave functions of so-called fundamental levels (i.e.

those with $v = 1$) always belong to the same symmetry species as the respective vibrational modes themselves.

20.7. Symmetry species of the wave functions of anharmonic vibrations

In Sections 20.5 and 20.6 we have made use of the special forms of simple-harmonic wave functions, the symmetry properties of which depend essentially upon those of the relevant Hermite polynomials. However, it is of great importance to realize that the conclusions reached transcend the limitations of the special simple-harmonic assumption, and apply with equal force to the corresponding wave functions of an anharmonic vibrator. The underlying reason for this can be seen as follows.

The introduction of a certain degree of anharmonicity into a previously simple-harmonic system means a change in the potential energy V. There will of course be consequent changes in the classical modes of vibration and in the associated wave functions. Nevertheless, the important symmetry property of V, namely, its invariance under any symmetry operation of the molecular point group, will be entirely unaffected. Consequently, when we solve the changed wave equation, the changed eigenfunctions will have exactly the same symmetry properties as did the corresponding simple-harmonic eigenfunctions.

Perhaps a more immediately convincing way of reaching the same conclusion is to start with the known mathematical fact that any eigenfunction for an anharmonic system can always be expressed as a linear combination of the complete set of simple-harmonic eigenfunctions. Moreover, all the simple-harmonic eigenfunctions which appear together in one such linear combination must transform in the same way as one another under every symmetry operation of the group. In real molecules, where the degree of anharmonicity is quite small, the anharmonic wave function will resemble rather closely what we have called the corresponding simple-harmonic function, and the latter will undoubtedly make a major contribution to the relevant linear combination. Thus we see that the anharmonic wave function in question must have the same symmetry properties (i.e. belong to the same species) as does the corresponding simple-harmonic one.

In dealing with real molecules, therefore, it will be permissible to invoke the symmetry species of the corresponding harmonic wave functions in all discussions where symmetry properties play a part. This is indeed a most convenient circumstance, for the simple-harmonic functions are of known mathematical form, whereas in order to find the form of the anharmonic functions it would be necessary to determine the exact nature of the potential field and to solve the relevant wave equation—a difficult, if not impossible, undertaking.

21 The general selection rule for fundamentals in infra-red absorption

21.1. Introduction

THE total intensity of a transition between quantum states with the eigenfunctions $\psi^{(n)}$ and $\psi^{(m)}$ is proportional to the square of the magnitude of the transition moment $\int \psi^{(n)} \boldsymbol{\mu} \psi^{(m)} \, d\tau$, where $\boldsymbol{\mu}$ is the molecular electric dipole moment. In greater detail, it is proportional to the sum of the squares of the components $\int \psi^{(n)} \mu_i \psi^{(m)} \, d\tau$, where μ_i is a component of $\boldsymbol{\mu}$ and $i = x$, y, or z. A vibrational frequency is forbidden in infra-red absorption (or emission) if all three of the components of the transition moment vanish. Whether or not this happens depends upon the symmetry properties of the integrands, which are functions of the positional displacement coordinates of the molecule—in the simple-harmonic case functions of the normal vibrational coordinates. In order to discuss this further, and so to arrive at general selection rules, we must first consider the symmetry properties of the three factors which make up the integrand $\psi^{(n)} \mu_i \psi^{(m)}$, each of which is in itself a function of the displacement coordinates. In the preceding chapter we have already discussed the symmetry properties of the simple-harmonic wave functions for levels with the vibrational numbers $v = 0$ and $v = 1$, i.e. for the levels involved in the infra-red absorption of fundamentals, where the transition is $v = 0 \rightarrow v = 1$. We must now consider the symmetry properties of the components μ_i of the molecular electric dipole moment, and then discuss how the behaviour of the transition moment integral $\int \psi^{(n)} \mu_i \psi^{(m)} \, d\tau$ (where $\psi^{(n)}$ and $\psi^{(m)}$ are respectively the wave functions of the ground and fundamental levels) is determined by the over-all symmetry of the integrand. For the time being we will assume that the vibrational modes are simple-harmonic, but because the

304

selection rule we shall deduce will be based solely upon symmetry considerations, its validity will not be restricted to this particular type of vibration.

21.2. Symmetry species of the components of the molecular electric dipole moment

The electric dipole moment of a molecule is a vector in physical space. If e_x, e_y, and e_z are unit vectors along three Cartesian axes, the dipole moment vector is representable as

$$\mu_x\, e_x + \mu_y\, e_y + \mu_z\, e_z,$$

where μ_x, μ_y, and μ_z are functions of the displacement coordinates of all the nuclei. Thus we may write

$$\mu_x(P) = \sum_t e^{(t)} x^{(t)}, \tag{21.1}$$

where P is a point in configuration space, $e^{(t)}$ is the effective electric charge associated with the tth nucleus, $x^{(t)}$ is the x-displacement of this nucleus corresponding to the point P, and the summation covers all the nuclei in the molecule. Now consider the effect of the symmetry operation R of the molecular point group. This will transform the displacement coordinates of site t and in general will move them to a new site t'. After the operation, as before it, the two sites t and t' are necessarily associated with equivalent nuclei, and these will necessarily have the same effective charge, i.e. $e^{(t')} = e^{(t)}$. The transformed coordinates at the site t' and the original ones at site t are related by the matrix equation

$$\begin{bmatrix} x^{(t')} \\ y^{(t')} \\ z^{(t')} \end{bmatrix} = \begin{bmatrix} R_{xx} & R_{xy} & R_{xz} \\ R_{yx} & R_{yy} & R_{yz} \\ R_{zx} & R_{zy} & R_{zz} \end{bmatrix} \begin{bmatrix} x^{(t)} \\ y^{(t)} \\ z^{(t)} \end{bmatrix}, \tag{21.2}$$

where the 3×3 matrix represents the operation R in the group representation which is based upon unit vectors in the x-, y-, and z-directions. This is the representation to which the translations T_x, T_y, and T_z of the molecule belong. In consequence of the transformations of the coordinates of the individual nuclei, as given in eqn (21.2), suppose that the point P in configuration space is moved to the new point P', and that the function μ_x is transformed into the new function μ'_x. Then by the fundamental definition stated in eqn (20.3), we have

$$\mu'_x(P') = \mu_x(P). \tag{21.3}$$

Combining this with eqn (21.1), we obtain

$$\mu'_x(P') = \sum_t e^{(t)} x^{(t)}. \tag{21.4}$$

Now inverting eqn (21.2) and making use of the fact that the square matrix is orthogonal, we have

$$\begin{bmatrix} x^{(t)} \\ y^{(t)} \\ z^{(t)} \end{bmatrix} = \begin{bmatrix} R_{xx} & R_{yx} & R_{zx} \\ R_{xy} & R_{yy} & R_{zy} \\ R_{xz} & R_{yz} & R_{zz} \end{bmatrix} \begin{bmatrix} x^{(t')} \\ y^{(t')} \\ z^{(t')} \end{bmatrix} \tag{21.5}$$

and in particular

$$x^{(t)} = R_{xx} x^{(t')} + R_{yx} y^{(t')} + R_{zx} z^{(t')}. \tag{21.6}$$

Substituting eqn (21.6) into eqn (21.4) and bearing in mind that $e^{(t)} = e^{(t')}$, we obtain

$$\mu'_x(P') = \sum_{t'} (R_{xx} e^{(t')} x^{(t')} + R_{yx} e^{(t')} y^{(t')} + R_{zx} e^{(t')} z^{(t')})$$

$$= R_{xx} \mu_x(P') + R_{yx} \mu_y(P') + R_{zx} \mu_z(P'). \tag{21.7}$$

Since P' is a general point in configuration space, eqn (21.7) is equivalent to

$$\mu'_x = R_{xx} \mu_x + R_{yx} \mu_y + R_{zx} \mu_z. \tag{21.8}$$

Corresponding expressions for μ'_y and μ'_z can be similarly derived. We have thus shown that the functions $\mu_x, \mu_y,$ and μ_z form a basis for the group representation to which the translations $T_x, T_y,$ and T_z belong. Except in the cases of the molecular point groups T_d and O_h, where all three directions in space are symmetrically equivalent and all three translations are degenerate, the representation in question will in general be reducible. When it is, it may be reduced by choosing the conventional special directions for the x-, y-, and z-axes—for example, in the case of the PCl_3 molecule, by choosing the C_3 axis as the z-direction. In this way we ensure that $\mu_x, \mu_y,$ and μ_z belong to the same species as $T_x, T_y,$ and T_z respectively. These species are conventionally included in the group character tables, where they can easily be looked up in any particular case.

21.3. Species of direct-product group representations

Suppose that the set of m entities $X_1, X_2,..., X_m$ forms a basis for a representation $\Gamma^{(X)}$ of a certain group, and that a different set of n entities $Y_1, Y_2,..., Y_n$ forms a basis for the representation $\Gamma^{(Y)}$ of the same group. Neither of these representations needs to be irreducible. Under any symmetry operation R of the group, any member of the first set will of course be transformed into a linear combination of all the members of that set; a similar statement is true of any member of the second set. It follows that the product of any member (say X_i) of the first set with any member (say Y_p) of the second set will be transformed by R into a linear combination of all the products (mn in number) obtained by multiplying together one member of the X-set with one of the Y-set. In fact the complete set of such products forms a basis for a new group representation of dimension mn. It is referred to as the *direct-product representation* and is written as $\Gamma^{(X)} \times \Gamma^{(Y)}$.

In order to reduce this representation we shall be specially interested in the characters of the matrices comprising it. For the operation R we may write

$$X_i \xrightarrow{R} X'_i = \sum_j R^{(X)}_{ji} X_j \qquad (21.9)$$

and

$$Y_p \xrightarrow{R} Y'_p = \sum_q R^{(Y)}_{qp} Y_q. \qquad (21.10)$$

Hence

$$X_i Y_p \xrightarrow{R} (X_i Y_p)' = \sum_j \sum_q R^{(X)}_{ji} R^{(Y)}_{qp} X_j Y_q. \qquad (21.11)$$

In the expression for $(X_i Y_p)'$, the coefficient of the term in $X_i Y_p$ is seen to be $R^{(X)}_{ii} R^{(Y)}_{pp}$. Thus for the character of the direct-product matrix we have, at once,

$$\chi^{(\text{prod})}_R = \sum_i \sum_p R^{(X)}_{ii} R^{(Y)}_{pp} = \chi^{(X)}_R \chi^{(Y)}_R. \qquad (21.12)$$

In general a direct-product representation will be reducible. We may reduce it by the general method discussed in Section

307

9.4, according to which (as in eqn (9.19)) the number of times a_i that the ith irreducible representation occurs in the completely reduced form is given by

$$a_i = \frac{1}{g} \sum_R \chi_R^{(i)} \chi_R^{(\text{prod})}, \tag{21.13}$$

where g is the order of the group and $\chi_R^{(i)}$ is the known character for the operation R in the ith irreducible representation.

Let us consider as an example the group C_{3v}, and the particular case when the two factors of the direct product are different but both belong to the same doubly degenerate species E. Here the direct product $E \times E$ is of dimension 4, and the character of the matrix representing the symmetry operation R is given by

$$\chi_R^{(\text{prod})} = (\chi_R^{(E)})^2.$$

Making use of the character table (Table 8.4), we find by eqn (21.13) that

$$a_{A_1} = \tfrac{1}{6}\{(1 \times 4) + 2(1 \times 1) + 3(1 \times 0)\} = 1,$$
$$a_{A_2} = \tfrac{1}{6}\{(1 \times 4) + 2(1 \times 1) - 3(1 \times 0)\} = 1,$$
$$a_E = \tfrac{1}{6}\{(2 \times 4) - 2(1 \times 1) + 3(0 \times 0)\} = 1.$$

We may now express the direct-product representation as a direct sum,

$$E \times E = A_1 + A_2 + E. \tag{21.14}$$

The term 'direct product' may be extended to cases with more than two factors. For example, continuing with the group C_{3v}, we may write

$$E \times E \times E = (E \times E) \times E = (A_1 + A_2 + E) \times E$$
$$= (A_1 \times E) + (A_2 \times E) + (E \times E).$$

It is very easy to show that $A_1 \times E = E$ and $A_2 \times E = E$. Therefore

$$E \times E \times E = E + E + (A_1 + A_2 + E)$$
$$= A_1 + A_2 + 3E.$$

21.4. General condition for the vanishing of the transition moment integral

The integral in question has the form $\int \psi^{(n)} \mu_i \psi^{(m)} \, d\tau$, where $\psi^{(n)}$ and $\psi^{(m)}$ are the wave functions of the levels involved, and

μ_i is a component of the molecular electric dipole moment. The integration extends over the whole of configuration space, of which $d\tau$ is a volume element. Knowing the symmetry species of each of the three factors in the integrand, we can find the species of their product. As we have already pointed out more than once, selection rules are determined by symmetry considerations, and we shall now investigate in detail how the species of the integrand determines whether or not the transition moment integral will vanish.

The integrand is a function of the vibrational displacement coordinates. Let its value at a certain point P in configuration space be $f(P)$. The effect of a symmetry operation R will be to move the point P (or the infinitesimal volume element at P) to a new point P', and to transform the function f into a new function f'. According to the fundamental eqn (20.3), the transformed function is defined by

$$f'(P') = f(P). \tag{21.15}$$

Now because the integration with which we are concerned extends over all the infinitesimal volume elements (i.e. over all the points) in configuration space, the fact that R transfers each value of f from one volume element to another can have no effect upon it. We may therefore write

$$\int f'(P') \, d\tau' = \int f'(P) \, d\tau. \tag{21.16}$$

Substituting the value of $f'(P')$ from eqn (21.15) into the left-hand side of eqn (21.16), we obtain

$$\int f(P) \, d\tau = \int f'(P) \, d\tau. \tag{21.17}$$

If we write $f' = \mathscr{R}f$, where $\mathscr{R}$ is the operator for the symmetry operation R, eqn (21.17) assumes the entirely equivalent form

$$\int f \, d\tau = \int \mathscr{R}f \, d\tau. \tag{21.18}$$

Now let us consider the effect of applying to the integrand function f the projection operator for the totally symmetric species of the group. Denoting this particular species by the superscript unity, the projection operator $\mathscr{P}^{(1)}$ in question is defined, as in eqn (11.1), by

$$\mathscr{P}^{(1)} = \sum_R \chi_R^{(1)} \mathscr{R}.$$

Since $\chi_R^{(1)} = 1$ for every R, the expression for $\mathscr{P}^{(1)}$ assumes the simpler form

$$\mathscr{P}^{(1)} = \sum_R \mathscr{R}. \tag{21.19}$$

So far in this book we have only used projection operators in connection with ordinary vectors. It is clear, however, from the complete analogy between functions in a function space and ordinary vectors in a vector space (see Section 20.3) that we can proceed in an analogous manner to apply the projection operator $\mathscr{P}^{(1)}$ to the function f which is the integrand of the transition moment integral. As with an ordinary vector, the result will be to annihilate all the components of f that belong to any species other than the totally symmetric one. Thus if the structure of the direct-product representation to which the integrand function f belongs does not contain the totally symmetric irreducible representation, the result will be zero. Only if f contains the totally symmetric representation can $\mathscr{P}^{(1)}f$ be different from zero.

Now

$$\mathscr{P}^{(1)}f = \sum_R \mathscr{R}f,$$

so

$$\int \mathscr{P}^{(1)}f \, d\tau = \sum_R \int \mathscr{R}f \, d\tau. \tag{21.20}$$

But by eqn (21.18) $\int \mathscr{R}f \, d\tau = \int f \, d\tau$ for every operation R. Since the number of such operations is equal to g, the order of the group, we have from eqn (21.20)

$$\int \mathscr{P}^{(1)}f \, d\tau = g \int f \, d\tau. \tag{21.21}$$

We have already seen that the left-hand side of eqn (21.21) must vanish unless the representation to which f belongs contains the totally symmetric irreducible representation. It therefore follows that the right-hand side of eqn (21.21), and hence the transition moment itself, must also vanish unless this condition is satisfied.

This important conclusion can be re-expressed as follows.

A transition between the states characterized by the wave functions $\psi^{(n)}$ and $\psi^{(m)}$ is forbidden in infra-red absorption unless for at least one of the components μ_i of the molecular electric dipole moment $(i = x, y, \text{ or } z)$ the product $\psi^{(n)}\mu_i\psi^{(m)}$ belongs to a representation whose structure contains the totally symmetric species.

21.5. The general selection rule for fundamentals in infra-red absorption

For fundamental transitions one of the two states (say the one characterized by $\psi^{(m)}$) is the ground vibrational state $(v = 0)$. As we saw in Section 20.5, it is totally symmetric for all normal modes. The symmetry properties of the integrand in the transition moment integral are therefore simply those of the product $\psi^{(n)}\mu_i$. The species of $\psi^{(n)}$ (i.e. the state with $v = 1$) is known from Section 20.6 to be always the same as that of the vibrational mode itself. The species of the components μ_i of the dipole moment can be found from the relevant character table. We therefore have to inquire as to the condition that the totally symmetric species shall be contained in the structure of the direct-product representation for two functions whose species (say γ_1 and γ_2) are known.

According to eqn (21.13), the number of times a_1 that the totally symmetric species will occur in the structure is given by

$$a_1 = \frac{1}{g} \sum_R \chi_R^{(\text{prod})}\chi_R^{(1)} = \frac{1}{g} \sum_R \chi_R^{(\text{prod})}.$$

Hence, using eqn (21.12), we have

$$a_1 = \frac{1}{g} \sum_R \chi_R^{(\gamma_1)}\chi_R^{(\gamma_2)}.$$

Finally, because of the orthogonality properties of the characters—see eqn (9.8)—the value of a_1 becomes

$$a_1 = \delta_{\gamma_1\gamma_2}. \tag{21.22}$$

We have thus shown that the direct-product representation to which $\psi^{(n)}\mu_i$ belongs, and hence to which the whole integrand for the fundamental transition also belongs, can only contain the totally symmetric species in its structure provided that at least

one component of the dipole moment belongs to the same symmetry species as does the wave function of the fundamental
level ($v = 1$). Since we have already shown that the species of
this wave function is always the same as that of the vibrational mode, we arrive finally at the following form of the
*general selection rule. A vibrational fundamental is inactive in
infra-red absorption unless its species is the same as that of at least
one of the components of the electric dipole moment of the molecule,
i.e. the same as that of at least one of the translations T_x, T_y, and T_z.*

The description 'general' is used to distinguish this kind of
selection rule, which is based solely upon symmetry theory,
from the rules described as 'restricted', the derivation of which
assumes simple-harmonic vibration and the absence of terms of
higher than the first degree in the Taylor expansion of the dipole
moment as a function of the normal coordinates Q. Neither of
these assumptions is involved for the derivation of general rules.
It is true that in considering the integrand of the transition
integral we invoked the symmetry properties of the simple-
harmonic wave functions; however, as we saw in Section 20.7,
these symmetry properties are conserved even when a certain
degree of mechanical anharmonicity is present. Likewise the
symmetry properties of the components of the dipole moment are
in no way affected by the presence of electrical anharmonicity.
Thus the validity of the general rules transcends the limitations
to which the restricted rules are subject.

Of course in the anharmonic case the actual frequency absorbed by the fundamental transition will not be precisely the
same as that calculated with a simple-harmonic approximation
to the real intramolecular force field, but the difference will be
small. It is customary to use the same symbol (ν_k) in both cases.
It should also be noted that the general rule, like the restricted
rule, makes no statement as to the intensities with which
permitted fundamentals will actually occur in the spectrum. In
certain cases they may be so low as to be experimentally unobservable. In fact the intensity might even fortuitously be
zero, for the selection rule expresses a necessary condition, but
in principle not a sufficient one, for it to be non-zero.

On the other hand, when the general rule forbids a fundamental it would appear at first sight to express a condition that is both necessary and sufficient to ensure zero intensity. It is for this reason that rules of this kind are frequently described as 'rigorous'. Actually, however, this is too strong a word, for there remain reasons why some transitions that are thus 'rigorously' forbidden may nevertheless be active, although only feebly so. In this connection we have to remember that, although the general rules (as we prefer to call them) remain valid despite the presence of mechanical and electrical anharmonicities, they still depend upon the Born–Oppenheimer approximation and the assumption that all rotation–vibration interactions are negligible (see Chapter 2). The Born–Oppenheimer approximation (i.e. that nuclear motions can be separated from electronic motions) is such a very good one that it is not expected to produce any observable effects. But the same cannot be said for the assumption that rotation–vibration interactions (and especially Coriolis forces) are negligible. We shall therefore expect that, in certain rare cases, these interactions may cause a 'forbidden' transition to become active. In fact the possibility of Coriolis interactions between different vibrational modes is itself controlled by symmetry considerations, and it transpires that not all transitions that are forbidden by the general selection rules can be rendered active by such interactions. In certain cases, however, forbidden infra-red absorptions have been observed experimentally with very low intensity, and the breakdown of the general selection rules has been interpreted as a Coriolis-interaction effect.

The great superiority of the rules of the general kind, as compared with the restricted kind, is to be seen when we come to infra-red absorptions of frequencies other than simple fundamentals. All such absorptions (overtones and combination tones) are forbidden by the restricted rules, but by extension of the methods of the present chapter it will be possible (in Chapter 22) to deduce general rules governing their activity. In fact these rules are found to be in good agreement with experimental observation, which shows that overtones and combination tones do appear quite commonly, and with appreciable intensities, in infra-red absorption spectra. It is for this reason that we use the word 'general' to describe the rules and to emphasize their much wider scope, as compared with rules of the restricted type. The description 'general' must not, however, be taken to imply that the rules are absolutely rigorous in every case.

21.6. 'Hot' transitions

In the preceding section, when considering the integrand $\psi^{(n)}\mu_i\psi^{(m)}$ for the fundamental of a particular simple-harmonic mode (say the kth), we did not explicitly take into account the vibrational states of any of the other modes of the molecule. In fact we stated that the symmetry properties of the total vibrational wave functions $\psi^{(m)}$ and $\psi^{(n)}$ were simply those **of** ψ_k with the vibrational quantum numbers $v_k = 0$ and $v_k = 1$ respectively. We were perfectly justified in doing this, for the state of every other mode, no matter what its vibrational quantum number, remains unchanged in the transition under consideration, and so contributes to the integrand a factor of the type $\psi_j\psi_j$. On integration this factor obviously becomes of no account, as it merely gives the value unity.

In the simple-harmonic case the value of the frequency absorbed in the fundamental transition is also independent of the vibrational excitation of the other modes. Suppose, for example, that in both the lower and upper states of the transition

$$v_k = 0 \to v_k = 1,$$

all other modes except the jth are in their ground states while for the jth mode $v_j = 1$. The energies of the upper and lower states (relative to the zero-point energy) are respectively $h\nu_k + h\nu_j$ and $h\nu_j$, and so it is reasonable and instructive to write the corresponding frequency for the transition in the form $\nu_k + \nu_j - \nu_j$, which, of course, is simply equal to ν_k.

With real molecules, the situation is somewhat less simple, for the presence of anharmonicity introduces interactions between the otherwise distinct modes, and so affects the relative energies of the levels. Consider, for example, the upper level corresponding to the harmonic one with $v_k = 1$, $v_j = 1$, and $v = 0$ for all other normal modes, and the lower level corresponding to the harmonic one with $v_k = 0$, $v_j = 1$, and $v = 0$ for all other modes. The energy difference will not be precisely the same as that between the upper level corresponding to $v_k = 1$ and $v = 0$ for all other modes (including the jth) and the ground level corresponding to $v = 0$ for all modes. Thus although the frequency in question is still written conventionally as $v_k + v_j - v_j$, its value is no longer exactly equal to v_k.

As far as activity in infra-red absorption is concerned, the general selection rule that applies to the ordinary fundamental v_k (i.e. to the transition $v_k = 0 \rightarrow v_k = 1$ when all other modes are in their ground states) remains valid for all non-zero values of v_j. Thus for a permitted fundamental of a molecule with an appreciable degree of anharmonicity, one expects to find, in the infra-red absorption spectrum, not just one unique frequency, but a series of closely-spaced frequencies corresponding to different degrees of vibrational excitation of other modes. Transitions in which such excitation is present are often termed '*hot' transitions*.

For a polyatomic molecule the complexity is theoretically considerable. In practice, however, hot transitions are rarely observable with samples at normal temperatures. It has to be remembered that a spectroscopically observable intensity of absorption is always proportional to the molecular population of the initial state involved in the transition. This is determined by the Boltzmann distribution, according to which the ratio of the population of a state with $v_j = n$ to that of the state with $v_j = 0$ is $\exp(-nv_j/kT)$, where k is the Boltzmann constant and T the temperature. Even for $n = 1$, this is a very small ratio at room temperature, except when v_j is an unusually low frequency. In general, therefore, hot transitions will have small spectroscopic intensities as compared with the corresponding ordinary, 'cold' transitions. The difficulty of observing them in practice will be increased if the degree of anharmonicity is small; for then, in addition to being very feeble, they will have frequencies lying very near to those of the corresponding normal transitions. Their appearance, however, is a not uncommon feature of actual infra-red spectra.

21.7. Application of the general infra-red selection rule to the fundamentals of the PCl_3 and BF_3 molecules

As an example of the application of the general infra-red selection rule, consider the fundamentals of the pyramidal PCl_3 molecule which belongs to the point group C_{3v}. As we have seen in Section 10.2, the number of distinct fundamentals is four, of which two belong to the species A_1 and two to the species E. The character table for the molecular point group (Table 8.4) shows that both of these species are represented amongst those of the translations, T_z belonging to species A_1 and the pair T_x and T_y to the species E. The general selection rule of Section 21.5 therefore tells us that all four fundamentals will be permitted in infra-red absorption. This is in agreement with observation.

By way of contrast, consider now the BF_3 moledule which, being planar instead of pyramidal, belongs to the point group D_{3h}. We saw in Section 10.7 that the number of distinct fundamentals is again four, and that one of them belongs to the species A_1', one to the species A_2'', and the remaining two to the species E'. Reference to the appropriate character table (Table 10.2) shows that T_z belongs to species A_2'' and the pair T_x, T_y to the species E'. Thus only three out of the four fundamentals are permitted in infra-red absorption. The one that is forbidden is the totally symmetric one. It is obviously associated with the pure bond-stretch mode (i.e. the 'breathing' mode) in which the central boron atom remains at rest while the three fluorine atoms move synchronously in and out. Incidentally we may note that the restricted selection rule—discussed in Sections 18.1 and 18.2—agrees in forbidding this mode. The symmetry of the equilibrium configuration of the molecule ensures that its dipole moment is zero. In the totally symmetric breathing mode this symmetry is conserved and therefore the dipole moment must remain zero throughout. Thus for every component we have $(\partial \mu_i / \partial Q)_0 = 0$, which is the condition, according to the restricted rule, for the mode to be forbidden. From the general rule we see that, because of the symmetry of the mode, this conclusion remains valid even when (as in real molecules) mechanical and electrical anharmonicities are present. In fact the requirements of the general selection rule are in good agreement with observation.

22 Application of the general selection rule to overtones and combination tones in infra-red absorption

22.1. Overtones of non-degenerate vibrations

ALTHOUGH all overtones are forbidden by the restricted selection rule of Section 18.2, certain of them are in fact observed in infra-red absorption spectra. We shall now consider the general selection rule governing their activity. As when we discussed fundamentals (in Chapter 21) we shall be concerned with the species of the transition-moment integrands. For the determination of these species we must first consider the question of the symmetry properties of the simple-harmonic wave functions of states with vibrational quantum numbers v greater than unity. The general form of the functions, as given in eqn (17.22), is

$$\psi_v = N_v \exp(-\tfrac{1}{2}\gamma Q^2) H_v(\gamma^{\frac{1}{2}} Q), \tag{22.1}$$

where H_v is the appropriate Hermitian polynomial and γ (equal to $4\pi^2\nu/h$) is a constant for any particular mode. In the non-degenerate case now under consideration, the exponential factor on the right-hand side of eqn (22.1) is always totally symmetrical, and so ψ_v transforms in the same way as does $H_v(\gamma^{\frac{1}{2}} Q)$. We remarked in Section 17.7 that when v is even the polynomial contains only even powers of Q, whereas when v is odd it contains only odd powers. Thus when v is even, $\psi_v(-Q) = \psi_v(Q)$, whereas when v is odd, $\psi_v(-Q) = -\psi_v(Q)$. Hence it follows at once that ψ_v is totally symmetric for all states with even v, no matter what the symmetry species of the non-degenerate vibration. It also follows that, for all states with odd v, ψ_v belongs to the same non-degenerate species as does the vibration itself. This is, of course, an extension to higher v-values of what we found in Sections 20.5 and 20.6 for the ground level ($v = 0$) and the fundamental level ($v = 1$).

As with the fundamentals (see Section 21.5) the general selection rule for the overtones of non-degenerate vibrations (i.e. transitions of the type $v = 0 \rightarrow v = n$ with $n > 1$) follows at once: *such overtones are forbidden in infra-red absorption unless the species of the upper level is the same as that of at least one of the translations T_x, T_y, or T_z.* It is a matter of experience that the intensity of permitted overtones falls off steeply with increase in the quantum number of the upper level.

We may illustrate the application of this general rule in the case of the PCl_3 molecule (point group C_{3v}). Two of its four normal modes are non-degenerate. Both belong in fact to the totally symmetric species A_1, so all their excited states likewise belong to this species. Because one of the translations (T_z) is totally symmetric, it follows that all the overtones of the two modes are (like the fundamentals) permitted in infra-red absorption. In the case of the BF_3 molecule, two of its four normal modes are again non-degenerate, but only one of them is totally symmetric (species A_1') while the other belongs to the species A_2''. The character table of the molecular point group (D_{3h}) shows that there is no translation belonging to the totally symmetric species A_1', and therefore all overtones of the totally symmetric mode are (like the fundamental) forbidden. Because one of the translations (T_z) belongs to the species A_2'', however, the overtones of the A_2'' mode with odd v in the upper level are all permitted; but the overtones of this mode with even v in the upper level (including in particular the first overtone $v = 0 \rightarrow v = 2$) are all forbidden.

If the vibrations were simple-harmonic, the frequency of a first overtone would be exactly double that of the fundamental, but when anharmonicity is present it is slightly less than this. For convenience, however, it is still conventionally written as $2\nu_k$, where ν_k is the fundamental frequency. Similarly, the second overtone is written as $3\nu_k$, although its actual frequency-value is slightly less than three times that of the fundamental.

22.2. Symmetry species of the wave functions of the level with $v = 2$ for a degenerate vibration

When the vibrational mode in question is degenerate, the situation is more complicated than that considered in the preceding section. As an example we shall take the first overtone level ($v = 2$) for a doubly degenerate simple-harmonic vibration with the normal coordinates Q_a and Q_b. Let v_a and v_b be the vibrational quantum numbers associated with Q_a and Q_b respectively. It is $v_a + v_b$ that determines the energy of a level, so that for the first overtone level we must have $v_a + v_b = 2$. This requirement can obviously be satisfied in three ways: either by $v_a = 2$ with $v_b = 0$, or by $v_a = 1$ with $v_b = 1$, or by $v_a = 0$ with $v_b = 2$. The level therefore has three-fold degeneracy.

As an orthonormal basis for the associated function space we may choose the three eigenfunctions $\psi_{2,0}$, $\psi_{1,1}$, and $\psi_{0,2}$, defined as

$$\left.\begin{aligned}\psi_{2,0} &= N_{2,0}\,\psi_{a2}\psi_{b0}\\ \psi_{1,1} &= N_{1,1}\,\psi_{a1}\psi_{b1}\\ \psi_{0,2} &= N_{0,2}\,\psi_{a0}\psi_{b0}\end{aligned}\right\}. \tag{22.2}$$

Here the Ns are suitable normalizing factors, and the functions ψ_{av} and ψ_{bv} are harmonic wave functions of the respective degenerate modes with quantum number v, the general form of which is given in eqn (22.1). Inserting the values of the Hermite polynomials as given in eqns (17.24), we obtain

$$\psi_{2,0}(Q_a, Q_b) = N_{2,0}\exp\{-\tfrac{1}{2}\gamma(Q_a^2+Q_b^2)\}(4\gamma Q_a^2-2) = A_{2,0}(4\gamma Q_a^2-2), \tag{22.3}$$

$$\psi_{1,1}(Q_a, Q_b) = N_{1,1}\exp\{-\tfrac{1}{2}\gamma(Q_a^2+Q_b^2)\}(4\gamma Q_a Q_b) = A_{1,1}(4\gamma Q_a Q_b), \tag{22.4}$$

$$\psi_{0,2}(Q_a, Q_b) = N_{0,2}\exp\{-\tfrac{1}{2}\gamma(Q_a^2+Q_b^2)\}(4\gamma Q_b^2-2) = A_{0,2}(4\gamma Q_b^2-2). \tag{22.5}$$

The A-factors are all totally symmetric.

We wish to find how these basis functions transform under a symmetry operation R of the molecular point group. Suppose that R transforms the normal coordinates according to the matrix equation

$$\begin{bmatrix} Q_a' \\ Q_b' \end{bmatrix} = \begin{bmatrix} R_{aa} & R_{ab} \\ R_{ba} & R_{bb} \end{bmatrix}\begin{bmatrix} Q_a \\ Q_b \end{bmatrix}. \tag{22.6}$$

This means that in the irreducible representation E the operation is represented by an orthogonal matrix whose character is given by

$$\chi_R^{(E)} = R_{aa} + R_{bb}. \tag{22.7}$$

Its effect is to move the point P in Q-space characterized by the coordinates Q_a, Q_b to the new point P' characterized by the transformed coordinates Q'_a, Q'_b. Let its effect upon the basis function $\psi_{2,0}$ of our three-dimensional function space be to transform it into the new function $\psi'_{2,0}$. Then according to the fundamental definition in eqn (20.3), we have

$$\psi'_{2,0}(P') = \psi_{2,0}(P),$$

i.e.
$$\psi'_{2,0}(Q'_a, Q'_b) = \psi_{2,0}(Q_a, Q_b). \tag{22.8}$$

Using eqn (22.3), we may therefore write

$$\psi'_{2,0}(Q'_a, Q'_b) = A_{2,0}(4\gamma Q_a^2 - 2). \tag{22.9}$$

Now since the square matrix in eqn (22.6) is necessarily orthogonal, inversion of this equation gives

$$Q_a = R_{aa} Q'_a + R_{ba} Q'_b \tag{22.10}$$

and
$$Q_b = R_{ab} Q'_a + R_{bb} Q'_b. \tag{22.11}$$

Substituting the value of Q_a from eqn (22.10) into the right-hand side of eqn (22.9), we obtain

$$\psi'_{2,0}(Q'_a, Q'_b) = A_{2,0}\{4\gamma(R_{aa} Q'_a + R_{ba} Q'_b)^2 - 2\}. \tag{22.12}$$

We now have the same coordinates Q'_a, Q'_b (i.e. the same point P' of Q-space) on both sides of eqn (22.12). Since P' is a perfectly general point, we may dispense with the primes on the coordinate and write simply

$$\psi'_{2,0}(Q_a, Q_b) = A_{2,0}\{4\gamma(R_{aa} Q_a + R_{ba} Q_b)^2 - 2\}. \tag{22.13}$$

In exactly the same way we may obtain the following expressions for the transformed basis functions $\psi'_{1,1}$ and $\psi'_{2,0}$:

$$\psi'_{1,1}(Q_a, Q_b) = A_{1,1}\{4\gamma(R_{aa} Q_a + R_{ba} Q_b)(R_{ab} Q_a + R_{bb} Q_b)\}, \tag{22.14}$$

$$\psi'_{0,2}(Q_a, Q_b) = A_{0,2}\{4\gamma(R_{ab} Q_a + R_{bb} Q_b)^2 - 2\}. \tag{22.15}$$

Our object is to find the species contained in the structure of the reducible group representation for which the three functions

$\psi_{2,0}$, $\psi_{1,1}$, and $\psi_{0,2}$ form a basis. In order to achieve this we shall need to know, for each operation R, the characters of the 3×3 matrix which transforms them. We therefore proceed to pick out the diagonal elements of the matrix for the operation R and add them together. Each of the transformed functions in eqns (22.13), (22.14), and (22.15) must be a linear combination of all the original functions defined in eqns (22.3) (22.4), and (22.5). In the case of $\psi'_{2,0}$ we have to pick out the coefficient of $\psi_{2,0}$. Now $\psi_{2,0}$ is the only one of the three original basis functions that contains a term in Q_a^2, so we need only concern ourselves with terms of this kind. On the right-hand side of eqn (22.13) the term in Q_a^2 has the coefficient $A_{2,0}(4\gamma R_{aa}^2)$, whereas on the right-hand side of eqn (22.3) the corresponding coefficient is $A_{2,0}(4\gamma)$. The desired coefficient of $\psi_{2,0}$ in the linear combination which is $\psi'_{2,0}$ is obtained by dividing $A_{2,0}(4\gamma R_{aa}^2)$ by $A_{2,0}(4\gamma)$. The result is simply R_{aa}^2. This is the first contribution to the character of the matrix representing R in the representation in question. In the same way we note that $\psi_{1,1}$ is the only one of the original basis functions that contains a term in $Q_a Q_b$. Now the coefficients of $Q_a Q_b$ in $\psi'_{1,1}$ and $\psi_{1,1}$ are seen from eqns (22.14) and (22.4) to be $4A_{1,1}\gamma(R_{aa}R_{bb}+R_{ab}R_{ba})$ and $4A_{1,1}\gamma$ respectively. Hence the second contribution to the desired character is $R_{aa}R_{bb}+R_{ab}R_{ba}$. The third contribution is analogous to the first; its value is clearly R_{bb}^2.

As is customary, we shall denote by E^2 the group representation based upon the wave functions of the level with $v = 2$ for a mode belonging to the species E. The character of the matrix representing the symmetry operation R in the group representation E^2 will accordingly be denoted by the symbol $\chi_R^{(E^2)}$. Adding together the three contributions evaluated above, we obtain

$$\chi_R^{(E^2)} = R_{aa}^2 + R_{aa}R_{bb} + R_{ab}R_{ba} + R_{bb}^2. \qquad (22.16)$$

We shall wish to use the values of $\chi_R^{(E^2)}$ for all the operations of the group, in order to determine the structure of the representation E^2 by the use of the usual eqn (9.19). The expression on the right-hand side of eqn (22.16) is not convenient for this purpose, as it contains the individual elements of the matrix of eqn (22.6), whereas only characters are available from the character tables. We therefore have to express the right-hand side of eqn (22.16) in terms of available matrix characters. This can be done in the following way. Take first the product of $\chi_R^{(E)}$ with itself. From the value of $\chi_R^{(E)}$ in eqn (22.7) the result is seen to be

$$\chi_R^{(E)}\chi_R^{(E)} = R_{aa}^2 + 2R_{aa}R_{bb} + R_{bb}^2. \qquad (22.17)$$

Next consider the symmetry operation R^2 (i.e. two successive performances of R). The matrix representing it is easily found by multiplying the square matrix of eqn (22.6) by itself. Its character is thus found to have the following value:

$$\chi_{R^2}^{(E)} = R_{aa}^2 + 2R_{ab}R_{ba} + R_{bb}^2. \qquad (22.18)$$

From eqns (22.16), (22.17), and (22.18) we see that

$$\chi_R^{(E^2)} = \tfrac{1}{2}(\chi_R^{(E)}\chi_R^{(E)} + \chi_R^{(E)}). \qquad (22.19)$$

This is the desired form, for $\chi_R^{(E)}$ can be read off from the group character table, and, since R^2 is an easily ascertainable operation of the group, $\chi_{R^2}^{(E)}$ can be similarly read off. Knowing $\chi_R^{(E^2)}$ for all the operations R of the molecular point group, we can find the species that are present in the completely reduced structure of the representation E^2, i.e. the species of the upper level ($v = 2$) or the first overtone transition for the doubly degenerate mode. This will determine whether, in any particular case, the overtone is permitted of forbidden by the general selection rule governing infra-red absorption. In Section 22.4 we shall consider an actual example; but first, in Section 22.3, we shall discuss the symmetry species of the level with $v = 3$ for a degenerate vibration.

22.3. Symmetry species of the wave functions of the level with $v = 3$ for a degenerate vibration

Again considering the case of a doubly degenerate vibration with normal coordinates Q_a and Q_b, we proceed in the same way for the second overtone level ($v = 3$) as for the first. The requirement $v_a + v_b = 3$ can be satisfied in four ways—either by $v_a = 3$ with $v_b = 0$, by $v_a = 2$ with $v_b = 1$, by $v_a = 1$ with $v_b = 2$, or by $v_a = 0$ with $v_b = 3$; the level therefore has fourfold degeneracy. (In passing we may remark that, as is easily seen, when $v = n$ the degeneracy is $(n+1)$-fold.) As an orthonormal basis for the associated function space we may choose the four functions $\psi_{3,0}$, $\psi_{2,1}$, $\psi_{1,2}$, and $\psi_{0,3}$ defined—compare with eqns (22.2)—as

$$\left.\begin{array}{l} \psi_{3,0} = N_{3,0}\,\psi_{a3}\,\psi_{b0} \\ \psi_{2,1} = N_{2,1}\,\psi_{a2}\,\psi_{b1} \\ \psi_{1,2} = N_{1,2}\,\psi_{a1}\,\psi_{b2} \\ \psi_{0,3} = N_{0,3}\,\psi_{a0}\,\psi_{b3} \end{array}\right\}. \tag{22.20}$$

The general form of the harmonic wave functions is given in eqn (22.1). Inserting the appropriate values of the Hermite polynomials from eqn (17.24), we obtain

$$\psi_{3,0}(Q_a, Q_b) = A_{3,0}(8\gamma^{\frac{3}{2}}Q_a^3 - 12\gamma^{\frac{1}{2}}Q_a), \tag{22.21}$$

$$\psi_{2,1}(Q_a, Q_b) = A_{2,1}(4\gamma Q_a^2 - 2)(2\gamma^{\frac{1}{2}}Q_b), \tag{22.22}$$

$$\psi_{1,2}(Q_a, Q_b) = A_{1,2}(2\gamma^{\frac{1}{2}}Q_a)(4\gamma Q_b^2 - 2), \tag{22.23}$$

$$\psi_{0,3}(Q_a, Q_b) = A_{0,3}(8\gamma^{\frac{3}{2}}Q_b^3 - 12\gamma^{\frac{1}{2}}Q_b). \tag{22.24}$$

The A-factors are all totally symmetric.

The effect of the group operation R upon the basis function $\psi_{3,0}$ is to transform it into the new function $\psi'_{3,0}$ defined—compare with eqn (22.8)—by

$$\psi'_{3,0}(Q'_a, Q'_b) = \psi_{3,0}(Q_a, Q_b). \tag{22.25}$$

Using eqn (22.21) we may write

$$\psi'_{3,0}(Q'_a, Q'_b) = A_{3,0}(8\gamma^{\frac{3}{2}}Q_a^3 - 12\gamma^{\frac{1}{2}}Q_a). \tag{22.26}$$

Substituting the value of Q_a from eqn (22.10) into the right-hand side of eqn (22.26), and dispensing with the primes on the coordinates, as we did in deriving eqn (22.13) from eqn (22.12), we obtain

$$\psi'_{3,0}(Q_a, Q_b) = A_{3,0}\{8\gamma^{\frac{3}{2}}(R_{aa}Q_a + R_{ba}Q_b)^3 - 12\gamma^{\frac{1}{2}}(R_{aa}Q_a + R_{ba}Q_b)\}. \tag{22.27}$$

This must be a linear combination of the four base functions defined in eqns (22.21), (22.22), (22.23), and (22.24). We note that the only one of them that contains a term in Q_a^3 is $\psi_{3,0}$. Therefore, in order to discover the relevant diagonal element of the 4×4 matrix representing R in the representation E^3, we need only consider the coefficient of Q_a^3 on the right-hand side of eqn (22.27). This coefficient is seen to be $8A_{3,0}\gamma^{\frac{3}{2}}R_{aa}^3$. Since the corresponding coefficient in $\psi_{3,0}$ is $8A_{3,0}\gamma^{\frac{3}{2}}$, we deduce at once that the first of the four contributions to $\chi_R^{(E^3)}$ is simply R_{aa}^3. In picking out the second, third, and fourth contributions we need consider only the coefficients of the terms in $Q_a^2 Q_b$, $Q_a Q_b^2$, and Q_b^3 respectively. The reader will easily confirm that the respective contributions to $\chi_R^{(E^3)}$ are $R_{aa}^2 R_{bb} + 2R_{aa} R_{ab} R_{ba}$, $R_{aa} R_{bb}^2 + 2R_{ab} R_{ba} R_{bb}$, and R_{bb}^3. By addition we obtain

$$\chi_R^{(E^3)} = R_{aa}^3 + (R_{aa} + R_{bb})(R_{aa} R_{bb} + 2R_{ab} R_{ba}) + R_{bb}^3. \quad (22.28)$$

As with $\chi_R^{(E^2)}$ in eqns (22.16) and (22.19), it is possible to express $\chi_R^{(E^3)}$ in a form which does not involve the individual elements of the matrix representing R in the group representation E. Taking first the product of $\chi_R^{(E)}$ with $\chi_R^{(E^2)}$, and using the values given respectively in eqns (22.7) and (22.16), we find

$$\chi_R^{(E)}\chi_R^{(E^2)} = R_{aa}^3 + 2R_{aa}^2 R_{bb} + 2R_{aa} R_{bb}^2 +$$
$$+ R_{aa} R_{ab} R_{ba} + R_{ab} R_{ba} R_{bb} + R_{bb}^3. \quad (22.29)$$

Next, if we consider the operation R^3 (i.e. three successive performances of R), the character of the matrix representing it is easily found to have the value

$$\chi_{R^3}^{(E)} = R_{aa}^3 + 3R_{aa} R_{ab} R_{ba} + 3R_{ab} R_{ba} R_{bb} + R_{bb}^3. \quad (22.30)$$

From eqns (22.28), (22.29), and (22.30) it can be seen that

$$\chi_R^{(E^3)} = \tfrac{1}{2}(\chi_R^{(E)}\chi_R^{(E^2)} + \chi_{R^3}^{(E)}). \quad (22.31)$$

This is the desired form, for R^3 is an easily ascertainable operation of the group, and so both $\chi_R^{(E)}$ and $\chi_{R^3}^{(E)}$ can be read off from the appropriate character table. The value of $\chi_R^{(E^2)}$ has already been given in eqn (22.19).

The analogy of eqn (22.31) with eqn (22.19) is obvious. In fact both are special cases of the general equation

$$\chi_R^{(E^n)} = \tfrac{1}{2}(\chi_R^{(E)}\chi_R^{(E^{n-1})} + \chi_{R^n}^{(E)}) \quad (22.32)$$

which may be shown to hold for all overtone levels of a doubly degenerate vibration. The general proof will not be given here, for in practice the intensities of overtones higher than the second are so very low as to be only rarely observable. Thus for $n > 3$ the general equation (22.32) is of little practical value.

Having evaluated $\chi_R^{(E^3)}$ in any particular case for all the symmetry operations R, we can employ the usual method to find the structure of the reducible 4-dimensional representation E^3. The general selection rule can then be applied to determine whether the second overtone is permitted or forbidden in infrared absorption.

In this and the preceding section we have considered the overtone levels of a doubly degenerate vibration. The case of a triply degenerate vibration can be dealt with in an analogous fashion.

22.4. Application of the general selection rule to the overtones of the doubly degenerate vibrations of the PCl$_3$ molecule

The PCl$_3$ molecule (point group C_{3v}) will again be used to illustrate the application of the conclusions of the two preceding sections. Two of the normal modes belong to the doubly

TABLE 22.1

Data for the point group C_{3v}

R	I	C_3	C_3^2	σ_{v_1}	σ_{v_2}	σ_{v_3}
R^2	I	C_3^2	C_3	I	I	I
R^3	I	I	I	σ_{v_1}	σ_{v_2}	σ_{v_3}
$\chi_R^{(E)}$	2	-1	-1	0	0	0
$\chi_{R^2}^{(E)}$	2	-1	-1	2	2	2
$\chi_{R^3}^{(E)}$	2	2	2	0	0	0
$\chi_R^{(E^2)}$	3	0	0	1	1	1
$\chi_R^{(E^3)}$	4	1	1	0	0	0

degenerate species E. We wish to find whether their first and second overtones are active in infra-red absorption.

The first three rows of Table 22.1 give respectively the sets of symmetry operations R, R^2, and R^3 for the molecular point group (C_{3v}). The next three rows give the corresponding characters, as read off from the character table (Table 8.4). The bottom two rows give the values of $\chi_R^{(E^2)}$ and $\chi_R^{(E^3)}$ as calculated by eqns (22.19) and (22.31) respectively.

The structure of the 3×3 representation E^2 for the first overtone level ($v = 2$) can be found by using eqn (9.19), according to which the number of times a_γ that a particular irreducible representation γ appears in the completely reduced form of E^2 is

$$a_\gamma = \frac{1}{g} \sum_R \chi_R^{(\gamma)} \chi_R^{(E^2)}.$$

Substituting the appropriate values of the characters from Table 22.1, we find that

$$a_{A_1} = \tfrac{1}{6}\{(1 \times 3) + (1 \times 1) + (1 \times 1) + (1 \times 1)\} = 1,$$

$$a_{A_2} = \tfrac{1}{6}\{(1 \times 3) - (1 \times 1) - (1 \times 1) - (1 \times 1)\} = 0,$$

$$a_E = \tfrac{1}{6}(2 \times 3) = 1.$$

This result is conventionally expressed by writing

$$E^2 = A_1 + E. \tag{22.33}$$

As regards the activity of the overtone in infra-red absorption, we note that the translation T_z belongs to the species A_1, and the degenerate pair of translations T_x, T_y belong to the species E. Either of these facts would ensure, by virtue of the general rule, that the overtone is permitted.

By the same method the reader will have no difficulty in verifying that the structure of the second overtone level ($v = 3$) is given by

$$E^3 = A_1 + A_2 + E. \tag{22.34}$$

Like the first overtone, therefore, the second overtone is permitted in infra-red absorption.

Triply degenerate vibrations (species F) occur with certain point groups where all three directions in physical space are symmetrically equivalent. The species of their overtones, and consequently the predictions of the general selection rule regarding their infra-red activity, may be deduced in the same way as for those of doubly degenerate vibrations.

22.5. The effect of anharmonicity on degenerate overtone levels

We have shown that in the simple-harmonic case the first overtone level of a doubly degenerate normal mode is triply degenerate, and that for the particular point group C_{3v} the group representation based upon a set of wave functions has the structure A_1+E. The introduction of a small amount of anharmonicity will perturb the eigenvalues and so cause a small splitting of the overtone level. However, since the symmetry species involved must remain unaffected, it follows that the triple degeneracy will not be completely removed. In fact the level will be split into two only, one belonging to the non-degenerate species A_1 and the other to the doubly degenerate species E. The general selection rule (which is of course still valid when anharmonicity is present) tells us that transitions from the ground level to both these overtone component levels are permitted in infra-red absorption.

In the case of the corresponding second overtone level, we saw that (for the same point group C_{3v}) the fourfold degeneracy is associated with the structure A_1+A_2+E. A small degree of anharmonicity will therefore split the level into three distinct components belonging respectively to the species A_1, A_2, and E. Here the general selection rule tells us that transitions from the ground level to the two components belonging to the species A_1 and E will be permitted in infra-red absorption, but that transitions from the ground level to the component level of species A_2 will be forbidden.

The extra degeneracy in the harmonic case, as compared with the anharmonic, is not due to symmetry, but to the very special form of the harmonic potential. For this reason it is sometimes included under the general heading of 'accidental' degeneracy. However, this term is perhaps better reserved for those cases where, even with anharmonic fields, certain states happen quite fortuitously to have practically identical eigenvalues, although in no way required by symmetry to be degenerate with one another.

22.6. Combination tones in infra-red absorption

The restricted selection rules of Section 18.2 forbid all combination tones, i.e. transitions which involve simultaneous changes in the vibrational quantum numbers of two (or more) different vibrational modes. Nevertheless certain combination tones are in fact observed in infra-red absorption spectra (just as are certain overtones). Here again activity or inactivity is determined by the general selection rule.

The simplest combination tones are the binary ones which involve the simultaneous transitions $v_j = 0 \to v_j = 1$ and $v_k = 0 \to v_k = 1$ for two different modes j and k. The two modes need not necessarily belong to different symmetry species. In order to be able to apply the general selection rule we need to know the species of the upper level. The wave function in question will simply be the product of the individual wave functions for the two different modes. If γ_1 and γ_2 are the species of the two factors, their direct-product representation is such —see eqn (21.12)—that for any symmetry operation R

$$\chi_R^{(\text{prod})} = \chi_R^{(\gamma_1)} \chi_R^{(\gamma_2)}. \tag{22.35}$$

This is true both when the species γ_1 and γ_2 are different and also when they are the same. If the direct-product representation $\gamma_1 \times \gamma_2$ is reducible, its structure can be determined by the usual method. Having thus found the symmetry species of the combination level, we can apply the general selection rule and find out whether the combination is active or inactive in infra-red absorption. When it is active, its frequency is formally written as $\nu_j + \nu_k$, although on account of anharmonicity the actual frequency value will not be exactly equal to the sum of the two fundamentals.

As an example we may take the binary combination tone $\nu_3 + \nu_4$ of the PCl_3 molecule. The two vibrational modes concerned both belong to the doubly degenerate species E, and so we are here concerned with the direct product $E \times E$. For each symmetry operation R of the molecular point group (C_{3v}) the relevant character is simply $(\chi_R^{(E)})^2$. Using eqn (9.19), the structure is easily found to be

$$E \times E = A_1 + A_2 + E. \tag{22.36}$$

Since both A_1 and E are represented amongst the species of the translations, the combination tone $\nu_3 + \nu_4$ is permitted in infrared absorption.

The general selection rule can be applied without difficulty to higher binary combinations, involving the simultaneous transitions $v_j = 0 \to v_j = n_j$ for the mode j and $v_k = 0 \to v_k = n_k$ for the different mode k, either n_j or n_k (or both) being greater than unity. As an example we may take the case of the fundamental transition $(0 \to 1)$ for one of the doubly degenerate modes (ν_3) of PCl_3 combined with the first overtone transition $(0 \to 2)$ for the other (ν_4). The combination frequency is formally written as $\nu_3 + 2\nu_4$. The degeneracy of the upper level will clearly be $2 \times 3 = 6$. The character of the matrix representing the operation R in the direct-product representation $E \times E^2$ is $\chi_R^{(E)} \chi_R^{(E^2)}$. This may be evaluated for each R of the molecular point group (C_{3v}), using the values of $\chi_R^{(E)}$ from the character table (Table 8.4) and the values of $\chi_R^{(E^2)}$ from Table 22.1. Hence, using eqn (9.19), the reader may easily verify the structure:

$$E \times E^2 = A_1 + A_2 + 2E. \tag{22.37}$$

More conveniently, we may make use of the structures of E^2 and of $E \times E$, as previously derived and given in eqns (22.33) and (22.36) respectively. Thus we have

$$E \times E^2 = E \times (A_1 + E) = (E \times A_1) + (E \times E)$$
$$= E + (A_1 + A_2 + E) = A_1 + A_2 + 2E.$$

The result is of course identical with that in eqn (22.37). We see that this higher combination tone $\nu_3 + 2\nu_4$ is permitted by the general selection rule.

It will be appreciated that the same kind of treatment may be further extended to ternary combination tones, i.e. to transitions involving simultaneous changes in the vibrational quantum numbers of three different vibrational modes. However, experience shows that permitted overtones of this and more complicated kinds usually have very low spectroscopic intensities. The applications of the general selection rule to them are therefore of correspondingly little practical interest.

22.7. A note on overtone and combination-tone levels of degenerate vibrations

Nothing essentially new will be presented in this section, the sole object of which is to direct the reader's attention to a point about which some confusion has sometimes been known to arise. It concerns the essential differences of degeneracy and symmetry species between an overtone level of a single degenerate mode and a combination level of two different degenerate modes. As an example consider, on the one hand, the first overtone level ($v_i = 2$) of a doubly degenerate mode i, and, on the other hand, the combination-tone level ($v_j = 1$ and $v_k = 1$) of two doubly degenerate modes j and k. We suppose that the two modes j and k are different from one another, though one of them could be identical with the mode i. For both levels the total vibrational quantum number is 2, but for the overtone level ($v_i = 2$) we saw in Section 22.2 that the degeneracy is 3, whereas for the combination-tone level ($v_j = 1$ and $v_k = 1$) it is $2 \times 2 = 4$. In order to distinguish between the corresponding group representations, the one for the overtone level is written as E^2 and the one for the combination-tone level is written as the direct product $E \times E$. Confusion is liable to arise unless the essential difference between the meanings of these two types of symbols is appreciated and borne clearly in mind. Of course the two representations have quite different structures. In the particular case of the point group C_{3v} they are given in eqns (22.33) and (22.36) respectively.

It may perhaps be helpful to present the relevant vibrational quantum numbers of the two levels in tabular form, so as to show how the different degeneracies arise. With each of the doubly degenerate modes will be associated two normal co-ordinates; the two corresponding quantum numbers will be distinguished by the additional subscripts a and b respectively. Thus the statement that $v_i = 2$ for the overtone level means that

331

$v_{ia} + v_{ib} = 2$. Likewise the statement that $v_j = 1$ and $v_k = 1$ for the combination-tone level means that $v_{ja} + v_{jb} = 1$ and $v_{ka} + v_{kb} = 1$. Table 22.2 shows clearly the origins of the three-fold degeneracy of the overtone level and the fourfold degeneracy of the combination-tone level.

TABLE 22.2

Quantum numbers and degeneracies of an overtone level and a combination-tone level involving only doubly degenerate vibrations

First overtone level ($v_i = 2$) of a doubly degenerate vibration		Combination-tone level ($v_j = 1$, $v_k = 1$) of two different doubly degenerate vibrations			
v_{ia}	v_{ib}	v_{ja}	v_{jb}	v_{ka}	v_{kb}
2	0	1	0	1	0
1	1	1	0	0	1
0	2	0	1	1	0
		0	1	0	1
$\Gamma = E^2$ Degeneracy = 3		$\Gamma = E \times E$ Degeneracy = 4			

Similar considerations apply to higher overtone and combination-tone levels of doubly degenerate vibrations, and also to analogous cases for triply degenerate vibrations. Thus, as the reader may easily verify, the first overtone level of a vibration belonging to the triply degenerate species F has sixfold degeneracy, whereas the combination-tone level involving two such vibrations, each with $v = 1$, has the degeneracy $3 \times 3 = 9$. The respective group representations are written as F^2 and $F \times F$.

22.8. Difference tones

The combination tones so far considered have involved only simultaneous upward transitions, i.e. simultaneous increases in the quantum numbers of the vibrations concerned. Such combination tones are often called *sum tones*, because (with the exception of small effects due to anharmonicity) the spectroscopic frequency is the sum of those associated with the relevant transitions for the different modes considered individually. It is sometimes possible, however, to observe *difference tones* in absorption, i.e. transitions involving an increase in the quantum number of one mode (say $v_j = 0 \to v_j = 1$) and a simultaneous drop in the quantum number of another mode (say $v_k = 1 \to v_k = 0$). The absorbed frequency is then written as $\nu_j - \nu_k$. Obviously, if absorption is to occur at all, it is necessary that $\nu_j > \nu_k$. Slightly more complicated cases (involving for example the frequency $n\nu_j - \nu_k$) will also be possible.

No extra information is required to determine whether a difference tone is permitted in infra-red absorption, if it is known whether the corresponding sum tone is permitted by the general selection rule. All that distinguishes the two is an interchange of initial and final states for one of the vibrational modes involved. This merely interchanges the corresponding wave functions in the transition-moment integrand, leaving it effectively unchanged. Corresponding sum and difference tones are therefore either both active or both inactive.

An important consideration is that, owing to the Boltzmann distribution factor, the molecular population of the initial state will be smaller for the difference tone than for the corresponding sum tone. Therefore if these tones are permitted, the spectroscopically observable intensity of the difference tone must be correspondingly lower than that of the sum tone. Thus it is improper for an infra-red spectroscopist to assign an observed feature in the spectrum to a difference tone, unless he is able also to observe the necessarily more intense corresponding sum tone.

22.9. Possibility of determining forbidden fundamental frequencies from permitted combination tones and overtones

It is a matter of practical importance that a combination tone that is permitted in infra-red absorption may involve, as one member of the combination, a vibrational transition that would be forbidden if considered on its own. As an example we may take the totally symmetric vibration ν_1 (symmetry species A_1') of the BF_3 molecule. As we saw in Section 21.7, this fundamental is forbidden in infra-red absorption. However, the other three fundamentals ν_2 (A_2''), ν_3 (E'), and ν_4 (E') are all permitted; it therefore follows from the totally symmetric nature of ν_1, that the combination tones $\nu_1 + \nu_2$, $\nu_1 + \nu_3$, and $\nu_1 + \nu_4$ are likewise permitted. This makes it possible in principle to deduce indirectly the value of ν_1 from the infra-red spectrum, although the direct observation of ν_1 in absorption is forbidden. In other cases a fundamental may be forbidden as such, but its first overtone may be permitted. Thus the observation and correct assignment of combination tones and overtones may in favourable cases yield information about fundamental frequencies that could not otherwise have been obtained from infra-red absorption spectra.

22.10. Considerations concerning intensities in infra-red absorption spectra

We have previously remarked that the general selection rule merely states the necessary condition for a transition to be permitted. It does not tell us anything about the intensity with which a permitted transition will in fact appear in the infra-red spectrum. However, in so far as the special assumptions of mechanical and electrical harmonicity are good approximations, we shall expect the restricted selection rules to be correspondingly close to the truth. The fact that these rules forbid all overtones and combination tones leads us, therefore, to anticipate that permitted features of these kinds will tend to have relatively low intensities, as compared with active fundamentals. Broadly speaking, this is borne out by experience—indeed many permitted overtones and combination tones are so weak as to escape observation. With decreasing degrees of mechanical and electrical anharmonicity the intensity must in all cases decrease, reaching zero in the limit. Thus an overtone or combination tone, despite being permitted by the general selection rule, could in fact have zero intensity because of the 'accidental' circumstance of its harmonic character.

This apparent paradox is resolved when it is realized that the condition which permits a frequency, according to the general rule, is a necessary condition for zero intensity, but not a sufficient one. Absence of anharmonicity is just a special circumstance under which its insufficiency becomes apparent. The logical situation is thus quite different from that which arises when a vibration is forbidden by the general rule; for then no special circumstance concerned with the presence or absence of anharmonicity could ever lead to a non-zero intensity. This is not to say, however, that no special circumstance whatever can have this effect, for, as pointed out in Section 21.5, the general selection rule is not completely rigorous. Its validity may be impaired by failure of the underlying assumption of all purely vibrational theory, that rotation–vibration interactions are negligible. In fact Coriolis forces may in certain cases lead to the appearance in the spectrum (albeit with very low intensities) of transitions that are forbidden by the general selection rule.

22.11. Fermi resonance

There is one 'accidental' circumstance in which a permitted overtone or combination tone may have an unexpected high intensity in infra-red absorption. It arises when two conditions are simultaneously fulfilled: first, the frequency of the overtone or combination tone in question must happen by chance to be nearly equal to that of a permitted fundamental; and secondly, the species of the fundamental must be the same as that of the overtone or combination tone (or one of its components). The effect of anharmonicity is to cause a mixing of the respective wave functions and a perturbation of the respective energy levels. In fact the higher level is displaced upwards and the lower one is displaced downwards—an effect often expressed by saying that the two 'repel' each other. The smaller the energy difference between the unperturbed levels, and the greater the degree of anharmonicity, the greater will be the mixing of the wave functions and the greater the energy perturbation. In any case it is no longer proper to speak of the one level as being an overtone or combination level and the other a fundamental level, for both the displaced levels will be partly of the one kind and partly of the other. This is known as *Fermi resonance*.

From our point of view, the important aspect is that the mixing of the wave functions results in a sharing of spectroscopic intensity. In general, the overtone or combination would normally appear with low intensity, as compared with the fundamental; but in consequence of the Fermi resonance it may 'borrow' intensity from the fundamental, and the two may appear as a so-called Fermi doublet with components of comparable intensities.

A well-known example occurs in the spectrum of the carbon tetrachloride molecule. Here the sum of the two fundamental frequencies ν_1 (totally symmetric) and ν_4 (triply degenerate) is by chance nearly equal to that of the fundamental ν_3 (triply degenerate), which is permitted in infra-red absorption. In consequence the combination tone $\nu_1+\nu_4$, being of the same species as ν_3, enters into Fermi resonance with it. Instead of an intense fundamental and a relatively weak combination tone, what is observed in the spectrum is a doublet consisting of two components with practically equal intensities.

23 General rules of selection and polarization for Raman scattering

23.1. The basis of the general selection rule

THE derivation of the general selection rule for the Raman effect is similar to that for infra-red absorption, except that in the case of the Raman effect we are concerned with the induced electric dipole moment π instead of the intrinsic dipole moment μ of the molecule. As we saw in Section 18.4, the expression for the transition moment may be written in the condensed form

$$\int \psi^{(n)}\pi\psi^{(m)}\,\mathrm{d}\tau = \int \psi^{(n)}\alpha E\psi^{(m)}\,\mathrm{d}\tau = E\int \psi^{(n)}\alpha\psi^{(m)}\,\mathrm{d}\tau, \quad (23.1)$$

where α is the molecular polarizability and E the electric field strength of the incident light. The detailed form of course takes into account the facts that α is a symmetric tensor with six distinct components α_{ij} and that E is a vector with three components E_i (i or $j = x, y$, or z).

The transition moment is proportional to E, and consequently the intensity of Raman scattering is proportional to E^2, i.e. to the intensity of the incident light. In considering selection rules, however, this proportionality may be ignored. We are therefore concerned with integrals of the type $\int \psi^{(n)}\alpha_{ij}\psi^{(m)}\,\mathrm{d}\tau$. By analogy with the case of infra-red absorption (discussed in Section 21.4), we may enunciate the basis of the general selection rule for Raman scattering in the following form. *A transition between states characterized by the wave functions $\psi^{(n)}$ and $\psi^{(m)}$ is forbidden in Raman scattering unless for at least one of the components α_{ij} of the molecular polarizability tensor (i or $j = x, y$, or z) the product $\psi^{(n)}\alpha_{ij}\psi^{(m)}$ belongs to a representation whose structure contains the totally symmetric species.*

23.2. The symmetry properties of the components of the molecular polarizability

Like the components of the intrinsic molecular electric dipole moment μ, the components of the polarizability tensor α are functions of the displacement coordinates of the nuclei. In Section 21.2 we saw that the dipole moment components μ_i belong to the same symmetry species as do the corresponding translations T_i. We shall now prove that the polarizability components α_{ij} are transformed in the same way as the products $T_i T_j$ of the corresponding translations. Thus we shall prove that α_{xx} is transformed in the same way as T_x^2 and that α_{xy} is transformed in the same way as $T_x T_y$, etc.

The effect of the group symmetry operation R upon the set of translational components may be expressed in the matrix form

$$\begin{bmatrix} T_x \\ T_y \\ T_z \end{bmatrix} \xrightarrow{R} \begin{bmatrix} T_x' \\ T_y' \\ T_z' \end{bmatrix} = \begin{bmatrix} R_{xx} & R_{xy} & R_{xz} \\ R_{yx} & R_{yy} & R_{yz} \\ R_{zx} & R_{zy} & R_{zz} \end{bmatrix} \begin{bmatrix} T_x \\ T_y \\ T_z \end{bmatrix}. \quad (23.2)$$

Hence we may write down expressions for the effects of R upon products of pairs of translations. For example,

$$\begin{aligned} T_x^2 \xrightarrow{R} (T_x^2)' &= (R_{xx} T_x + R_{xy} T_y + R_{xz} T_z)^2 \\ &= R_{xx}^2 T_x^2 + R_{xy}^2 T_y^2 + R_{xz}^2 T_z^2 + 2 R_{xx} R_{xy} T_x T_y + \\ &\quad + 2 R_{xy} R_{xz} T_y T_z + 2 R_{xz} R_{xx} T_z T_x, \quad (23.3) \end{aligned}$$

and

$$\begin{aligned} T_x T_y &\xrightarrow{R} (T_x T_y)' \\ &= (R_{xx} T_x + R_{xy} T_y + R_{xz} T_z)(R_{yx} T_x + R_{yy} T_y + R_{yz} T_z) \\ &= R_{xx} R_{yx} T_x^2 + R_{xy} R_{yy} T_y^2 + R_{xz} R_{yz} T_z^2 + \\ &\quad + (R_{xx} R_{yy} + R_{yx} R_{xy}) T_x T_y + (R_{xy} R_{yz} + R_{yy} R_{xz}) T_y T_z + \\ &\quad + (R_{xz} R_{yx} + R_{yz} R_{xx}) T_z T_x. \quad (23.4) \end{aligned}$$

Similar expressions can be written down for all the other binary products.

We now have to show that the polarizability components

α_{xx}, α_{xy}, etc., transform in the same ways as T_x^2, $T_x T_y$, etc. To this end we make use of the fact (previously noted in Section 18.3) that the relation between the induced dipole moment vector $\boldsymbol{\pi}$ and the electric field-strength vector $\boldsymbol{E}$ can be written in the matrix form

$$\begin{bmatrix} \pi_x \\ \pi_y \\ \pi_z \end{bmatrix} = \begin{bmatrix} \alpha_{xx} & \alpha_{xy} & \alpha_{xz} \\ \alpha_{yx} & \alpha_{yy} & \alpha_{yz} \\ \alpha_{zx} & \alpha_{zy} & \alpha_{zz} \end{bmatrix} \begin{bmatrix} E_x \\ E_y \\ E_z \end{bmatrix}$$

or, more briefly, $\qquad\qquad \boldsymbol{\pi} = [\alpha]\mathbf{E}.$ (23.5)

(Of course **E**, as used here, is not to be confused with a unit matrix.)

After the performance of the operation R we have similarly

$$\boldsymbol{\pi'} = [\alpha']\mathbf{E'}. \tag{23.6}$$

Now (just as is the case for the vector $\boldsymbol{\mu}$) the vectors $\boldsymbol{\pi}$ and $\boldsymbol{E}$ must transform in the same way as does the translational vector **T**. Therefore we can write

$$\mathbf{E'} = \mathbf{RE} \tag{23.7}$$

and $\qquad\qquad\qquad \boldsymbol{\pi'} = \mathbf{R}\boldsymbol{\pi},$ (23.8)

where **R** is the square matrix representing the operation R in eqn (23.2). Substituting eqns (23.7) and (23.8) into eqn (23.6), we obtain

$$\mathbf{R}\boldsymbol{\pi} = [\alpha']\mathbf{RE},$$

whence $\qquad\qquad\qquad \boldsymbol{\pi} = \mathbf{R}^{-1}[\alpha']\mathbf{RE}.$ (23.9)

Comparing eqn (23.9) with eqn (23.5) we see that

$$[\alpha] = \mathbf{R}^{-1}[\alpha']\mathbf{R}. \tag{23.10}$$

From eqn (23.10), bearing in mind that **R** is necessarily orthogonal, we finally conclude that

$$[\alpha'] = \mathbf{R}[\alpha]\mathbf{R}^\dagger. \tag{23.11}$$

339

By performing the matrix multiplications on the right-hand side of eqn (23.11) the reader may easily confirm that

$$\alpha'_{xx} = R^2_{xx}\alpha_{xx} + R^2_{xy}\alpha_{yy} + R^2_{xz}\alpha_{zz} + 2R_{xx}R_{xy}\alpha_{xy} +$$
$$+ 2R_{xy}R_{xz}\alpha_{yz} + 2R_{xz}R_{xx}\alpha_{zx} \quad (23.12)$$

and

$$\alpha'_{xy} = R_{xx}R_{yx}\alpha_{xx} + R_{xy}R_{yy}\alpha_{yy} + R_{xz}R_{yz}\alpha_{zz} +$$
$$+ (R_{xx}R_{yy} + R_{yx}R_{xy})\alpha_{xy} + (R_{xy}R_{yz} + R_{yy}R_{xz})\alpha_{yz} +$$
$$+ (R_{xz}R_{yx} + R_{yz}R_{xx})\alpha_{zx}. \quad (23.13)$$

Similar expressions can be written down for all the other tensor components. In obtaining them account is taken of the symmetric nature of the tensor α (and hence of the matrix $[\alpha]$), by virtue of which $\alpha_{ij} = \alpha_{ji}$. Comparison of eqn (23.12) with eqn (23.3), and of eqn (23.13) with eqn (23.4), shows that the components α_{ij} of the polarizability transform in exactly the same way as do the corresponding products $T_i T_j$ of the translations. This is what we set out to prove.

23.3. The symmetry species of the polarizability components for the point group C_{3v}

As an example of the determination of the symmetry species to which the components of the polarizability belong, we shall consider the point group C_{3v}. Reference to the conventional character table (Table 8.4) shows that T_z belongs to the totally symmetric species A_1. It follows that A_1 is the species of T^2_z and also, therefore, of α_{zz}. From the character table we also find that the pair T_x, T_y belongs to the doubly degenerate species E. Because of the totally symmetric nature of T_z, it follows that the products $T_z T_x$ and $T_y T_z$ form a doubly degenerate pair, and that therefore the polarizability components α_{xz} and α_{yz} belong to the species E.

We are left with the task of determining the species of the remaining three products T^2_x, T_{xy}, and T^2_y. Because, under any symmetry operation, the individual translation components T_x and T_y are transformed into linear combinations of both of them, each of the three products in question must be transformed into a linear combination of all three of them. In fact they must generate a 3-dimensional group representation. The situation here is exactly analogous to that which we encountered in the case of the three wave functions of the first overtone level ($v = 2$) of a doubly degenerate vibration (see Section 22.2), which involved respectively Q_a^2, $Q_a Q_b$, and Q_b^2. For the same point group (C_{3v}) as we are now considering, the relevant group representation was found (see eqn (22.33)) to have the completely reduced structure $A_1 + E$. It must therefore be possible to find three linearly independent linear combinations of the three products T_x^2, $T_x T_y$, and T_y^2, one of which remains unchanged under all the symmetry operations of the group (i.e. belongs to the symmetry species A_1), while the other two form a pair belonging to the doubly degenerate species E. In fact it is easy to see that the A_1 combination must be $T_x^2 + T_y^2$, for this is merely the square of the magnitude of the translation vector of the xy plane, which clearly remains unchanged under all symmetry operations. However, not only this combination but also those of species E (which are not so immediately obvious) can be found by application of the projection operator method.

TABLE 23.1

Values of $\mathscr{R}(T_x^2)$ for the point group C_{3v}

$\mathscr{R}$	I	C_3	C_3^2	σ_{v_1}	σ_{v_2}	σ_{v_3}
$\mathscr{R}(T_x^2)$	T_x^2	$(-\tfrac{1}{2}T_x - \tfrac{1}{2}\sqrt{3}T_y)^2$	$(-\tfrac{1}{2}T_x + \tfrac{1}{2}\sqrt{3}T_y)^2$	T_x^2	$(-\tfrac{1}{2}T_x - \tfrac{1}{2}\sqrt{3}T_y)^2$	$(-\tfrac{1}{2}T_x + \tfrac{1}{2}\sqrt{3}T_y)^2$

The procedure is analogous to that exemplified in Section 11.3. In order to produce from T_x^2, $T_x T_y$, and T_y^2 a linear combination belonging to the species A_1, we apply the projection operator $\mathscr{P}^{(A_1)}$ to any one of the three, say to T_x^2. The definition of $\mathscr{P}^{(A_1)}$, as given in eqn (11.1), is $\sum_R \chi_R^{(A_1)} \mathscr{R}$, which (since $\chi_R^{(A_1)} = 1$ for every R) reduces to $\sum_R \mathscr{R}$. The matrices of an irreducible representation of the species E, to which T_x and T_y belong, are given in Table 8.2. The corresponding values of $\mathscr{R}(T_x^2)$ for each operation R are given in Table 23.1.

Hence we find that $\sum_R \mathscr{R}(T_x^2)$ is equal to $3(T_x^2 + T_y^2)$. This then is a combination belonging to the species A_1. The numerical factor is of course irrelevant. The result agrees with what we saw intuitively above. For the species E the relevant combination $\sum_R \chi_R^{(E)} \mathcal{R}(T_x^2)$ is easily found to have the value $\frac{3}{2}(T_x^2 - T_y^2)$. Ignoring the numerical factor, the desired combination is thus $T_x^2 - T_y^2$. This is one member of a linearly independent pair belonging to the species E. In order to derive the other member we make use of the fact that the matrix $\mathbf{W}$ which, by a change of basis, effects the complete reduction of the 3-dimensional representation, must be an orthogonal one. We accordingly normalize to unity the combinations $T_x^2 + T_y^2$ and $T_x^2 - T_y^2$ by multiplying each by the factor $1/\sqrt{2}$, and write the transformation by the matrix $\mathbf{W}$ in the form

$$
\begin{bmatrix} (T_x^2)' \\[2mm] (T_x T_y)' \\[2mm] (T_y^2)' \end{bmatrix} = \begin{bmatrix} \dfrac{1}{\sqrt{2}} & 0 & \dfrac{1}{\sqrt{2}} \\[2mm] \dfrac{1}{\sqrt{2}} & 0 & -\dfrac{1}{\sqrt{2}} \\[2mm] a & b & c \end{bmatrix} \begin{bmatrix} T_x^2 \\[2mm] T_x T_y \\[2mm] T_y^2 \end{bmatrix}, \qquad (23.14)
$$

where a, b, and c are elements to be determined. From the orthogonal properties of the matrix $\mathbf{W}$ we have

$$
\left. \begin{array}{l} \dfrac{1}{\sqrt{2}} a + \dfrac{1}{\sqrt{2}} c = 0 \\[4mm] \dfrac{1}{\sqrt{2}} a - \dfrac{1}{\sqrt{2}} c = 0 \end{array} \right\}. \qquad (23.15)
$$

and

It therefore follows at once that

$$
a = c = 0. \qquad (23.16)
$$

Also, since $a^2 + b^2 + c^2 = 1$, we have

$$
b = 1. \qquad (23.17)
$$

Eqns (23.16) and (23.17) show that the desired linear combination belonging to the species E is simply $T_x T_y$.

If we return now to the components α_{ij} of the polarizability tensor, which (we remember) transform in exactly the same ways as the respective products $T_i T_j$, we see that the above discussion has shown that $\alpha_{xx} + \alpha_{yy}$ belongs to the species A_1 and that

$\alpha_{xx} - \alpha_{yy}$ and α_{xy} form a degenerate pair belonging to the species E. It is customary to include this sort of information in the group character tables used by vibrational spectroscopists. As an example we now give in full (Table 23.2) the character table of this kind for the point group C_{3v}. This is the group to which belongs the PCl_3 molecule, which we have used so often for illustrative purposes in this book, and which we shall presently use again in connection with the general rules of selection and polarization for Raman scattering.

TABLE 23.2

Conventional character table for the point group C_{3v}

C_{3v}	I	$2C_3$	$3\sigma_v$		
A_1	1	1	1	T_z	$\alpha_{xx} + \alpha_{yy}$; α_{zz}
A_2	1	1	-1	R_z	
E	2	-1	0	(T_x, T_y) ; (R_x, R_y)	$(\alpha_{xx} - \alpha_{yy}, \alpha_{xy})$; $(\alpha_{yz}, \alpha_{zx})$

TABLE 23.3

Conventional character table for the point group D_{3h}

D_{3h}	I	$2C_3$	$3C_2$	σ_h	$2S_3$	$3\sigma_v$		
A_1'	1	1	1	1	1	1		$\alpha_{xx} + \alpha_{yy}$; α_{zz}
A_2'	1	1	-1	1	1	-1	R_z	
E'	2	-1	0	2	-1	0	(T_x, T_y)	$(\alpha_{xx} - \alpha_{yy}, \alpha_{xy})$
A_1''	1	1	1	-1	-1	-1		
A_2''	1	1	-1	-1	-1	1	T_z	
E''	2	-1	0	-2	1	0	(R_x, R_y)	$(\alpha_{yz}, \alpha_{zx})$

The species of the polarizability components (or appropriate linear combinations of them) are given in the last column. The remainder of the table has been previously discussed, and is the same as was shown in Table 8.4.

It will be instructive also to consider the rules of selection and polarization for the molecule BF_3. Accordingly we give here the full character table, in conventional form, for the point group D_{3h} to which this molecule belongs (Table 23.3). This is to be compared with the previous incomplete version in Table 10.2.

23.4. General selection rule for fundamentals in Raman scattering

The application of the basic conclusion of Section 23.1 is particularly simple in the case of fundamental transitions. The wave function of the ground level being always totally symmetric, the necessary condition for Raman activity is that the product of the wave function of the upper level and the polarizability component α_{ij} shall be totally symmetric or contain in its structure the totally symmetric species. As we showed in Section 21.4, this can only be so provided that the two factors belong to the same species. The general selection rule for fundamentals in the Raman effect is therefore as follows. *A fundamental is inactive in the Raman effect unless its species is the same as that of at least one of the components of the polarizability tensor (or one combination of these components).*

From character tables of the kind represented by Tables 23.2 and 23.3 it is thus an easy matter to read off which species of fundamentals are permitted and which forbidden. It should not be forgotten, however, that the mere knowledge that a particular fundamental is permitted by the selection rule is no guarantee that it will actually appear in the Raman spectrum with easily observable intensity. Nor is it absolutely certain that a fundamental that is forbidden by the rule will in fact have an exactly zero intensity; for (as in the case of the infra-red absorption selection rule of Section 21.5) there is a possibility (although in practice a very small one) that the validity of the Raman rule may be slightly impaired by the presence of appreciable rotation–vibration interaction. For the same reasons, therefore, as in the infra-red case, we prefer to describe the rule as 'general' rather than 'rigorous'.

As the reader will realize, the application of the general rule could be extended to overtones and combination tones in the Raman effect, in the same sort of way as for these types of transitions in infra-red absorption. However, it is a matter of experience that in Raman scattering overtones and combination tones are generally of very low intensity indeed, as compared with fundamentals. In fact they are only seldom observed. The application of the general selection rule to them is therefore of correspondingly little practical significance, and we shall not discuss it further here. In this connection we recall that the restricted selection rule forbids overtones and combination tones in Raman scattering. The situation is thus parallel to that for infra-red absorption, but the observed intensities indicate that the assumptions underlying the restricted rules represent a better approximation to reality in the Raman case than in the infra-red case.

23.5. Application of the general selection rule to the fundamentals of the PCl₃ and BF₃ molecules

The PCl$_3$ molecule (point group C_{3v}) has four distinct fundamental frequencies, two of the modes belonging to species A_1 and the other two to species E. Reference to the group character table (Table 23.2) shows that there are polarizability components that belong to both of these species. It follows that all four fundamentals must be permitted in the Raman effect. In this particular example, for which we found in Section 21.7 that all four fundamentals are also permitted in the infra-red, it happens that the application of the respective general selection rules gives the same result for both kinds of vibrational spectrum. This, however, is by no means always the case. The two phenomena, Raman scattering and absorption, are physically quite different from one another, and activity (or inactivity) in the one does not necessarily imply activity (or inactivity) in the other.

For instance, consider the fundamentals of the BF$_3$ molecule (point group D_{3h}). Like PCl$_3$, it has four distinct fundamental frequencies. One mode is of species A_1', one of species A_2'', and the remaining two of species E'. In this case reference to the group character table (Table 23.3) shows that there are polarizability components belonging to the species A_1' and E', but none belonging to the species A_2''. It therefore follows that only three of the four fundamentals are permitted in the Raman effect. Here the result is not the same as for infra-red absorption (see Section 21.7). It is true that we also found that only three of the four fundamentals are permitted in the infra-red, but the striking fact is that they are not the same three as for the Raman effect. Actually, the one that is forbidden in infra-red absorption (the totally symmetric one) is not only permitted in the Raman effect but is found by experiment to have a very high Raman intensity. This is not unexpected when we reflect that it is the 'breathing frequency', for which the bonds are all extended in one phase and all compressed in the other. In contrast to the dipole moment, which remains zero throughout, the molecular polarizability must evidently change strongly in the course of a vibration, and this will cause the Raman intensity to be high.

The BF_3 molecule provides an example of the essentially complementary nature of the Raman and infra-red methods, for in order to observe all the four fundamentals, neither is alone sufficient. In certain cases, however, it may happen that a particular fundamental may be forbidden in both kinds of spectrum, and so may be inaccessible to direct observation. In this connection we recall (from Section 22.9) that a fundamental that is forbidden as such in infra-red absorption, may possibly be permitted as an overtone or as a combination with a permitted fundamental. If so, it becomes possible in principle (and perhaps also in practice) to determine its frequency indirectly. In Raman spectroscopy, however, the corresponding use of overtones and combination tones is rarely practicable on account of their extremely low intensities.

23.6. Fermi resonance

The nature of Fermi resonance has already been discussed in Section 22.11 in connection with infra-red absorption. It produces similar effects in Raman spectra. When an overtone or combination-tone frequency happens to be nearly coincident with that of a fundamental which is permitted in the Raman effect, then (provided that the symmetry species of the fundamental is the same as that of the overtone or combination or one of its components) a Fermi doublet may appear in the Raman spectrum. The components of the doublet will have comparable intensities, and we may roughly describe the effect by saying that the overtone or combination (which would normally be very weak) has not only had its frequency somewhat altered, but has 'borrowed' intensity from the permitted fundamental. In the case of the CCl_4 molecule, for example, the Fermi resonance between $\nu_3(E)$ and $\nu_1(A_1) + \nu_4(E)$, which was quoted in Section 22.11 in connection with infra-red absorption, produces an analogous effect in the Raman spectrum. This is due to the circumstance that the fundamental ν_3 happens to be active in the Raman effect as well as in the infra-red absorption. The Fermi doublet (component frequencies 762 and 790 cm^{-1}), which appears instead of the expected single fundamental ν_3, is a familiar feature to all Raman spectroscopists, who tend to use carbon tetrachloride as a convenient substance for testing the performance of Raman spectroscopic apparatus. It so happens that both $\nu_1(A_1)$ and $\nu_4(E)$, the members of the combination involved in the Fermi resonance, are separately permitted as fundamentals in the Raman spectrum (though not in the infra-red). Their observed frequencies are 459 and 314 cm^{-1} respectively. The sum tone is therefore expected to lie close to $459 + 314 = 773$ cm^{-1}. We note that this anticipated frequency is in between the actual frequencies of the Fermi doublet components.

Another very striking example of Fermi degeneracy is encountered in the Raman spectrum of gaseous carbon dioxide. The CO_2 molecule is linear and symmetrical. Of its three distinct fundamental frequencies, shown in Fig. 2.5 (linear model), only the totally symmetric 'breathing' frequency ν_1 is permitted in Raman scattering. The spectrum would accordingly be expected to consist (on the Stokes side) of just one intense line. In fact an intense doublet is observed. Its components lie at 1 285 and 1 388 cm^{-1} and are practically equal in intensity. The very large frequency difference (over 100 cm^{-1}, which amounts to between 7 and 8 per cent of the mean frequency) indicates that we are here concerned with a very strong Fermi resonance, resulting from a very near coincidence between the true value of ν_1 and some overtone or combination of the appropriate species. In fact it turns out that it is the first overtone $2\nu_2$ of the doubly degenerate bending mode (see Fig. 2.5) which is involved. As a fundamental, ν_2 is forbidden in the Raman effect, but it is permitted in the infra-red absorption spectrum, where its frequency is observed to be 667 cm^{-1}. Its overtone $2\nu_2$ will be expected to have a value near to $2 \times 667 = 1\,334$ cm^{-1}. It will be noted that this anticipated value is very nearly midway between the values for the two components of the observed Fermi doublet.

23.7. General polarization rule for Raman scattering

Symmetry considerations lead to an important general rule concerning the states of polarization of the Raman scattering for normal vibrational modes belonging to different species, as characterized by the degree of depolarization ρ_n. We recall from Section 19.4 that this quantity refers to the scattering of light at right angles to the direction of the incident beam when the incident light is natural.

For Rayleigh scattering we found in Section 19.7 that for samples containing free molecules the value of the degree of depolarization is

$$\rho_n = \frac{6\gamma_0^2}{45(\overline{\alpha_0})^2 + 7\gamma_0^2},$$ (23.18)

where α_0 and γ_0 are respectively the mean-value invariant and the anisotropy invariant of the molecular polarizability tensor α_0. Now the essential difference between Rayleigh and Raman scattering is that the former leaves the vibrational state of the scattering molecule unchanged, whereas the latter involves a transition from an initial vibrational state m to a different final vibrational state n. We shall here be concerned with fundamentals, i.e. cases in which m is the ground state and n the state in which $v_k = 1$ for one particular normal mode k. However, the argument which follows is not restricted to fundamentals, but (*mutatis mutandis*) applies equally well to overtones and combination tones.

In going over from the consideration of Rayleigh scattering to the consideration of Raman scattering, we have to replace the actual polarizability tensor α_0 by what we may call the transition polarizability tensor α_{nm}. This is related to α by an expression which is analogous to that which expresses the relation of the transition moment $\mathbf{\mu}_{nm}$ to the actual dipole moment $\mathbf{\mu}$ in the infra-red case. We thus have

$$\alpha_{nm} = \int \psi^{(n)} \alpha \psi^{(m)} \, \mathrm{d}\tau. \tag{23.19}$$

This is the convenient condensed form, and is to be understood as implying a similar expression relating each individual component of α_{nm} to the corresponding component of α. For the degree of depolarization ρ_n of a Raman line we thus obtain an expression analogous to eqn (23.18), i.e.

$$\rho_n = \frac{6\gamma_{nm}^2}{45(\overline{\alpha_{nm}})^2 + 7\gamma_{nm}^2}. \tag{23.20}$$

Here $\overline{\alpha_{nm}}$ and γ_{nm} are respectively the mean-value invariant and the anisotropy invariant of α_{nm}. In particular

$$\overline{\alpha_{nm}} = \int \psi^{(n)} \overline{\alpha} \psi^{(m)} \, \mathrm{d}\tau = \tfrac{1}{3} \int \psi^{(n)} (\alpha_{xx} + \alpha_{yy} + \alpha_{zz}) \psi^{(m)} \, \mathrm{d}\tau. \tag{23.21}$$

A Raman line is said to be polarized when $\rho_n < \tfrac{6}{7}$. From eqn (23.20) we see that the condition for this is that $\overline{\alpha_{nm}}$ is different from zero. If $\overline{\alpha_{nm}} = 0$ for a Raman line, then $\rho_n = \tfrac{6}{7}$, and the line is said to be depolarized. Now whether or not $\overline{\alpha_{nm}}$ vanishes is clearly determined by the symmetry species of the integrand in eqn (23.21), and hence (the species of $\psi^{(n)}$ and $\psi^{(m)}$ being known) by the species of $\overline{\alpha}$. For fundamentals, considerations of the kind we used in Section 21.4 lead to the conclusion that $\overline{\alpha_{nm}}$ can only be different from zero (i.e. the Raman fundamental can only be polarized) provided that $\overline{\alpha}$ belongs to the same symmetry species as does the vibration in question. We therefore have to find the species to which $\overline{\alpha}$ belongs. In fact this is a very simple matter.

By definition $\overline{\alpha} = (\alpha_{xx} + \alpha_{yy} + \alpha_{zz})/3$ and therefore, under the symmetry operations of the molecular point group, it must transform—see Section 23.2—in the same way as does $(T_x^2 + T_y^2 + T_z^2)$. But $(T_x^2 + T_y^2 + T_z^2)$ is equal simply to the square of the magnitude of the translation vector, and so must remain unchanged under all the symmetry operations. We have thus shown that, no matter what the molecular point group, $\overline{\alpha}$ always belongs to the totally symmetric species.

We are now in a position to enunciate the general rule of polarization for fundamentals in Raman scattering: *only fundamentals belonging to the totally symmetric species can be polarized (i.e. have $\rho_n < \frac{6}{7}$). All fundamentals belonging to other symmetry species must be depolarized (i.e. have $\rho_n = \frac{6}{7}$).*

23.8. Application of the general rule of polarization to the PCl_3 and BF_3 molecules

The general rule of polarization is of great value, because it is in principle possible to determine experimentally which lines in a Raman spectrum are polarized and which depolarized. By virtue of the rule we are thus able to find out which frequencies must be assigned to totally symmetric vibrational modes and which to modes of other species. In the case of the PCl_3 molecule we saw in Section 23.5 that all four fundamentals are permitted in Raman scattering. We can now go further and say that two of them (the ones belonging to species A_1) will be polarized, and two depolarized. This is, in fact, what is observed experimentally. The observation is important in that it enables us, with complete assurance, to assign values to the two totally symmetric frequencies, ν_1 and ν_2. Indeed, in this simple and favourable case, it enables us to make the full assignment of all the four fundamentals, for the remaining two, ν_3 and ν_4, both belong to the species E. Such a correct complete assignment is, of course, a necessary prerequisite for the use of observed frequencies as a basis for the calculation of force constants.

The BF_3 molecule has only one fundamental belonging to the totally symmetric species, and so only one of the three permitted Raman lines will be polarized. This, again, is what is observed experimentally.

23.9. A note on observed degrees of depolarization of Raman lines

The rule of polarization states that only fundamentals of totally symmetric vibrations can have $\rho_n < \frac{6}{7}$; but in general it tells us nothing about how much less than $\frac{6}{7}$ the value of ρ_n will

be in any particular case. The only exceptions arise with isotropic molecules, for which $\gamma = 0$ and likewise (for a totally
symmetric vibration) $\gamma_{nm} = 0$. In such very exceptional cases
(for instance the symmetric 'breathing' vibration of the CCl_4
molecule) it follows from eqn (23.20) that $\rho_n = 0$. For all other
totally symmetric Raman lines, however, we know only that
$\rho_n < \frac{6}{7}$. It may happen that the value in some particular case
is so little less than $\frac{6}{7}$ that the precision of the experimental
method may not be adequate to detect the difference. In these
circumstances the assignment of the line to a totally symmetric
species (rather than to some other) is open to doubt. The situation is thus analogous to that for the selection rule, where it
may occasionally happen that a frequency, although permitted,
actually has so low an intensity as to escape observation and
seem to be forbidden.

24 The use of vibrational spectroscopy to determine molecular symmetry

24.1. Principles of the method

As we have seen in Chapters 21, 22, and 23, it is molecular symmetry that determines the activity of vibrations in infra-red absorption, and their activity and states of polarization in Raman scattering. By observation of the infra-red and Raman spectra of a molecular species, therefore, it is possible to obtain information about its symmetry.

Suppose we are interested in a molecule containing known numbers of nuclei of known kinds. If we assume a certain nuclear arrangement in the equilibrium molecular configuration, we can deduce the number of vibrational modes belonging to each symmetry species, and hence the number of distinct normal frequencies (see Chapter 10). We can also make use of the general selection rules and the general rule of Raman polarization (stated in Sections 21.5, 23.4, and 23.7 respectively) in order to make the following predictions about the vibrational spectra:

(1) The number of fundamental frequencies permitted in infra-red absorption.
(2) The number of fundamental frequencies permitted in the Raman effect.
(3) The number of 'coincidences', i.e. frequencies permitted in both types of spectrum.
(4) The number of permitted Raman lines that are polarized.

These predictions, which are all concerned with fundamentals, are the most important in practice because fundamentals are the most intense and easily observable spectral features. In addition we can predict the activity of all overtones and combination tones. This is likely to be more significant for the infra-red than for the Raman spectrum.

Suppose now that the real structure and symmetry of the molecule are unknown, and that it is possible to propose two or more models belonging to different point groups. A simple example would be a molecule of the type XY_3, which might reasonably be expected to be either pyramidal (C_{3v}) or planar (D_{3h}). For each of the alternative models in any particular case, we can make predictions of the kinds enumerated above; in general the predictions for the different models will not be alike. Experimental investigation of the actual vibrational spectra (both infra-red and Raman) will therefore enable us in principle to discriminate between the proposed structures, and in favourable cases to decide which is the correct one.

24.2. Application to molecules of type XY_3

For a molecule whose formula is known to be of the type XY_3 it is natural to suggest that its structure will be either pyramidal (point group C_{3v}) or planar (point group D_{3h}). Predictions for each of the two rival models have been made earlier in this book

TABLE 24.1

Predictions for fundamentals of a molecule of type XY_3

Structure	Point group	Number active in infra-red absorption	Number active in the Raman effect	Number of coincidences	Number of polarized Raman lines
Pyramidal	C_{3v}	4	4	4	2
Planar	D_{3h}	3	3	2	1

in the course of discussions of the general rules of selection and Raman polarization (see Sections 21.7, 23.5, and 23.8). For convenience of comparison, the predictions for fundamentals are collected in Table 24.1.

Now let us imagine for a moment that we do not know the structure of the PCl_3 molecule and that we wish to discover, from its vibrational spectra, whether it is pyramidal or planar.

The Raman spectrum is found to consist of four lines with the measured frequency shifts of 511, 484, 258, and 190 cm^{-1}. The first and third of these are observed to be quite strongly polarized (i.e. to have ρ_n values very much less than $\frac{6}{7}$). The other two are observed to be depolarized ($\rho_n = \frac{6}{7}$). There is no reason at all to doubt that all four frequencies are fundamentals. This Raman evidence alone rules out the planar structure. Confirmation comes from the observation of all four fundamentals in the infra-red absorption spectrum. The experimental findings are

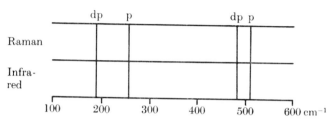

F$_{IG}$. 24.1. Observed vibrational spectra of the PCl$_3$ molecule. (p = polarized Raman line; dp = depolarized Raman line.)

n fact entirely consistent with the symmetrical pyramidal structure. As already pointed out in Section 23.8, the Raman polarization evidence enables us to assign confidently all the four fundamental frequencies. The fact that 511 cm^{-1} and 258 cm^{-1} are polarized makes it certain they are $\nu_1(A_1)$ and $\nu_2(A_1)$ respectively; it then follows that 484 cm^{-1} and 190 cm^{-1} must be $\nu_3(E)$ and $\nu_4(E)$ respectively. We have here followed the usual convention of numbering fundamentals of the same species in order of descending frequency.

Not only are the numbers of active fundamentals in accordance with expectation, but also (as shown at the end of Section 10.3) their spectral 'pattern', i.e. the fact that two are of relatively high frequency and two of relatively low frequency (see Fig. 24.1).

By way of contrast, let us examine the experimental evidence for BF$_3$, again imagining for the moment that we are ignorant of its structure. Here we encounter the special circumstance that a natural sample of the compound contains two isotopes of boron with appreciably different nuclear masses (10 and 11) and

354

with an abundance ratio of approximately $1:4$. The observed vibrational spectra of such a sample will thus consist of superpositions of the corresponding spectra of the two species $^{10}BF_3$ and $^{11}BF_3$. To avoid confusion, we shall here confine our attention to the spectra attributable to the more abundant species $^{11}BF_3$. This molecule (like PCl_3) has four distinct fundamental frequencies, but the Raman spectrum is found to consist of only three lines with the measured frequency shifts 1 454, 888, and 480 cm^{-1}. Of these the second is very intense and is the only one that is polarized. These findings are of a different kind from those for the PCl_3 molecule, and are in agreement with the expectations for the planar structure rather than the pyramidal (see Table 24.1). The infra-red spectrum provides further evidence pointing to the same conclusion. Here again only three fundamentals are observed, but the number of coincidences (see Table 24.1) is only two. Most striking is the fact that the fundamental frequency which is missing from the infra-red spectrum is 888 cm^{-1}—the one which is intense and polarized in the Raman spectrum. This is very strong evidence in favour of the planar model; for the Raman polarization shows that the mode in question is totally symmetric, and the totally symmetric 'breathing' frequency of a planar symmetrical XY_3 molecule obviously involves no change of molecular dipole moment and must be forbidden in infra-red absorption. The frequency which is observed in the infra-red spectrum, but which does not appear in the Raman spectrum, is found to be 691 cm^{-1}. Accepting the evidence for the planar structure, we conclude that this frequency must be assigned to the out-of-plane deformation mode of species A_2''. Thus the spectroscopic evidence again enables us (as for PCl_3) to make a complete assignment of the fundamentals. The observed spectra are shown diagrammatically in Fig. 24.2. The contrast with Fig. 24.1 for the PCl_3 molecule is obvious.

Of course, the structures of PCl_3 and BF_3 are in fact both well known. What we have done is to confirm that their vibrational spectra show characteristic differences which are in accordance with the predictions for the respective point groups. This gives us a basis for hoping that, with other molecules of the type XY_3, the study of their vibrational spectra will provide evidence as to structure in cases where this is unknown. A general discussion of the method, which is applicable in principle to any type of molecular species where alternative point groups can be proposed, will be given in Section 24.5.

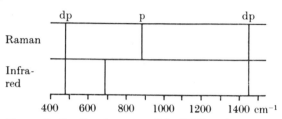

FIG. 24.2. Observed vibrational spectra of the $^{11}BF_3$ molecule. (p = polarized Raman line; dp = depolarized Raman line.)

24.3. The ClF_3 molecule

Throughout this book extensive use has been made of the PCl_3 and BF_3 molecules in order to illustrate the usual alternative point groups, C_{3v} and D_{3h} respectively, of molecules of the type XY_3. The reader will probably find it refreshing, therefore, to consider a very unusual XY_3 molecule, the spectroscopic evidence for which does not conform to the predictions for either of these two symmetries. The molecule in question is that of chlorine trifluoride, ClF_3. Its infra-red absorption spectrum is found to show no less than six frequencies, all of which must be fundamentals. Their measured values are 752, 703, 528, 434, 364, and 326 cm^{-1}. The spectrum also contains a number of higher frequencies which can be satisfactorily accounted for as overtones or combination tones of the six fundamentals. Clearly the point group cannot be either C_{3v} or D_{3h}. The observation that all the $3N-6 = 6$ fundamental frequencies are distinct, i.e. that no degeneracy is present, rules out the possibility of a three-fold axis. The infra-red spectrum is in fact in harmony with the very unusual molecular structure which has been established by other methods. The molecular point group is actually C_{2v}, the molecule being planar, but not symmetrical like BF_3. Its shape has been shown to be that of a slightly distorted T, with the Cl nucleus at the junction of the upright and horizontal members, and the F nuclei at the three extremities. Thus instead of all three F nuclei being symmetrically equivalent (as in BF_3), two are equivalent and the third (the one at the foot of the vertical member of the T) is different. Application of the general selection rule leads to the prediction that all six distinct fundamentals will be permitted in infra-red absorption. The general Raman selection rule also permits all six in the Raman spectrum. The compound is not an easy one for the Raman spectroscopist, and the spectra so far reported are doubtless far from complete. However, two of the fundamentals (750 and 529 cm^{-1}) have been observed.

24.4. Isotopic effects in the vibrational spectra of boron trifluoride

Table 24.2 shows the observed fundamental frequencies for the isotopic species $^{10}BF_3$ and $^{11}BF_3$, and the isotopic shifts in each case.

TABLE 24.2

Isotopic frequency shifts for boron trifluoride (values in cm^{-1})

	$\nu_1(A_1')$	$\nu_2(A_2'')$	$\nu_3(E')$	$\nu_4(E')$
$^{10}BF_3$	888	719	1505	482
$^{11}BF_3$	888	691	1454	480
Isotopic shift	0	28	51	2

The assignments, the correctness of which is beyond doubt, have already been discussed. Especially noticeable is the fact that the totally symmetric frequency 888 cm^{-1}, unlike all the others, shows no isotopic shift. This is entirely in accordance with expectation; for in this 'breathing' mode all three F nuclei move in phase, while the central B nucleus necessarily remains at rest. As a result the mass of the B nucleus cannot influence the frequency. If the molecule were pyramidal, like PCl_3, the corresponding totally symmetric bond-stretching mode would necessarily involve a motion of the apical nucleus in order that the centre of mass of the molecule should remain at rest, and in

357

this case an isotopic shift would occur. Thus the single observation, from the Raman spectrum, that ν_1 is the same for both isotopic species, provides in itself strong evidence in favour of the planar symmetrical structure. The additional observation that this frequency is inactive in infra-red absorption makes the total evidence very strong indeed.

Still further support is forthcoming from application of the product rules (see Chapter 14) to the other vibrational species A_2'' and E'. Assuming the point group D_{3h}, the group representation Γ generated by a complete set of $3N$ nuclear displacement coordinates was found (see eqn (10.14)) to have the structure

$$\Gamma = A_1' + A_2' + 3E' + 2A_2'' + E''. \tag{24.1}$$

Each of the isotopic molecules possesses two sets of symmetrically equivalent nuclei—the first consists simply of the central B nucleus, the second consists of the three F nuclei. As far as the product rules are concerned, only the first set need be taken into account, since the masses of the F nuclei are the same in both isotopic species. By the same method as we used for PCl_3 in Section 9.6, the reader may easily confirm that the first set generates the representation Γ_B with the structure

$$\Gamma_B = E' + A_2''. \tag{24.2}$$

Reference to the group character table (Table 23.3) shows that the only non-genuine vibrations belonging to any of the species A_1', A_2', and E' with which we shall be concerned in applying the vibrational product rules, are T_z (which belongs to the species A_2'') and the degenerate pair T_x, T_y (which belongs to the species E').

Thus using eqn (14.15) and letting primes indicate quantities for the isotopic molecule $^{11}BF_3$, the product rules are easily found to be as follows:

for species A_1', $\dfrac{\nu_1}{\nu_1'} = 1;$ $\tag{24.3}$

for species A_2'', $\dfrac{\nu_2}{\nu_2'} = \sqrt{\left\{ \dfrac{(m_B + 3m_F)m_B'}{(m_B' + 3m_F)m_B} \right\}};$ $\tag{24.4}$

for species E', $\dfrac{\nu_3 \nu_4}{\nu_3' \nu_4'} = \sqrt{\left\{ \dfrac{(m_B + 3m_F)m_B'}{(m_B' + 3m_F)m_B} \right\}}.$ $\tag{24.5}$

We have already discussed the very simple case of the totally symmetric species A_1', and have noted the absence of any isotopic frequency shift (in Table 24.2). For the other species A_2'' and E', using $m_B = 10$, $m_B' = 11$, and $m_F = 19$, we find that the right-hand side of both eqn (24.4) and eqn (24.5) has the value 1·041. Reference to Table 24.2 shows that the pair of corresponding frequencies 719 and 691 comes closest to satisfying the product-rule requirement for the species A_2''. The value 1·040 which this pair gives for the left-hand side of eqn (24.4) is of course subject to a slight uncertainty on account of possible errors in the measurements. As with ν_1, the application of the product rule confirms the ν_2 assignments given in Table 24.2. For the species E' the observed frequencies give the value 1·039 for the left-hand side of eqn (24.5). This is again in satisfactory agreement with the value 1·041 required by the product rule. Altogether the application of the product rules provides support for the conclusion that the molecular point group is indeed D_{3h}.

24.5. Practical limitations in the use of the general rules of selection and polarization for the determination of molecular symmetry

When different point groups can be proposed for a molecular species of unknown structure, it is in principle possible to discriminate between them by comparing the observed vibrational spectra with the respective predictions based upon the general rules of selection and Raman polarization. The evidence so obtained is often of very considerable value, and vibrational spectroscopy has a recognized place amongst the various methods used for the investigation of molecular structure.

However, it is subject to certain practical limitations. These arise from the fact (which we have previously noted) that the general rules merely tell which frequencies may have non-zero intensities, and which Raman lines may have values of ρ_n less than $\frac{6}{7}$. Because they are silent on the question of how much the intensity of a permitted feature will differ from zero, it is always possible that the feature will in fact have an intensity that is so low as to escape observation. Similarly, for a polarized line the

ρ_n-value may be so little less than $\frac{6}{7}$ that the difference may be too small to be detected. Thus the investigator may get the impression that a permitted feature is forbidden, or that a polarized Raman line is depolarized. Such impressions may in some cases lead him to erroneous conclusions as to molecular symmetry.

For the sake of argument let us consider the case of a molecular species like $^{11}BF_3$ (see Fig. 24.2) and try to put the evidence in the worst possible light. We could argue that the observation of only three Raman lines conceivably represents a failure to observe the fourth, not because it is forbidden but because, being permitted, it happens to have an unobservably low intensity. We could argue in the same way about the observation of only three frequencies in infra-red absorption. In fact we could point out that all the evidence in favour of the planar symmetrical structure, rather than the pyramidal one, is essentially *negative* in character, and that negative evidence can never be convincing. Even the reported absence of an isotopic shift for ν_1 on substituting ^{10}B in place of ^{11}B could be discredited as a failure to observe an actual shift because of its smallness. From a purely logical point of view these comments must be acknowledged to be justified, though it remains true that the circumstantial evidence in favour of the planar structure is remarkably strong.

It is probable that failures to observe fundamentals under favourable experimental conditions will be due to the operation of symmetry in forbidding them, in accordance with the selection rules. The conclusions from such failures will then be well founded. But it should always be borne in mind that failure can arise from the inadequacy of experimental techniques. Clearly the greater the number of pieces of evidence that point in the same direction, the more likely is that direction to be the correct one. For this reason it is very advisable, wherever this is possible, to investigate both the infra-red spectrum and the Raman spectrum, and to investigate the states of polarization of the Raman lines.

Generally speaking, the effects of the general rules of selection and polarization will be more restrictive for a higher symmetry than for a lower. For molecules like PCl_3 and BF_3, the higher symmetry is D_{3h}, and the lower is C_{3v}; accordingly, reference to Table 24.1 shows that the numbers of permitted fundamentals and polarized Raman lines are smaller for the former of these point groups. It follows that, in the case of a molecule which possesses the *lower* of two proposed symmetries, it will in principle be possible to establish this fact unequivocally from spectroscopic evidence of an essentially positive kind. On the other hand, in the case of a molecule which possesses the *higher* of two proposed symmetries, the evidence must necessarily be essentially negative. Often such negative evidence is very strong; but it can never be wholly convincing. Indeed, cases have arisen in which evidence of this kind, though circumstantially very impressive, has turned out to be seriously misleading.

It is worth noting that a determination of molecular symmetry which is based solely upon the general rules of selection and polarization (i.e. solely upon symmetry considerations) is quite independent of the nature of the force field of the molecule. The force field, with the nuclear masses, determines the actual values of the frequencies. The rules of selection and polarization, however, are not concerned with these values, but only with the symmetry properties of the associated vibrational modes. This is an important circumstance, for, as we saw in Chapter 15, force fields cannot in general be fully determined. Because it is not subject to this limitation, the vibrational spectroscopic method of determining molecular symmetry is one of great power and elegance.

361

Index

REFERENCES are to page numbers, which will be found at the *bottom* outer corners of the pages.

The Index includes some important symbols. Those consisting of a single letter (subscripts and superscripts being ignored) are placed before all other entries beginning with that letter. Greek letters are listed according to the initial letters of their English names, e.g. π and ψ are placed under P. Greek letters follow English letters, e.g. σ follows S.

Index

Index

Kronecker delta δ_{ij}, 44, 55–6, 59, 133, 311

L, matrix relating internal coordinates D with normal coordinates Q, 238

L, matrix relating internal symmetry coordinates S with normal coordinates Q, 233
  columns as eigenvectors of **GF** matrix, 234–5
  normalization, 236–7
  relation to **F** and **G** matrices, 234
  relation to **L** matrix, 238
  rows of inverse as eigenvectors of **FG** matrix, 235
linear molecules, number of vibrational normal coordinates, 31–3
  See also XY_2 molecules, symmetrical linear

μ, *see* dipole moment
matrices, 13, 15–16, 34–5
  addition, 39
  character χ, 69–70
  characteristic equation, 231
  diagonal, 42, 50
  diagonal-block form, 40–1, 117, 118–19
  eigenvalues and eigenvectors, 231
  elements, 35, 59
  equivalent, 69
  inverse, 44–7
  null, 43
  orthogonal, 47–8, 63
  partitioned, 39–40, 73–4, 76–7
  products, 35–7
    inverse of, 46
    transpose of, 41–2
  quadratic forms expressed by, 49–50
  single-column, 38, 42
    representing a vector, 53, 58, 64–5
  single-row, 38, 42
    representing a set of basis vectors, 60, 65
  singular, 44, 45–6
  symmetric, 42–3
  transpose, 41, 46–7
  unit, 43–4
  See also representations of a group

non-genuine vibrations, 8, 9, 28, 201–3
  and product rules, 211–12
  See also rotation *and* translation
normal coordinates Q, 13, 233
  in terms of internal coordinates D, 238
  in terms of internal symmetry coordinates S, 237
  and kinetic energy, 13–14, 233
  in terms of mass-weighted Cartesian coordinates q, 15
  and potential energy, 14, 233
  and separability of wave equation, 244–5
  symmetry species, 147–50
  unnecessary for determination of number of distinct normal frequencies and their species, 151
  unnecessary for setting up secular equation, 82
normalization:
  of basis vectors, 172, 174–5
  of **L** matrix, 236–7
  of vibrational amplitudes, 12
  of wave functions, 248–9
normal modes, 6
  vibrational:
    number of, 30, 32–3
    See also BF_3 *and* PCl_3
  of zero frequency, *see* non-genuine vibrations
normal vibrational frequencies:
  conventional numbering, 152
  equality of classical and spectroscopic, 4, 252
  number of distinct values, 150–4, 162–6

O_h, point group, 123
operators, 64. *See also* Hamiltonian operator, projection operators, *and* symmetry operators
orthogonality properties:
  of characters of matrices in irreducible representations, 134–6, 141, 169
  of elements of matrices in irreducible representations, 130–4
  of wave functions, 245–8
orthogonal transformations, 48
  symmetry operations as, 63
orthonormal basis, *see* function space *and* vector space